Counterparty Risk and Funding

A Tale of Two Puzzles

CHAPMAN & HALL/CRC
Financial Mathematics Series

Aims and scope:
The field of financial mathematics forms an ever-expanding slice of the financial sector. This series aims to capture new developments and summarize what is known over the whole spectrum of this field. It will include a broad range of textbooks, reference works and handbooks that are meant to appeal to both academics and practitioners. The inclusion of numerical code and concrete real-world examples is highly encouraged.

Published Titles

American-Style Derivatives; Valuation and Computation, *Jerome Detemple*
Analysis, Geometry, and Modeling in Finance: Advanced Methods in Option Pricing, *Pierre Henry-Labordère*
Computational Methods in Finance, *Ali Hirsa*
Counterparty Risk and Funding: A Tale of Two Puzzles, *Stéphane Crépey and Tomasz R. Bielecki, With an Introductory Dialogue by Damiano Brigo*
Credit Risk: Models, Derivatives, and Management, *Niklas Wagner*
Engineering BGM, *Alan Brace*
Financial Mathematics: A Comprehensive Treatment, *Giuseppe Campolieti and Roman N. Makarov*
Financial Modelling with Jump Processes, *Rama Cont and Peter Tankov*
Interest Rate Modeling: Theory and Practice, *Lixin Wu*
Introduction to Credit Risk Modeling, Second Edition, *Christian Bluhm, Ludger Overbeck, and Christoph Wagner*
An Introduction to Exotic Option Pricing, *Peter Buchen*
Introduction to Risk Parity and Budgeting, *Thierry Roncalli*
Introduction to Stochastic Calculus Applied to Finance, Second Edition, *Damien Lamberton and Bernard Lapeyre*
Monte Carlo Methods and Models in Finance and Insurance, *Ralf Korn, Elke Korn, and Gerald Kroisandt*
Monte Carlo Simulation with Applications to Finance, *Hui Wang*
Nonlinear Option Pricing, *Julien Guyon and Pierre Henry-Labordère*
Numerical Methods for Finance, *John A. D. Appleby, David C. Edelman, and John J. H. Miller*

Option Valuation: A First Course in Financial Mathematics, *Hugo D. Junghenn*

Portfolio Optimization and Performance Analysis, *Jean-Luc Prigent*

Quantitative Finance: An Object-Oriented Approach in C++, *Erik Schlögl*

Quantitative Fund Management, *M. A. H. Dempster, Georg Pflug, and Gautam Mitra*

Risk Analysis in Finance and Insurance, Second Edition, *Alexander Melnikov*

Robust Libor Modelling and Pricing of Derivative Products, *John Schoenmakers*

Stochastic Finance: An Introduction with Market Examples, *Nicolas Privault*

Stochastic Finance: A Numeraire Approach, *Jan Vecer*

Stochastic Financial Models, *Douglas Kennedy*

Stochastic Processes with Applications to Finance, Second Edition, *Masaaki Kijima*

Structured Credit Portfolio Analysis, Baskets & CDOs, *Christian Bluhm and Ludger Overbeck*

Understanding Risk: The Theory and Practice of Financial Risk Management, *David Murphy*

Unravelling the Credit Crunch, *David Murphy*

Proposals for the series should be submitted to one of the series editors above or directly to:
CRC Press, Taylor & Francis Group
3 Park Square, Milton Park
Abingdon, Oxfordshire OX14 4RN
UK

Chapman & Hall/CRC FINANCIAL MATHEMATICS SERIES

Counterparty Risk and Funding

A Tale of Two Puzzles

Stéphane Crépey

University of Evry Val d'Essonne, Evry, France

and

Tomasz R. Bielecki

Illinois Institute of Technology, Chicago, USA

With an Introductory Dialogue by

Damiano Brigo

Imperial College, London, UK

CRC Press
Taylor & Francis Group
Boca Raton London New York

CRC Press is an imprint of the
Taylor & Francis Group, an **informa** business
A CHAPMAN & HALL BOOK

CRC Press
Taylor & Francis Group
6000 Broken Sound Parkway NW, Suite 300
Boca Raton, FL 33487-2742

Printed on acid-free paper
Version Date: 20140305

Printed and bound in India by Replika Press Pvt. Ltd.

International Standard Book Number-13: 978-1-4665-1645-8 (Hardback)

Contents

Preface

Introduction

This book is concerned with studying risk embedded in financial transactions between two parties, called the bank and its counterparty. This terminology is a useful convention used throughout the text, as, of course, the bank itself is the counterparty in the transaction relative to the bank's counterparty in the transaction. The term counterparty risk thus applies to both parties in the transaction. It will be analyzed from the perspective of the bank throughout the book.

The credit crisis and the ongoing European sovereign debt crisis have highlighted the native form of credit risk, namely counterparty risk. This is the risk of non-payment of promised cash flows due to the default by a party in an over the counter (OTC) derivative transaction. Thus, this is the risk born by a party in an OTC transaction, that its counterparty defaults on delivery of the promised payments. The value (price) of this risk was initially called credit valuation adjustment (CVA), and the volatility of this value is of interest. A key related issue, especially with credit derivatives, is the wrong-way risk, i.e. the risk that occurs when exposure to a counterparty is adversely correlated with the credit quality of that counterparty. Moreover, as banks themselves have become risky, counterparty risk must be analyzed from the bilateral perspective: not only CVA needs to be accounted for, but also debt valuation adjustment (DVA), so that the counterparty risk of the two parties to the OTC contract are jointly accounted for in the modeling. In this context the classical assumption of a locally risk-free asset which is used for financing purposes (lending and borrowing as needed) is no longer sustainable, which raises the companion issue of proper accounting for the funding costs of a position, namely the funding valuation adjustment (FVA), also called in this book liquidity valuation adjustment (LVA) when considered net of the credit spread. Another related issue is that of replacement cost (RC) corresponding to the fact that, at the default time of a party, the valuation of the contract by the liquidator may differ from the economic value of the contract (cost of the hedge) right before that time. Also since August 2007 one saw the emergence of a systemic counterparty risk, referring to various significant spreads between quantities that were very similar before, such as OIS rates versus Libor. Through its relation with the concept of discounting, this systemic component of counterparty risk has impacted on all derivatives markets. And the list goes on since people now talk about KVA for capital valuation adjustment (in reference to the capital cost of CVA volatility), or AVA for additional valuation adjustment (to reach a notion of "prudent value", integrating model risk and credit spreads including own, recently opposed to fair value by Basel). In the book we also consider RVA for a rating valuation adjustment, accounting for rating triggers. An acronym XVA was even introduced to generically refer to this increasing list of adjustments (see Carver (2013)).

All these adjustments, which are interdependent and must be computed jointly, are today among the main P&L generators of investment banks. As it is seen, dealing with these adjustments involves many areas: modeling, computation, pricing, risk management, regulation, economics, legal, lobbying, politics (even geopolitics, through varying legislative and legal practices), areas that often have conflicting objectives. The current trend of regulation is to push participants to negotiate centrally via clearing houses and to guarantee their failure through collateralization, posing a serious liquidity constraint on the market, for margin calls abound.

The basic counterparty risk mitigation tool is a Credit Support Annex (CSA) specifying the valuation scheme which will be applied by the liquidator in case of default of a party before a certain horizon T, including the netting rules which will be applied for valuing the different OTC derivatives between the two parties at that time. This "CSA value process", i.e. the value process of an OTC contract subject to the CSA specification, is also used to define a collateralization scheme between the two parties, similar to the margining procedure in futures contracts (except that the collateral posted through a CSA typically bears some interest). However, wrong-way risk and gap risk (slippage of the value of the portfolio between the time of default, where the collateral is frozen, and the time of liquidation of the portfolio) imply that collateralization cannot be a panacea (and it also poses liquidity problems as indicated above). Therefore counterparty risk cannot be simply mitigated through collateral; it also needs to be hedged against default and/or market risk. Eventually, the collateralized and hedged portfolio needs to be funded, which takes us to the related funding issue, with the related controversial issues of windfall benefit at own default and DVA/FVA overlap or double counting.

In this book we provide an analytical basis for the quantitative methodology of dynamic valuation, mitigation and hedging of bilateral counterparty risk on OTC derivative contracts under funding constraints. We review counterparty risk in its different aspects such as CVA, DVA, FVA, LVA, RC, RVA, wrong-way risk, multiple funding curves and collateral. The book is intended primarily for researchers and graduate students in financial mathematics. In addition, it is addressed to financial quants and managers in banks, CVA desks, as well as members of the supervisory bodies.

Outline

Within the bank, every particular business trading desk has an accurate picture of only its own activity. It usually lacks the global view of the bank activities and, specifically, the aggregated data that are required to properly value the counterparty risk and funding cash flows. Therefore in major investment banks the trend is to have central desks in charge of collecting the global information and of valuing and hedging counterparty risk, also accounting for any excess funding costs involved. We will call total valuation adjustment (TVA) the aggregate value of all the adjustments which are required in order to account for bilateral counterparty risk under funding constraints. Accordingly, we generically call TVA desk a central desk in charge of the corresponding risks (in practice it is typically split between a unit called CVA desk in charge of the counterparty risk adjustments, and the ALM (or treasury desk), which takes care of the funding issues). A "clean" price and hedge indicate a price (cost of the hedge) and hedge of promised cash flows ignoring counterparty risk and assuming that the trading strategy is funded at the OIS rate (as a typical rate of remuneration of the collateral, also the best market proxy for a risk-free rate). The price-and-hedge of a contract is then obtained as the difference between the clean price-and-hedge provided by the related business trading desk and a price-and-hedge adjustment provided by the TVA desk. Note that, since market quotes typically reflect prices of fully collateralized transactions which, as we will see in the book, can be considered as clean prices, the clean price is also the relevant notion of valuation at the stage of model calibration.

This allocation of tasks between the various business trading desks of an investment bank and a central TVA desk motivates a mathematical TVA approach to the problem of valuing and hedging counterparty risk and funding constraints. Moreover, the TVA will emerge in the book not only as a very important financial concept, but also as a useful tool in the mathematical analysis of counterparty risk subject to funding constraints. Indeed, the mathematical analysis reveals that the TVA process Θ can be interpreted as the price process of an option on the clean price P; this option, which is called the contingent credit default swap (CCDS), pays so-called exposure at default and funding dividends.

Part I sets the financial landscape, starting with a "Galilean dialogue" touching upon most of the topics of the book. We then provide an economic analysis of the post-crisis multi-curve reality of financial markets, which is an important feature underlying several aspects of the financial and mathematical analysis of counterparty risk and funding costs. In Part II we describe in mathematical but model-free terms all the basic elements of this pricing and hedging framework. Though nice theoretically, these do not immediately lend themselves to any concrete computations. To make them practical, we therefore specify in Part III the setup for a reduced-form framework with respect to a reference filtration, in which the risk of default of the two parties only shows up through their default intensities. The above-mentioned reduced-form modeling approach results in good tractability, but is obtained at the cost of a standard immersion hypothesis between the reference filtration and the full model filtration, which is the reference filtration progressively enlarged by the default times of the two parties. From the financial point of view, such a standard immersion setup implies a weak or indirect dependence between the counterparty risk and the underlying contract exposure. In particular, it excludes the possibility of a promised dividend of the contract at the default time of a party. This is acceptable in standard cases, such as counterparty risk on interest rate derivatives. But it is too restrictive for cases of strong wrong-way risk, such as with counterparty risk on credit derivatives, which is treated separately in Part IV by means of dynamic copula models. However, due to the high-dimensional context of credit portfolio models, we only deal here with counterparty risk, ignoring the nonlinear funding issue. Moreover, the credit portfolio models of this part only admit two possible states for each obligor: default and non-default. In Part V, we first present a credit migrations model which allows one to account for rating dependent CSA clauses. The concluding chapter then sheds a unified perspective on the different kinds of modeling employed in this book (except for the above-mentioned credit migrations model), showing how a reduced-form approach can be developed in the models of Part IV, which also opens the door to nonlinear FVA computations in credit portfolio models. Part VI is a mathematical appendix covering classical tools from stochastic analysis and a brief introduction to the theory of Markov copulas.

About the title of the book

The two "puzzles" alluded to in the title of the book[1] are (rather than counterparty risk and funding in general):

- the DVA/FVA puzzle regarding the interaction and possible overlap between DVA (own credit) and FVA (funding) terms;
- the top-down versus bottom-up portfolio credit modeling puzzle, which the CDO industry had been trying to solve for a long time and has basically failed, that also needs to be addressed in order to properly deal with counterparty risk on credit derivatives.

Regarding the first puzzle, it is shown in Part III how, once the dependence structure of the problem is understood and suitably formalized in terms of pre-defaut BSDEs, the DVA/FVA overlap issue can be solved. Regarding the second puzzle, we propose in Part IV dynamic copula models of portfolio credit risk and, in particular, the Markov copula common-shock model of Chapters 8 through 10. These are, in our opinion, the two major contributions of the work presented in this book. A bridge between the approaches of Part III and IV is made in the concluding chapter of Part V, where a unified perspective is given in terms of marked default times, under the so-called condition (A).

[1]Cf. "A Tale of Two Cities" by Charles Dickens.

Standing Notation, Terminology and Assumptions

A time dependence is denoted in functional form by (t) when it is deterministic and as a subscript by $_t$ for a stochastic process. Somewhat unconventionally, we will frequently use notation X_t, rather than just X or $X_\cdot$, for a stochastic process $(X_t)_{t\geq 0}$. Regarding operators, a subscript (if any) indicates the variable in which the action takes place, e.g. ∂_t for time derivation, or $\mathcal{A}_x$ for an operation acting in the direction of x.[2] By default, a real valued function of real arguments is assumed Borel measurable; all price processes are assumed to be special semimartingales and all semimartingales (including finite variation processes) are taken in a càdlàg version; "martingale" means local martingale, but the strict martingale property is assumed whenever necessary; all inequalities between random quantities are to be understood almost surely or almost everywhere, as suitable; all the cash flows are assumed to be integrable; all the market quotes (calibration data) are denoted with a star to distinguish them from their model counterparts (model clean prices as explained in the Outline above), which are denoted by the same letter without a star.

The symbols τ_b and τ_c represent the default times of the bank and of the counterparty respectively; $\mathbb{H}^b = (\mathcal{H}_t^b)_{t\in[0,T]}$ and $\mathbb{H}^c = (\mathcal{H}_t^c)_{t\in[0,T]}$ stand for the natural filtrations of τ_b and τ_c, that is the filtrations generated by the corresponding indicator processes, i.e. $\mathcal{H}_t^i = \sigma(\tau_i \wedge t) \vee \sigma(\tau_i > t)$, for any time t and $i = b$ or c; $|S|$ is the cardinality of a finite set S; $\mathbb{R}_+$ denotes the nonnegative half-line $[0, +\infty)$; $\int_a^b$ is to be understood as $\int_{(a,b]}$ (so in particular $\int_a^b = 0$ whenever $a \geq b$); $x^+ = \max(x, 0)$ and $x^- = \max(-x, 0) = (-x)^+$ are the positive and the negative parts of x a real number x (sometimes, in case there is a superscript as in x^e, we write $x^{e,\pm}$ instead of $(x^e)^\pm$); $\boldsymbol{\delta}$ represents a Dirac measure; $^\mathsf{T}$ stands for transpose of a matrix.

Bibliographic Guidelines

Here are the main sources for this book:

- for Part I: Crépey and Douady (2013a,2013b) (Chapter 2); the Galilean dialogue in Chapter 1 is essentially original material renewing the Platonic dialogue of Brigo, Morini, and Pallavicini (2013);
- for Part II: Bielecki and Crépey (2013) (Chapter 3); Crépey (2012a) (Chapter 4);
- for Part III: Crépey (2012b) (Chapter 5); Crépey, Gerboud, Grbac, and Ngor (2013) (Chapter 6);

[2] In this we depart from the standard operator theory notation where a subscript t typically emphasises a time-inhomogeneity of the operator $\mathcal{A}$. So, in our case, any such time-inhomogeneity is implicit in $\mathcal{A}$ (or $\mathcal{A}_x$).

- for Part IV: Crépey, Jeanblanc, and Wu (2013) (Chapter 7); Bielecki, Cousin, Crépey, and Herbertsson (2013c,2013a, 2013b) and Bielecki and Crépey (2013) (Chapter 8); Crépey, Jeanblanc, and Zargari (2010) and Bielecki, Crépey, Jeanblanc, and Zargari (2012) (Chapter 9); Assefa, Bielecki, Crépey, and Jeanblanc (2011) and Crépey and Rahal (2013) (Chapter 10);
- for Part V: Bielecki, Cialenco, and Iyigunler (2013) (Chapter 11); Crépey (2014) (Chapter 12);
- for Part VI: Sect. 3.5, 4.2 and 4.3.2 in Crépey (2013) (Chapter 13); Bielecki, Jakubowski, and Nieweglowski (2013,2012) and Bielecki, Jakubowski, Vidozzi, and Vidozzi (2008) (Chapter 14).

All the numerical calculations were done using Matlab.

Acknowledgements

We would like to address warm thanks to:

- all our co-authors cited in the above references;
- students and course participants at Global Derivatives April 2012 in Barcelona, at the Bachelier world congress pre-conference practitioner workshop "quantitative aspects of counterparty risk" in June 2012, at the AIMS financial mathematics summer school 2013 in February 2013, at different quantitative finance master programs throughout the world and different WBS courses (most recently October 2013 in Munich);
- institutions: our home universities as well as the University of Gothenburg which hosted us at the time we were starting to work on this book in July 2012. Here we would like to extend our special thanks to Dr. Alex Herbertsson, from the Economics Department in the University of Gothenburg, whose help and hospitality is greatly appreciated;
- grants: the research of S. Crépey has benefited from the support of the "Chair Markets in Transition" under the aegis of Louis Bachelier laboratory, a joint initiative of École polytechnique, Université d'Évry Val d'Essonne and Fédération Bancaire Française. Research of Tomasz R. Bielecki has been supported by the NSF grants DMS-0908099 and DMS-1211256;
- people with whom we had the opportunity to exchange or interact in one way or another on the topics of this book. In particular: Claudio Albanese (Global Valuation Limited and King's College London), Antonio Castagna (Iason consulting),

Giovanni Cesari (UBS London), Nicole El Karoui (Université Paris 6 Jussieu), Jean-Pierre Lardy (JPLC Credit Risks Advisory), Jean-Paul Laurent (Université Paris 1 Sorbonne and BNP-Paribas London), Alex Lipton (Bank of America Merrill Lynch), Marek Musiela (Oxford-Man Institute of Quantitative Finance), Andrea Pallavicini (Banca IMI, Milan, and Imperial College, London) and Marek Rutkowski (University of Sydney);

- Gary Wong (Ipotecs Ltd, London), who allowed us to use his original idea and design for the cover of the book (see Wong (2013)).

Part I

Financial Landscape

Chapter 1

A Galilean Dialogue on Counterparty Risk, CVA, DVA, Multiple Curves, Collateral and Funding

In this introductory chapter, which is written in the form of a Galilean dialogue renewing the Platonic dialogue of Brigo, Morini, and Pallavicini (2013), we present the financial landscape of the book, touching upon credit value at risk (credit VaR), potential future exposure (PFE), expected exposure (EE), expected positive exposure (EPE), credit valuation adjustment (CVA), debt valuation adjustment (DVA), CVA versus DVA hedging, close-out conventions, netting clauses, collateral modeling, gap risk, rehypothecation, wrong-way risk, Basel III, inclusion of funding costs, first-to-default risk, contingent credit default swaps (CCDS), CVA restructuring possibilities through margin lending, backward stochastic differential equations (BSDEs) and dynamic copulas (Markov copulas in particular).

The dialogue is in the form of a conversation among three friends respectively working as a quant (Salva), a trader (Sage) and a regulator (Simeone), in the manner of *Dialogo sopra i due massimi sistemi del mondo* (Galileo Galilei). Quoting Wikipedia: "The book is presented as a series of discussions, over a span of four days, among two philosophers and a layman: Salviati, from Florence, argues for the Copernican position and presents some of Galileo's views directly. [...] Sagredo, from Venice, is an intelligent layman who is initially neutral. [...] Simplicio, a dedicated follower of Ptolemy and Aristotle, presents the traditional views and the arguments against the Copernican position".

In our case:

Sage - Sagredo has been a trader for most of his career, and recently joined the top management of a tier–1 investment bank. He has a general and utilitarian grasp of maths. He believes in global growth, developed economy and he is a capitalist at heart. He is a derivatives and more generally a market entusiast. He has a master's from Chicago. He is based in London.

Simeone - Simplicio is a US Central Bank employee, he is very attentive to regulation. He has a very prudent stance on financial innovation. He is in line with Chicago school economics, he worked briefly as model validator previously. He has a good knowledge of academic literature in economics and finance. He has a master's from Paris, following education in a grand école. He is momentarily based in Paris at the time of the dialogue but often travels to London by train.

Salva - Salviati is a quant who worked in a variety of roles, from Front Office (rates, Credit) of investment banks to Risk Management of commercial banks. She is open minded and gently opinionated. She tries to mediate the extreme positions of the other two and is somehow a skeptic. She has a PhD in a university of a marginal (from the corporate/anglo saxon centric point of view) country. She is also based in London.

The three characters were in touch at the time of their undergraduate studies.

At this point it is fair to stress that the characters do not have an exact correspondence with Galileo's characters. In Galileo's dialogue, roughly speaking, SIM is the bad guy, SAL is the good guy, and SAG is an enlightened layman/billionaire siding with SAL. Here we are more in a mixed situation, and while we retained the three-person scheme and a formal name mapping, we would discourage3 a full identification. For example, you'll notice that our SIM comes out much better than Galileo's SIM, whereas our SAG comes out definitely worse than Galileo's. A similar remark applies to the following section, "to the discerning reader", that should not be interpreted literally.

1.1 To the Discerning Reader

Several years ago[1], there was published in Basel a powerful edict which, in order to obviate the dangerous tendencies of our present age, imposed a seasonable silence upon the Pythagorean opinion that counterparty risk, collateral and funding pricing cannot be standardized. There were those who asserted that this decree had its origin not in judicious inquire, but in passion none too well informed. Complaints were to be heard that advisers who were totally unskilled at financial observations ought not to clip the wings of reflective intellects by means of rash standardization and regulation. In reaction to the Pythagorean view, the Basel supporters impudently accused the standardization critics of lack of a global vision.

Upon hearing such carping insolence, my zeal could not be contained. Being thoroughly informed about that prudent determination, I decided to appear openly in the theater of the world as a witness of the sober truth.

I was often to be found in the marvelous city of London, in discussions with Signore Sage, a man of notable wealth and trenchant wit. From London came also Signora Salva, a sublime intellect who fed no more hungrily upon good wine than it did upon fine meditations. I often talked with these two of such matters in the presence of Signor Simeone, a Central Banker whose interpretations of Basel I, II and III always struck me by their depth.

Now, since diverging destinies have deprived London of those three great intellects, I have resolved to make their fame live on in these pages, so far as my poor abilities will permit, by introducing them as interlocutors in the present argument.

It happened that several discussions had taken place casually at various times among them and had rather whetted than satisfied their thirst for learning. Hence very wisely they resolved to meet together on certain days during which, setting aside all other business,

[1]Refashioned after the version translated by Stillman Drake and annotated and condensed by S. E. Sciortino.

they might apply themselves more methodically to the contemplation of the economics wonders of the markets. They met in a pub in London; and, after the customary but brief exchange of compliments, Salva commenced as follows.

1.2 The First Day

1.2.1 General Introduction, Size of Derivatives Markets, Exposures, Credit Var, Basel

SALVA. Good to see you, Sage, good to meet you again, Simeone. It's good to meet after such a long time.

SAGE. Indeed, it has been a while. I find you well, you have not lost an ounce of acumen, beauty and charme.

SALVA. Flattery will get you anywhere. [smiles]

SIMEONE. You both look well.

SAGE. What have you been doing since university? After my master in Chicago I started a career as a junior quant in a trading desk, working on rates and FX. Then I moved on to trading, first with a junior role and then more and more senior. I traded in the CDO correlation desk of Top Bank, then moved to Super Bank and in the last four years I have taken charge of the CVA trading desk of Platinman Bank.

SALVA. You must have made quite a lot of money.

SAGE. Not as much as I'd love to [laughs] and I lost a lot of it too .... You know the saying, if you still need to work it means you haven't made enough money.

SALVA. Well, the last few years have been tough on a lot of people. I also started as a quant, but four years later than you, since I had to complete a PhD. I have stayed in the quant space, as I have worked on modeling of credit risk and derivatives pricing and hedging more generally, first for the front office of an investment bank and then for the risk department of a commercial bank. Now I am taken between departments as I coordinate efforts for a consistent CVA modeling platform for the bank.

SAGE. Not a bad idea, since I heard you had two babies? Congratulations, by the way. I guess the risk office is becoming quite an environment for quants, although I noticed the front office keeps being the most sought-after destination for quants, especially young ones.

SIMEONE. These quants would work on derivatives... that business that in 2011 reached a notional of more than 700 trillion USD$?

SAGE. Well that is according to the Bank of International Settlements, that bunch of bureaucrats

SALVA. But you have to admit these numbers can dwarf imagination. 700 trillion... 7 10^{14}, or more prosaically 700 000 000 000 000 USD.

SAGE. Never traded on those notionals, though; I wish I had. [laughs]

SIMEONE. You are not funny. You know what was the nominal GDP of US in 2011? About 15 trillion... and Europe was slightly above 17 trillion. The world GDP was about 70 trillions. Keeping this in mind, 700 trillion for derivatives sounds like madness, even when discounting double counting etc. Maybe we bureaucrats sometimes look at the big picture rather than being obsessed with our personal fortune.

SAGE. Now you can't say that. I eat what I kill! You don't.

SIMEONE. Strictly speaking you don't kill the cows whose meat you eat. Nor do you produce the electricity you consume. Nor...

SALVA. Now, slow down you both!! Testosterone is overrated anyway. I still think that derivatives are useful, ok? and are not bad in themselves. As many other objects, they can be used well or abused, their use is not implicit in their definition.

SAGE. Well spoken.

SIMEONE. I am not convinced. Do you find it sensible that the derivatives market can grow to be 10 times as large as the GDP of the planet? It seems madness to me.

SAGE. Why? Nobody pointed a gun and forced the banks to trade derivatives.

SALVA. Mmmhhhh I don't know about that... it looks like a game that got out of hand at some point and nobody could stop...

SAGE. You can't stop progress, you need to keep growing and accelerating, going back does not help.

SALVA. Slowing down can be helpful occasionally, you know [smiles]. And I am not sure about this "perennial economic growth" idea anymore. On the other hand, you studied at Chicago and I recently became a responsible mum [smiles].

SIMEONE. Well, there is also the economic cycle; it is not always growth, only in the long run. You can't deny that the industrial revolution and technological progress and free trading lead to growth.

SALVA. ... and to depletion of natural resources, pollution, global warming and a lot of developed economies on the verge of default, if you look at the data.

SAGE. I think the average person lives better and healthier today than 100 years ago.

SALVA. The real question is how long is this sustainable? All developed economies are quite in trouble with this "crisis"... is it really a crisis? A crisis has a turning point, it starts looking more like a chronic syndrome.

SIMEONE. This is not unprecedented, look at the great depression.

SAGE. You look at it! It is already bad enough to hear all the bad news...

SALVA. Ok that's quite enough, you two!! [laughs gently] Simeone, what have you been up to? You seem to know the big picture pretty well.

SIMEONE. After I finished my master's in London I worked in model validation for a bank and then moved to regulation. I joined the Central Reserve Bank in Metropolis, in the States, five years ago, to work on counterparty risk.

SAGE. That is definitely one of the safest places where you can be now. It's funny that now we are all connected to counterparty risk.

SALVA. I guess that has not made you popular with fellow traders in other desks!!

SAGE. You don't know the half of it... these xxxxxxx don't have a clue of the risks we cover for them and don't want to be charged a penny.

SIMEONE. They may feel that you are not really helping them. You charge them a fee that should then be used to hedge counterparty risk, funding costs, etc, but can you really hedge such risks effectively? I don't think you can, so perhaps these colleagues of yours feel they are wasting their money and that if the bad day comes when a counterparty to their deal defaults, you won't really protect them in any relevant way.

SALVA. That's a little harsh. I have seen the same attitude but this might be more a reluctance to give up part of the P&L rather than lack of trust of the CVA desk. After all, what trader is happy to be charged additional costs?

SAGE. Yeah, that's right.

SALVA. That's not the whole story, though ... in fact, difficulties in hedging CVA and DVA are objective. You are an experienced trader. Do you think CVA can be hedged? And DVA? Can you hedge CVA and DVA just with sensitivities just because you can hedge an equity call option in Black- Scholes like that? How do you hedge wrong-way risk, recovery risk, jump to default risk? And what about even credit risk, for those counterparties that do not have a liquidly traded CDS? You can't really hedge any of those risks properly. What is the point of implementing cross-gamma hedging if you have an uncertainty of 20 percent on the recovery? And the fun one would be proxy hedging... who mentioned that? Goldman Sachs I think.

SAGE. Whose side are you on??? Goldman Sachs are simply the best, so if they say that...

SALVA. Right, right. As the resident anti-establishment nuts, I need to remind you of: Greek debt camouflage operations, going long or short California bonds depending on what is more convenient (not always to the client, understatement of the year), the resignation letter of Greg Smith, continued conflict of interest and public/government positions: Henry Paulson, Mario Draghi, Mario Monti, Mark Patterson.... you are a big boy, so you read the newspapers, don't you? [smiles]

SAGE. [shakes head] Give me a break, next you will start a conspiracy theory... you need to read more credible sources.

SALVA. You know, I didn't find this in some obscure revolutionary magazine. As I said, it is in mainstream newspapers (New York Times anyone?) and even in wikipedia. [laughs]

SAGE. Ah, then it must be right. [laughs]

SALVA. I am simply trying to be intellectually honest here! [pretends to be offended]

SAGE. [pretends to be offensive] Are you sure?

SIMEONE. If you two are quite finished with the cabaret, I would like to say something as well. Going back to valuation and hedging of CVA and proxy hedging in particular, you should both make up your mind: is CVA a proper concept or not? Should it be pursued? Can it be hedged? If not, does it make sense as a price? I personally think it is not helpful; it should not be traded and it cannot be hedged.

SAGE. [confident] Of course it is.

SALVA. You cannot always pretend to have easy and simple answers. CVA-DVA-FVA is a very complex concept that requires very sophisticated computational techniques to be addressed properly and even defining it properly is a headache [runs hand along luminous light-brown hair]

SIMEONE. [didactically] Quantities and products that cannot be managed properly and responsibly should be banned. Imagine if you cannot even define them properly...

SAGE. [rolling eyes] Come on guys, CVA is relatively simple: your counterparty may default, thus causing you a loss if you were a creditor at the time.

SALVA. There are a lot of choices to be made when computing CVA, both on the models to be used and on the type of CVA to be computed. There are choices to be made on whether it is unilateral or bilateral (does DVA make sense?), on the closeout formulation, on how you account for collateral and rehypothecation, on whether you include first to default and on how you account for funding costs and so on. Due to the variety of possible different definitions of CVA and of modeling choices, there appears to be material discrepancies in CVA valuation across financial institutions, as pointed out in the article by Watt (2011). And you think it is simple? I do not go as far as Simeone, but I am definitely not as optimistic as you are.

SAGE. [shrugs] If you let complexity paralyze you, you will never do anything in your life. We must go as far as we can with the analysis in a reasonable time, but then we should take action. That is what a CVA desk is about.

SALVA. I think we are looking at different time horizons. Your philosophy may be ok for very short terms, but over long terms it can be quite different.

SIMEONE. Salva is right. Even your performance as a trader is measured annually with a bonus and then you may be gone. However, the consequences of your trades may affect the bank over the next 10 or 20 years.

SAGE. You know too well that if I trade poorly my reputation is ruined and I will have difficulties in finding a good position again. [mimicks gun with fingers] So it's not like bad performances are instantly forgotten.

SALVA. I am not sure about that. [laughs]

SIMEONE. As a regulator I have seen all cases happen, but there is indeed memory in the market.

SALVA. Most of the traders I know are gossip oriented, present company excepted of course, so you may be right. Gossip, however, is hardly the most reliable source of information.

SAGE. As a trader your P&L speaks for you, and I thought women were more gossip oriented. [grinning]

SIMEONE. As if mark to market were an exact science.

SALVA. [higher voice pitch] "Book this with flat volatilities, book this with bid, book this with ask, book that with the smile, book that with..."

SAGE. Yeah, I bet you read that too in Wikipedia... [shakes head]

SALVA. No I actually heard it from our traders. [laughs]

SIMEONE. Would you mind remaining serious for a minute?

SAGE. Why don't you talk to us about your regulator's job? And do you think all this exposure / PD / LGD / ED and Credit VaR stuff from Basel I, II and now III has been helping really?

SIMEONE. [didactically] Let us be careful here, as we are mixing a number of different notions. Let me cite Canabarro and Duffie (2004) from my graduate studies. This paper contains basic prototype definitions for exposure that are not always in line with the market, but can be used as a reference. First, counterparty exposure at any future time is the larger between zero and the market value of the portfolio at that time. Current exposure (CE) is the current value (say now, at time 0) of the exposure to a counterparty, so it is the current value of the portfolio if positive, and zero otherwise. It is the expectation under the pricing measure $\mathbb{Q}$ of future cash flows, each discounted back to time 0 (now) and added up, if positive and zero otherwise. [loosens tie] Potential future exposure (PFE) at a given time is the maximum of exposure at that time under a high degree of statistical confidence. For example, the 99% PFE is the level of potential exposure that is exceeded with only 1% $\mathbb{P}$-probability. PFE plotted against time is the potential exposure profile, up to the final maturity of the portfolio of trades with the counterparty. I think our friend Salva can explain to us this measures issue ($\mathbb{P}$ and $\mathbb{Q}$); during my master's I studied it in terms of risk premia, but maths people managed to do it much more complicated. [drinks]

SALVA. [adjusting skirt] Is that so? Well even among "quants" there is a lot of mileage there... some think there is only $\mathbb{Q}$, some think there is only $\mathbb{P}$ and so on... But let me comment on the following first: PFE is typically obtained through simulation. For each future time, the price of the netting set of trades (the relevant portfolio) with a given counterparty is simulated. A $\mathbb{P}$-percentile of the distribution of exposures is chosen and this is the PFE at the future time. The maximum of PFE over the life of the portfolio is called maximum potential future exposure (MPFE).

SAGE. Bla bla bla... do we really need all this technobubble?

SALVA. [flashing an ironic look] It depends on what you are trying to achieve of course.

SIMEONE. As a regulator I can tell you that banks care about such quantities because PFE and MPFE are usually compared with credit limits in the process of deciding whether a trade is permissible or not. I am sure you know about this. We also have expected exposure ($\text{EE}(t)$), the expectation of the exposure at a future date t under the $\mathbb{P}$-measure conditioned on the first to default occurring at time t. The curve of $\text{EE}(t)$ plotted against time t is the expected exposure profile. Expected positive exposure (EPE) is the time average of EE up to a given future time.[2] Finally, we have exposure at default (ED). This is naturally defined as the exposure value at the (random future) time of default for the counterparty.

SAGE. All right, but we are talking about default risk here and I haven't heard anything about default yet, although you mentioned twenty different definitions of exposure... [stifles a yawn] Why are we not mentioning any default probabilities, recoveries, etc?

SIMEONE. Indeed, in exposure there is no default simulation involved. The future portfolio value is simulated, but not the default of the counterparty. With exposure we deal with this matter: IF default happens, what is going to be our loss due to that default? So in a way, assuming default happens, we check what would be the loss in that case.

SAGE. [removing jacket] But this is not Credit VaR, is it?

SALVA. No, indeed. With Credit VaR we address the problem: what is the loss level that is not exceeded with a given $\mathbb{P}$ probability, over a given time horizon? This does involve the inclusion of the default event of the counterparty as we generate the loss and the default probability of the counterparty and also the recovery rates are key here. But I would let our regulator colleague talk about Credit VaR

SIMEONE. [in a didactical tone, removing jacket, then pinching chin] Credit VaR is basically an attempt to measure counterparty default risk for capital requirements. It tries to measure the risk that one bank faces in order to be able to lend money or invest towards a counterparty with relevant default probability. The bank has to cover that risk by re-

[2]In Chapter 3 we discuss a somewhat different notion of EPE which, in particular, is considered under the pricing measure $\mathbb{Q}$. See also the remark 3.2.11 in this regard.

serving capital, and how much capital can be decided after the risk has been measured. A popular such measure is precisely Value at Risk (VaR), which is a quantile on the loss distribution associated with the position held by the bank, over a given time horizon. More precisely, it is a percentile (say the 99.9 percentile) of the difference between the initial value of the position and the future value of the position at the risk horizon, across scenarios. When applied to default risk, leading to Credit VaR, the horizon is usually one year. If one chooses the 99.9-th percentile, we have the loss that is exceeded in only 1 case out of 1000. There are a few different definitions, more precisely. For example, Credit VaR can be either the difference of the percentile from the mean, or the percentile itself. There is more than one possible definition. [sweats slightly, drinks, looks at Salva]

SALVA. [adjusting long hair with left hand] We may also mention that in general VaR is not a universally good measure. It has often been criticized for not being sub-additive. In a way it does not always acknowledge the benefits of diversification. In some peculiar situations the risk of a total portfolio can be larger than the total of the risks in single positions. Another measure used to be preferred, namely expected shortfall, also known as tail VaR, conditional VaR, etc. This is the expected value of the losses beyond the estimated VaR percentile. However, expected shortfall is not elicitable, as discussed in Ziegel (2013), whereas VaR is. There are other good reasons why expected shortfall doesn't work and also why coherent risk measures, ten years later, turned out not to be such a great idea after all, so the choice is not so clearcut[3] [pauses a minute, sips drink] However, a large part of the industry still uses Credit VaR. This is obtained through a simulation of the basic financial variables affecting the portfolio and of the default of the counterparty before one year. The simulation is done under the historical probability measure, the $\mathbb{P}$ we mentioned earlier, up to the risk horizon. This simulation does include the default of the counterparties. At the risk horizon, the portfolio is re-priced in every single simulated scenario. For example, if the risk horizon is one year, we then have a number of scenarios for the value of the portfolio in one year. In each scenario, if the counterparty has defaulted, we check the value of the portfolio. If this is positive, all is lost except a recovery and this enters our loss. If it is negative or zero, nothing is lost. If there has been no default up to one year in that scenario, again nothing is lost (due to default risk). After we analyze this in every scenario at one year, a distribution for the portfolio losses at the horizon is assembled based on these scenarios of portfolio values and defaults. [leans forward] Notice that by saying "priced" above, we are stating that the discounted future cash flows of the portfolio after the risk horizon are averaged conditional on each scenario at the risk horizon itself. However, this averaging happens under the pricing probability measure $\mathbb{Q}$ and not under the historical measure. [pauses, sips drink]

SAGE. What is it with this confusion of probability measures?

[3] A risk measure is elicitable if it is possible to verify and compare competing estimation procedures for this measure.

SIMEONE. I warned you...

SALVA. [sighs, sits back] OK, suppose your portfolio is an equity forward that is traded with a client, with a final maturity of five years. To obtain the Credit-VaR you simulate equity under the $\mathbb{P}$ measure up to one year, obtain several one year equity scenarios. You also simulate the default of the counterparty up to one year, so that you know in each scenario whether the counterparty has defaulted or not. This is again under the measure $\mathbb{P}$. It may be important to include "correlation" between counterparty default and equity, leading to wrong-way risk (WWR). [leans forward] Summing up: in the first part we simulate under $\mathbb{P}$ because we need the risk statistics in the real world, under the physical probability measure. After that, in the second part, in each one-year scenario, if the counterparty has defaulted then there is a recovery value and all else is lost. Otherwise, we price the equity forward over the remaining year. However, this price is like taking the expected value of the forward in five years, conditional on each equity scenario in one year. Since we are doing valuation, these last expected values will be taken under the pricing measure $\mathbb{Q}$ and not under $\mathbb{P}$. [pauses, sips drink, checks mobile] Excuse me, just checking the baby sitter messages.

SAGE. [sympathetically] Take your time.

SALVA. All right. In the first part, it may be difficult to obtain the statistics needed for the simulation. For example, for default probabilities, often one uses probabilities obtained through aggregation methods or proxying, such as the probability associated to the rating of the counterparty. This is not accurate. But by definition default of a single name happens only once, so its $\mathbb{P}$ probability cannot be estimated in any direct way. About $\mathbb{P}$ and $\mathbb{Q}$, typically default probabilities under $\mathbb{Q}$ (which are usually obtained from CDS or corporate bond prices) are larger than those under $\mathbb{P}$. For example, a comparison of the $\mathbb{P}$ and $\mathbb{Q}$ loss distributions involved in Collateralized Debt Obligations (CDO tranches) is considered in Torresetti, Brigo, and Pallavicini (2009).

SAGE. I need a refresher, I studied VaR way back in Chicago but I am not up to speed.

SALVA. A basic intro is the book Jorion (2006), whereas for a higher technical level you have books like McNeil, Frey, and Embrechts (2005), which may, however, be too advanced for you, no offence meant [smiles]

SAGE. [fascinated] None taken, my beautiful professor!

SALVA. [smiling] From an historical perspective it is interesting to look at the original Credit VaR framework in the original "Credit Metrics Technical Document" by Gupton, Finger, and Bathia (1997). I am sure you all came across that at some point.

SAGE. I suddenly realized we haven't talked about CVA, but only Credit VaR. Why don't we discuss CVA, since that seems to be our main activity, even if from different points of view. Why don't we have an open minded discussion on this subject from different angles in front of a few drinks? We can do that tomorrow evening if you are ok with that, it would be fun I think.

SALVA. Yeah, why not?

SIMEONE. I will be there, I can prepare an agenda

SAGE. Again the bureaucrat... Why don't we keep it informal?

SALVA. I guess an agenda will help us stay focused and avoid more personalized confrontation (look at today!).

SAGE. You think we have been confrontational? And you said you worked with traders? Man, you don't know the half of it.

SALVA. You said that already. [smiles feigning boredom]

SAGE. [magnanimously] "Repetita iuvant".

SIMEONE. Here, you see why we need an agenda? On one side I feel like hugging you both after all these years, on the other hand I would like to kick Sage out of this pub, push him down the stairs, kick him again, shoot and kiss Salva and then go all together for a beer anyway.

SALVA. All right, all right. Thanks guys; see you tomorrow. Same time, ok? [adjusts skirt, pushes chair back, stands up and takes purse]

SAGE. Good for me. [checks mobile, updates calendar, stands up, kisses Salva's cheeks, shakes hands with Simeone] See you tomorrow.

SIMEONE. Excellent. [stands up, shakes hands with Sage, kisses Salva's cheeks]

1.3 The Second Day

1.3.1 CVA, DVA, Pricing, Arbitrage Free Theory, Closeout. And the Data? Ratings?

SALVA. Hello Sage, let's sit outside. The weather is quite pleasant with this sunny spring we are having.

SAGE. Hi, good idea Salva. Let me get a couple of drinks while we wait for Simeone. What will you have?

SALVA. A glass of Valpolicella please.

SAGE. Here you are, Salva.

SALVA. Thank you. So how did the trading go today?

SAGE. Today I mostly struggled through meetings with the treasury and the risk management. They are all pushing for a different system for calculation and charge of funding costs. A political nightmare.

SALVA. I imagine, since you are dealing with something that is quite P&L sensitive. [rolls the chalice in left hand]

SIMEONE. Hello Colleagues, let me get a drink and I will join you.

SAGE. Hi Simeone, great, we'll wait for you before saving the world. [laughs]

SALVA. [smiling] Yes we have big plans.

SIMEONE. Uff...

SAGE. First of all let me celebrate our reunion: Cheers! [raises wine glass]

SIMEONE. Cheers!

SALVA. Salute! [sipping the wine] So Sage was telling me the sort of political nightmare CVA is for his bank. It is the same where I work, but I am less impacted, being a quant. However, let's start by checking if we agree on what is CVA.

SAGE. All right, best way to explain is like this. Suppose you can trade a product either with a default-free counterparty or with the defaultable one. Having a choice, you would always take the default-free one. This means that in order to do the trade with the risky one you require additional compensation. In other terms, you require a reduction in the product price. This reduction is exactly the credit valuation adjustment, or CVA.

SIMEONE. In other words we can define it as the difference between the price of the product without default risk and the price of the product with default risk of the counterparty included.

SAGE. That's what I said!

SALVA. Good enough for a start. So the key point here is that this is a reduction in price, or in other terms it is itself a price. As a price it is a risk neutral or pricing measure expectation of future discounted cash flows. This is an expected value at valuation time, typically now, i.e. 0, of the discounted future cash flows involved with the default risk on the specific portfolio. While Credit VaR was a $\mathbb{P}$ measure percentile of the loss due to counterparty default at the risk horizon, this is a price at time 0, i.e. a $\mathbb{Q}$ expectation given the information today.

SAGE. Again with this $\mathbb{P}$ and $\mathbb{Q}$, we can't let you get away with this anymore. Please explain. [lights a cigar]

SALVA. [leaning back disapprovingly] You took up smoking cigars now?

SAGE. [innocently] After you do your first million it becomes compulsory, you know. It's written in Basel. [laughs] What, I hope you are not a health nut!

SIMEONE. You can smoke as much as you like as long as you puff that smelly smoke away from me. If you want to harm yourself it is none of my business.

SAGE. Pfffff.

SALVA. ...and from me, my hair and my clothes. I am sure you know how persistent that smoke smell is.

SAGE. [feigning offense, pushes chair back a little] Pfffff. This is perfume, not smell. And we are in the open. Why can't a guy smoke his cigar in peace after a hard day of work?

SALVA. [smiling ironically] All right, forgive us for not being in the unhealthy millionaires club...

SAGE. Ahahah, you know you are the only friends I have who are under one million? [puffs his cigar further]

SALVA. It would be fun if this conversation was intercepted and published by a journalist.

SIMEONE. Stop the cabaret and focus on the topic at hand.

SAGE. "Focus on the topic at hand"?? What ridiculous nonsense is that? Don't talk to me like that. I have known you too long. [puffs]

SALVA. All right, all right. $\mathbb{P}$ and $\mathbb{Q}$ then. Where do I start from? Ok, statistics of random objects such as future time losses depend on the probability measure being used. Clearly, under two different probability measures the same random variable will usually have two different expected values, variances, percentiles, etc. The probability $\mathbb{P}$, the historical or physical probability measure (also known as real world measure), is the probability measure we use to do historical estimation of financial variables, econometrics, historical volatility calculations, historical correlations, autocorrelations, maximum likelihood estimation, etc. As we compute VaR, for example, when we simulate the financial variables up to the risk horizon, we do it under $\mathbb{P}$. Also, when we try to make a prediction of future market variables through economic forecast, technical analysis etc we do it implicitly under $\mathbb{P}$. This is because risk measurement and prediction are interesting under the statistics of the observed world. However, if we are rather trying to price an option or a structured product in a no-arbitrage framework, the no-arbitrage theory states that expected values of future discounted cash flows are to be taken under a different probability measure, namely $\mathbb{Q}$. These two measures are connected by a mathematical relationship depending on risk aversion, or market price of risk. In simple models, the real measure $\mathbb{P}$ expected rate of return is given by the risk-free rate plus the market price of risk times the volatility. Under $\mathbb{P}$ the expected value of the rate of return of an asset is hard to estimate, whereas under $\mathbb{Q}$ one knows that the rate of return is the risk-free rate. Arbitrage-free theory says that dependence on the $\mathbb{P}$ rate of return can be hedged away through replication techniques. And if you think about it, this is why derivatives markets have exploded. The price of a derivative does not depend on your perception of the actual ($\mathbb{P}$) expected future return. Actual expected returns of assets are very difficult to estimate or predict and rightly so; otherwise we would all be very rich. But to price a derivative depending on the growth of such assets you do not need to know such ($\mathbb{P}$) returns, in principle.

SIMEONE. You seem to imply that the creation of derivatives markets was a good development.

SAGE. Again with this story? Derivatives are here and are here to stay. How can an airline hedge fuel risk without an oil swap? Give up this nerd attitude, take note and move on! [puff]

SIMEONE. [indignant] What if you stopped and thought about something for more than three seconds?

SALVA. [waving hands] Ok, ok, calm down you guys, I told you that testosterone is overrated. Let's go back to CVA. [adjusting hair] But before we do that we need a detour through a city in Switzerland.

SAGE. And I bet it's not Zurich and starts by "B". [laughs, puffs]

SALVA. And it's not Bern [anticipating Simeone's protest] ok? Basel: 1 2 3...

Ok, seriously, ... "Basel" is a set of recommendations on banking regulation issued by the Basel Committee on Banking Supervision. We are mostly interested in the second critical set of such recommendation, termed Basel II, and in the third one – Basel III. Basel II was first issued in 2004 and updated later on, to create a standard that regulators can use to establish how much capital a bank has to set aside in order to cover financial and operational risks connected to its lending and investing activities. Often banks tend to be willing to reserve as little capital as possible when covering risks, so as to be able to use the remaining funds for their activity and to have more liquidity available. The capital requirements concern overall the three areas of credit - or counterparty - risk, market risk and operational risks. The counterparty risk component of capital requirements can be measured at three different levels of increasing complexity, the "standardized approach", the foundation internal rating-based approach (IRBA) and the advanced IRBA. The standardized approach is more conservative and is based on simple calculations and quantities, so that if a bank follows that approach it is likely to find higher capital requirements than with the IRBA's. Obviously this is an important incentive for banks to develop internal models for counterparty risk and credit rating. Still, the ongoing credit crisis is generating a lot of doubt and debate on the effectiveness of Basel II and of banking regulation more generally. Basel regulation is currently under revision in view of a new set of rules commonly referred to as Basel III (later). Basel I, II and III have been heavily criticized also from a methodological point of view; see for example Blundell-Wignall and Atkinson (2010).

SIMEONE. So, summarizing, we mentioned above two main areas: (i) counterparty risk measurement for capital requirements, following Basel II, and the related Credit VaR risk measure, or (ii) counterparty risk from a pricing point of view. Basel II deals with (i) mostly and hence with concepts such as Credit VaR. I have been quite happy with that and with the efforts to improve the framework. Where I have problems is with (ii). In (ii) we are updating the value of a specific instrument or portfolio, traded with a counterparty, by altering the price to be charged to the counterparty. This reduction in price accounts for the default risk of the counterparty. As Sage said earlier, all things being equal, we would always prefer entering a trade with a default-free counterparty than with a default risky one. Hence we charge the default risky one a supplementary amount besides the default-free cost of the contract, that is thus reduced to us exactly of that amount. This is often called credit valuation adjustment, or CVA. As Salva explained earlier, since it is a price, it is computed entirely under the $\mathbb{Q}$ probability measure, the pricing measure. $\mathbb{P}$ does not play a role here. We are computing a price, not measuring risk statistics. As I mentioned earlier, [looks around skeptically] I have doubts this notion is helpful or even appropriate. Despite this, it has been there for a while; see for example Duffie and Huang (1996), Bielecki and Rutkowski (2001), Brigo and Masetti (2005).

SALVA. However, it became more and more important after the 2008 defaults. I recall that in 2002 not many people cared about this.

SIMEONE. Now perhaps Salva could summarize how this CVA term looks from a modeling point of view, so that we can discuss its drawbacks?

SAGE. My old friend, the glass can be half full as well, you know. [puffs and drinks wine]

SIMEONE. Your wine glass is almost empty. If you keep drinking at this rate the discussion will become impossible.

SALVA. Or more funny. [smiles] Back to CVA, boys. CVA looks like an option on the residual value of the traded portfolio of netting sets, with a random maturity given by the default time of the counterparty. Why an option, you may ask? To answer this question you need to look at cash flows. If the counterparty defaults before the final maturity and the present value of the portfolio at default is positive, then the surviving party only gets a recovery. If, however, the present value is negative, the surviving party has to pay it in full to the liquidators of the defaulted entity. Once we have done all calculations by netting these cash flows, we have that the value of the deal under counterparty risk is the value with no counterparty risk minus a positive adjustment, called CVA. This adjustment is the option price in the above sense. See for example Brigo and Masetti (2005) for details and a discussion. Of course, an option with random maturity is a complicated object... [looks at Sage and Simeone, sighs] It is complicated because it generates model dependence even in products that were model independent without counterparty risk.

SAGE. Let's use examples, so we avoid getting lost in the technobabble. [orders another glass of wine]

SIMEONE. Let us take for example a portfolio of plain vanilla swaps.

SALVA. [checks watch, adjusts skirt] You have to price an option on the residual value of the vanilla swap portfolio at default of the counterparty. You need an interest rate option model and, by adding counterparty risk, your valuation has become model dependent. Quick fixes to pricing libraries are quite difficult to obtain. And model dependence of course means that volatilities and "correlations" would impact this calculation and similarly for dynamics features more generally.

SIMEONE. And then my problems start. It is very hard to measure implied volatilities and implied correlations, i.e. under $\mathbb{Q}$ rather than $\mathbb{P}$.

SAGE. You do what you must, take some shortcuts, proxying, etc.

SALVA. But in any case no, they are not easy to measure. We are pricing and we are under $\mathbb{Q}$. We need volatilities and correlations extracted from traded prices of products that depend on such parameters. But from what market products do I imply information on the correlation between a specific corporate counterparty default and the underlying of the trade, for example gold, or a specific FX rate? And where do I extract credit spread volatilities from? Single name CDS options are not liquid.

SIMEONE. I agree, but it's actually worse than that. For some small/medium and sometimes even large counterparties it is even difficult to imply default probabilities, not to mention expected recoveries. From what I understand [looks at Salva] $\mathbb{Q}$ default probabilities can be deduced from credit default swap or corporate bond counterparty data. But how many small and medium enterprises have no reliable CDS or Bond quotes? Many counterparties do not have a liquid CDS or even an issued bond that are traded. What if the counterparty is the port of Mouseton? Where can one imply default probabilities from, leave alone credit volatilities and credit-underlying "correlations"? Not to mention the often overlooked recoveries (and do not get me started on forty percent).

SAGE. Puffff... you are making a fuss for nothing. Default probabilities, when not available under $\mathbb{Q}$ as you would say, may be considered under $\mathbb{P}$. You may then ajust them for an aggregate estimate of credit risk premia obtained from credit index data. For example, rating information can yield rough aggregate default probabilities for the port of Mouseton if we have a rating system for small medium enterprises (these are available by credit agencies). One may also consider $\mathbb{P}$-statistics like historical credit spread volatility and historical correlation between underlying portfolio and credit spread of the counterparty. That allows you to model even wrong-way risk to some extent.

SALVA. [amused/sarcastic] I have to admit I am fascinated by the phenomenal speed you show in solving all problems. They really should consider you for the Nobel award in economics. [leaning forward, murmuring] Do you realize that in your answer above there are a number of very fundamental problems? Do you trust rating agencies? Do you think their ratings for all SME[4] are reliable? Do you think most financial institutions have suitable internal rating systems? Do you think you can hedge a price you obtained through historically estimated statistics easily? Do you...

[4]small and medium-sized enterprises.

SAGE. Wait, isn't your Girsanov theorem saying that instantaneous vols and correlations for diffusions are the same under the two measures?

SALVA. [looking at Sage carefully] As usual, you pretend to be a jerk but you are quite smart. Yes, you would be right. But it's not that simple. Parameterizing the Radon-Nikodym...

SAGE. [finishes drinking the second glass] Please, Salva, stop. I told you, this is all about action. You should not be paralyzed by too much analysis and too many questions. [orders a third glass]

SALVA. This starts looking like "shoot first and ask questions later". Do you realize we are just scratching the surface of all the problems CVA has?

SIMEONE. I agree again. Are we having an honest debate or are we here to pretend everything is easy and doable and get drunk?

SAGE. You two are ganging up against me. It's not fair two against one. [feigns helplessness] And I am under the spell of this wonderful lady who has completely enchanted me.

SALVA. [ironic] Indeed, poor boy, look how we cornered you into a fatal position. You have been outsmarted by the femme fatale quant and by the Machiavellian Byzantine regulator. Checkmate. [laughs]

SAGE. Ahahaha, you are fantastic, we need to get out more often. [drinks more wine] All right [puffs], I will concede that the data may be a problem. But not an impossible one. And let me paraphrase Jules Verne: I am ignorant, sure, but I am so ignorant that I even ignore difficulties. They do not stop me. [winks]

SIMEONE. You should work as a comedian, not as a trader. Can we be serious for a minute? Let us talk about wrong-way risk. There is a lot of attention to that in the regulators space and we are not doing well there. Basel is having a particularly hard time with wrong-way risk.

SAGE. I don't see all this fuss about wrong-way risk. This is simply the additional risk we have when the underlying portfolio and the default of the counterparty are "correlated" in the worst possible way. Suppose, for example, that you are trading an equity forward with a corporate and you will receive fixed and paying (variable) floating equity at maturity. Suppose the equity of the forward and the equity of the counterparty to be positively correlated, for example the counterparty is Nokia and the underlying is Vodafone. We may have a negative correlation between the default likelihood of Nokia and Vodafone's equity, since higher prices of Vodafone will lead to higher values of Nokia's equity, which in turn implies lower default likelihood for Nokia. On the opposite side, lower values of the Vodafone equity will lead to higher default probabilities for Nokia. When Vodafone's equity decreases importantly, the default probability of Nokia will increase a lot due to wrong-way correlation and the value of the residual receiver equity forward will increase as well. This means that the embedded CVA option term will be more in the money precisely in those situations where the default probability of the counterparty Nokia is larger, thus causing much more damage than in the case with low correlation. This is an example of wrong-way risk. The opposite case, with negative correlation between Nokia and Vodafone equities would be right-way risk.

SIMEONE. That is a good example.

SALVA. A good one, indeed. More generally, Wrong Way Risk has been studied, for example, in the following references in different asset classes: Redon (2006), Brigo and Masetti (2005), Brigo, Morini, and Tarenghi (2011), Brigo and Tarenghi (2004), Brigo and Tarenghi (2005) for equity, Brigo and Pallavicini (2007), Brigo and Pallavicini (2008) for interest rates, Brigo, Chourdakis, and Bakkar (2008) for commodities (Oil), Brigo and Chourdakis (2008) for Credit (CDS).

SAGE. To our resident librarian! [bows, raises wine chalice, drinks] On the other hand, who needs to read all those papers? It's just as I told you.

SALVA. [smiling] "Also sprach Zarathustra"

SIMEONE. Look...

SALVA. Ok ok. I think another quite difficult problem is how to correctly frame the debt valuation adjustment, DVA. It has to do with the possibility that both parties in a deal agree on the counterparty risk charge. Maybe Sage, who is so good with examples, could start with some introduction?

SAGE. [kills cigar on the tray] I will tell you what I understood. [sips wine] Let's say that we are default-free and all recognize this. We are pricing the risk that the counterparty defaults before the final maturity of the deal on a given netting set. This is the CVA to us and as we have seen it is a positive adjustment to be subtracted from the default-risk-free price to us. Again, as we said above, having the choice and all things being equal, we prefer trading with a default risk-free counterparty rather than with a risky one. So we understand the risk-free price to us needs to be diminished by subtracting a positive quantity called CVA, so as to compensate us for the additional default risk. Now ask yourself: what happens from the point of view of the other party?

SIMEONE. You mean the other party looks at our default probability in the deal?

SAGE. No, or at least not yet. I mean the same setup as above, where we are still default free and the counterparty retains the default probability, but now it is the counterparty doing the analysis, not us.

SALVA. [looking at Sage admiringly] Very smart.

SAGE. The counterparty will mark a corresponding positive adjustment (the opposite of our negative one) to the risk-free price to her. This way both parties will agree on the price, because we agree we have to pay less for the contract (the price to us has decreased by CVA) and the counterparty agrees it has to pay more (the price to it has increased by the same amount, CVA). This makes sense. The adjustment for the counterparty client is positive because the counterparty needs to compensate us for its default risk. So finally we have gotten to DVA: the adjustment seen from the point of view of the counterparty is positive and is called debt valuation adjustment, DVA. It is positive because the early default of the counterparty herself would imply a discount on its payment obligations towards us, and this means a gain for her. So the counterparty marks a positive adjustment over the risk-free price by adding the positive amount called DVA. In this case, where we are default free, the DVA computed by the counterparty is also called Unilateral DVA, UDVA, since only the default risk of the counterparty is included. Similarly, the adjustment marked by us by subtraction is called Unilateral CVA, UCVA. In this case UCVA(us) = UDVA(counterparty), i.e. the adjustment to the risk-free price is the same, but it is added by the counterparty and subtracted by us. Notice also that, since we are default-free, UDVA(us) = UCVA(counterparty) = 0.

SIMEONE. Very clear. But this is not yet a bilateral case really, since there is only the counterparty that may default. So this DVA is not the general one.

SAGE. You have the general one when the two firms do not agree on one of them being default free. Say that in your example the counterparty does not accept that we are default free (a reasonable objection, after the eight credit events that happened to Financial Institutions in one month of 2008).

SALVA. [finishing the first glass of wine] Allow me to step in. In this case then the only possibility for both parties to agree on a price is for both parties to consistently include their defaults into the valuation. Now both parties will mark a positive (bilateral) CVA to be subtracted and a positive (bilateral) DVA to be added to the default risk-free price (MtM) of the deal. The CVA of one party will be the DVA of the other one and viceversa. Both parties will compute the final price as follows:

MtM - CVA + DVA

and, in Sage's example,

Price To Us = MtM to Us + DVA(us) - CVA(us),

whereas when the counterparty does the calculation she gets a completely analogous formula

Price To Counterparty = MtM to Counterparty + DVA(Counterparty) - CVA(Counterparty)

and, recalling that

MtM to Us = - MtM to Counterparty,

DVA(us) = CVA(Counterparty), DVA(Counterparty) = CVA(us),

we get that eventually

Price To Us= - Price To Counterparty,

so that both parties agree on the price. We could call Total (bilateral) Valuation Adjustment (TVA) to one party the difference CVA - DVA as seen from that party,

TVA = CVA - DVA.

Clearly TVA to us = - TVA to counterparty. We need to pay attention to terminology. By "bilateral CVA" the market refers both to TVA and to the CVA component of TVA. Mostly the industry uses the term to denote TVA and we will do so similarly, except when explicitly stated otherwise.[5]

SIMEONE. What is the technical literature on Bilateral CVA and on DVA?

SAGE. Yaaawnn

SALVA. [ignoring Sage] The first calculations are probably Duffie and Huang (1996), but Wrong Way Risk is hard to model in their framework. Furthermore, that paper deals mostly with swaps. Again, swaps with bilateral default risk are dealt with in Bielecki and Rutkowski (2001), but the paper where bilateral risk is examined in detail and DVA derived is Brigo and Capponi (2008a). In that paper bilateral risk is introduced in general and then analyzed for CDS. In Brigo, Pallavicini, and Papatheodorou (2011), Brigo, Capponi, and Pallavicini (2014) and Brigo, Capponi, Pallavicini, and Papatheodorou (2011) other aspects of bilateral risk are carefully examined, also in relationship with wrong-way risk, collateral and extreme contagion and gap risk. Those works analyze what happens when default happens between margining dates and a relevant mark to market change for worse has occurred. Brigo, Capponi, and Pallavicini (2014) considers a case of an underlying CDS with strong default contagion where even frequent margining in collateralization is quite ineffective.

SIMEONE. All right, thank you Salva. So we can summarize this by saying that Total bilateral Valuation adjustment is the difference between CVA and DVA as seen by the party doing the calculation.

SALVA. Yes but be careful. TVA is not just the difference of CVA and DVA in a universe where only one name can default. In computing DVA and CVA in the difference, you

[5] As it is frequent in the counterparty risk literature, the same terminology is used by different authors and users with different meaning. So is the case with TVA.

need to account for both defaults in both terms. There is thus a first-to-default check: If we are doing the calculation, in scenarios where we default first the DVA term will be activated and the CVA term vanishes, whereas, in scenarios where the counterparty defaults first, our DVA vanishes and our CVA payoff activates. So we need to check who defaults first. However, some practitioners implemented a version of TVA that ignores first to default times. If we compute

TVA(us) = CVA(us) - DVA(us)

(see for example Picoult (2005)) we are computing DVA(us) in a world where only we may default and then compute CVA(us) in a world where only the counterparty may default. But we do not eliminate the other term as soon as there is a first default. So in a sense we are double counting. The appropriate TVA contains a first to default check. The difference between the two approximations has been considered in Brigo, Buescu, and Morini (2012). The error in neglecting the first to default term can be quite sizeable even in seemingly harmless examples.

SIMEONE. Understood. From the conversations I had with our regulated entities, it looks like the industry prefers to leave the first to default out because this avoids the need to model default correlation between the parties involved in the deal.

SAGE. That is not very important, since if the underlying netting set is credit sensitive (for example contains CDS) then you need to model default correlation anyway or you have no wrong-way risk.

SIMEONE. But I understand the large majority of deals are interest rate swaps, so for those with the simplified formula without first to default you could indeed use your Unilateral CVA libraries to compute the bilateral adjustment without bothering with default correlation.

SAGE. Sure. But I still don't understand all this fuss everyone is making about DVA.

SIMEONE. Well, it's not hard to see why. DVA is a reduction on my debt due to the fact that I may default, thus not paying all my debt and a missed debt payment is like a gain, but I only can realize this gain as a cash flow if I default. Do you find that natural or straightforward?

SAGE. Yes, it is quite natural actually.

SIMEONE. There is more: if your credit quality worsens and you recompute your DVA, you mark a gain.

SAGE. Which, again, is perfectly natural. What is wrong? Everyone coming up with this story.

SIMEONE. You are kidding, right? Let us look at some numbers. Citigroup, in its press release on the first quarter revenues of 2009, reported:

> "Revenues also included [...] a net 2.5$ billion positive CVA on derivative positions, excluding monolines, mainly due to the widening of Citi's CDS spreads".

More recently, from the Wall Street Journal:

> october 18, 2011, 3:59 PM ET. Goldman Sachs Hedges Its Way to Less Volatile Earnings. "Goldman's DVA gains in the third quarter totaled $450 million, about $300 million of which was recorded under its fixed-income, currency and commodities trading segment and another $150 million recorded under equities trading. That amount is comparatively smaller than the $1.9 billion in DVA gains that J.P. Morgan Chase and Citigroup each recorded for the third quarter. Bank of America reported $1.7 billion of DVA gains in its investment bank. Analysts estimated that Morgan Stanley will record $1.5 billion of net DVA gains when it reports earnings on Wednesday"

So it is a sizeable effect. Now you think this is not relevant? Let me ask you a question then. How could DVA be hedged? Because, you will agree with me, a price that is not backed by a hedging strategy is hard to implement and will hardly prevail over time. To make DVA real, one should have a hedging strategy, i.e. one should sell protection on oneself, a very difficult feat, unless one buys back bonds that he had issued earlier. This may be hard to implement.

SAGE. Come on, proxy hedging is not hard. Most times DVA is hedged by proxying. Instead of selling protection on oneself, one sells protection on a number of names that are highly correlated to oneself.

SIMEONE. I heard of this. The WSJ article reported:

"[...] Goldman Sachs CFO David Viniar said Tuesday that the company attempts to hedge [DVA] using a basket of different financials. A Goldman spokesman confirmed that the company did this by selling CDS on a range of financial firms. [...] Goldman wouldn't say what specific financials were in the basket, but Viniar confirmed [...] that the basket contained "a peer group". Most would consider peers to Goldman to be other large banks with big investment-banking divisions, including Morgan Stanley, J.P. Morgan Chase, Bank of America, Citigroup and others. The performance of these companies' bonds would be highly correlated to Goldman's".

Now let me object to this. Proxying can be misleading. It can approximately hedge the spread risk of DVA, assuming the spread correlation is strong, but not the jump to default risk. Morgan hedging DVA risk by selling protection on Lehman would not have been a good idea. In fact this can worsen systemic risk when you look at jump-to-default risk. If I sell protection on a firm that is correlated to me and then that firm not only has its credit quality worsen (which would hedge my DVA changes due to spread movements) but actually defaults, then I have to make the protection payments and that could push me into default. Do you think this is an unrealistic scenario?

SAGE. Protection proxying is not traded on a single name, but on a basket of names. So there is a diversification.

SALVA. Again let me step in. There is a contradiction here. On one hand you want the names to be as much correlated as possible with yourself, or the hedge won't be effective. On the other hand, to avoid systemic risk, you hope to lower such correlations through diversification.

SAGE. Well I am not saying that diversification lowers all correlation, but only systemic risk correlation.

SALVA. And what other correlation is there? Are you thinking about a factor structure? I am sceptical. Think about the eight credit events that happened in one month of 2008 to Financials.

SIMEONE. That's right, my sentiments exactly.

SAGE. All right, I will concede again that it is probably not as simple as I think. I have not considered the big picture carefully, I need to think about it more.

SALVA. That would be a first. [smiles, orders a bottle of water, looks at Sage a little mysteriously] Now do not feel alone in this. Regulators are having a very hard time in deciding what to do with DVA. [leans back, adjusts skirt] Simeone, why don't you tell us what is happening with DVA in the regulatory space?

SIMEONE. [didactically] Basel III recognizes CVA risk but does not recognize DVA risk. This is creating a gap between CVA calculations for capital adequacy and CVA calculations for accounting and mark to market. For example (I can be a bit of librarian too):

"This CVA loss is calculated without taking into account any offsetting debt valuation adjustments which have been deducted from capital under paragraph 75". (Basel III, page 37, July 2011 release)

"Because nonperformance risk (the risk that the obligation will not be fulfilled) includes the reporting entity's credit risk, the reporting entity should consider the effect of its credit risk (credit standing) on the fair value of the liability in all periods in which the liability is measured at fair value under other accounting pronouncements". (FAS 157)

SAGE. [rolls his shirt sleeves] I see that regulators have clear ideas.

SIMEONE. [ignoring Sage, pinching chin] And here is what the former president of the Basel Committee said:

"The potential for perverse incentives resulting from profit being linked to decreasing creditworthiness means capital requirements cannot recognize it, says Stefan Walter, secretary-general of the Basel Committee: The main reason for not recognizing DVA as an offset is that it would be inconsistent with the overarching supervisory prudence principle under which we do not give credit for increases in regulatory capital arising from a deterioration in the firm's own credit quality".

SALVA. See? You are not alone in your DVA perplexity. [crosses legs, adjusts skirt]

SIMEONE. [frowning, joining hands] Another problem where we are trying to obtain some clarity from ISDA is closeout.

SALVA. Yes, that is interesting. I worked on that a little, so let me summarize. [smiles at both] Closeout is what happens when the first of the two parties in the deal defaults. So suppose in our example the counterparty defaults. Closeout proceedings are then started according to regulations and ISDA[6] documentation. The closeout procedure fixes the residual value of the contract to us and establishes how much of that is going to be paid to us. If, however, it is negative then we will have to pay the whole amount to the counterparty. So far this seems the standard definition we have seen above. But think carefully: at the default time of the counterparty, do we value the remaining contract by taking into account our own residual credit risk (namely by including our now unilateral DVA, "replacement closeout") or just by using a default risk-free valuation ("risk-free closeout")? The replacement closeout maintains that if we are now going to re-open the deal with a risk-free party, the risk-free party will charge us a unilateral CVA, which, seen from our point of view, is our unilateral DVA. Hence in computing the replacement value we should include our DVA to avoid discontinuity in valuation. If we always included DVA to value the deal prior to default, we should not stop doing so at default. So we should use replacement closeout. On the other hand, since we are liquidating the position now, why bother about residual credit risk? So we should use risk-free closeout.

SAGE. [looking at Salva admiringly] Interesting, does the valuation change strongly?

SALVA. In Brigo and Morini (2011), Brigo and Morini (2010a) and Brigo and Morini (2010b) it is shown that a risk-free closeout has implications that are very different from what we are used to expecting in case of a default in standardized markets such as the bond or loan markets. Let us take a case of TVA where the valuation is always in the same direction, e.g. a loan or a bond. Suppose we own the bond that we bought at time 0

---

[6]International Swaps and Derivatives Association.

from the counterparty. We are waiting for the payment of the final notional and we paid in the beginning, so we are like a lender in a loan. If we default, i.e. default of the lender, this should imply no losses to the bond issuer or to the borrower (the counterparty in our example). Instead, if the risk-free default closeout is adopted, if we default then the value of the liability of the counterparty will suddenly jump up. In fact, before the default, the liability of the counterparty (net debtor) had a mark-to-market that took into account the risk of default of the counterparty itself. In other words, it had a DVA term as seen from the counterparty. As a negative cash flow, it was smaller, in absolute value, than if DVA had not been there. After our default, if a risk-free closeout applies, this mark-to-market transforms into a risk-free one, surely larger in absolute value than the pre-default mark-to-market, because DVA is no longer there. The increase will be larger the larger the credit spread. This is a dramatic surprise for the counterparty that will soon have to pay this increased amount – no longer offset by DVA – to our liquidators. This effect does not exist in the bond or loan market. Clearly, net debtors at default (the counterparty in our example) will not like a risk-free closeout. They will prefer a replacement closeout, which does not imply a necessary increase of the liabilities since it continues taking into account the credit-worthiness of the debtor also after the default of the creditor.

SAGE. [whistles] I hadn't thought about this. But it does make sense...

SALVA. One could then decide to use replacement closeout. However, the replacement closeout has shortcomings opposite to those of the risk-free closeout. The risk-free closeout will be preferred by the creditors. The more money debtors pay, the higher the recovery will be. The replacement closeout, while protecting debtors, can in some situations worryingly penalize the creditors by abating the recovery. This for example happens when the defaulted entity is a company with high systemic impact, so that when it defaults the credit spreads of its counterparties are expected to jump high.

SAGE. It seems unbelievable that no clear regulation was available for this issue. What are you waiting for, Simeone??

SIMEONE. There are other issues, it's not over yet with the list of problems. Closeout is the least of our worries.

SALVA. So what are the other issues that are keeping you regulators busy?

SIMEONE. Collateral modeling, especially rehypothecation. Capital requirements on CVA for Basel III. Deciding what types of CVA restructuring we should allow. Consistent inclusion of funding costs. Proper conservative accounting of Gap Risk. Central Clearing. But I think we covered enough for today. I have to say that I enjoyed it more than I had thought. Same time tomorrow?

SALVA. [stands up, straightens skirt, picks up purse] Sure.

SAGE. You have been wonderful, Salva [kisses his cheeks]

SIMEONE. Salva, it must have been similar to kissing an ahstray [kisses Salva's cheeks in turn, shakes hands with Sage]

SAGE. Ahahah very funny. See you tomorrow, guys.

1.4 The Third Day

1.4.1 FVA, Hard Maths with No Data? CVA VaR, Basel III Problems, Collateral and Gap Risk

SAGE. Hi Simeone, let's sit outside. This year we are very lucky with the weather.

SIMEONE. Hi, Sage. Indeed we have wonderful weather this year. There, Salva is joining us.

SALVA. Hi Boys. How are you today? Why, what is this bouquet of roses on the table? [smiling, amused]

SAGE. I don't know. It was there. Look, there is a tag.

SALVA. "To the beautiful charming and intelligent lady at the outside table, from a secret admirer". [bows head slightly, looks penetratingly at Sage and then at Simeone] Mmmmm. [smiles] Anyway, my turn to buy drinks. Perhaps I"ll find a secret admirer inside the pub. What would you like? [...]

SALVA. So how did the day go?

SAGE. More and more meetings. I am spending days meeting people who are mostly bureaucrats. Not really exciting. I am itching for action.

SIMEONE. It's funny that despite being in quite different institutions, we end up doing very similar activities. I have been doing that too.

SALVA. That's three of us then. Today I beat the record of "useless meetings". [laughs musically] All right, gentlemen, let us restart from where we ended up yesterday. [pensively runs left hand along long hair] We covered a lot of issues in the first two days and I believe the brainstorming was quite helpful. [leans forward, counting on fingers] We tried to clarify the difference between risk measures like Credit VaR and pricing quantities like CVA, we tried to explain $\mathbb{P}$ and $\mathbb{Q}$, we tried to reason on CVA and DVA hedging, on the conflicting regulations on DVA and then we talked about closeout, analyzing a little the risk-free vs replication closeouts and we talked about first-to-default risk. We also commented on the difficulties of finding data for the calculation of CVA.

SIMEONE. We have an interesting picture emerging from this debate. CVA and DVA seem to be so full of choices. I heard this interesting comment at an industry panel recently: five banks may be computing CVA in 15 different ways. And it is no surprise: from all the choices you listed, one is not even sure about which payout should be implemented for CVA, let alone the models. So how can a bank price it while being sure that other banks price it consistently? This is an issue for regulation.

SAGE. [rolls shirt sleeves] It does not matter. I simply price it and charge the other desks. If they are not happy, we may discuss. When the market opens, they may be buying CVA protection somewhere else through CCDS or similar structures, and then it will be simply an offer and demand process.

SIMEONE. [pinching chin] What if bank B manages to sell protection to bank A, not because it is doing a better job than bank C, but because it is pricing CVA with a sloppy payout and thus is obtaining a lower price, but based on assumptions that do not reflect reality? We already commented on the article by Watt (2011), that implicitly seems to suggest this may happen.

SAGE. [waving hand] Look, if a bank is sloppy in its procedures, this is that bank's problem, and not a problem with CVA.

SIMEONE. [raising hand] But the problem stems precisely from doubts about the CVA definition and conventions. So it is not entirely the bank's responsibility.

SALVA. [leaning forward] Simeone, I think this is what we expect from regulators to some extent. ISDA in particular is working on this, although I must say I have not been impressed with their work in the CDS "Big Bang" story of Beumee, Brigo, Schiemert, and Stoyle (2010). But let's not digress.

SIMEONE. ISDA is not a regulator, strictly speaking. It is basically a trade organization for the over-the-counter derivatives market participants. It is not something like the BIS,[7] for example.

SALVA. You are right, I was using the term regulator in a loose sense. Anyway I think that regulation is quite tricky. Especially because it would require the best people, a little

[7]BIS stands for the Bank of International Settlements.

like politics and police work. However, compensation levels often drive the most energetic and talented professionals to the industry rather than regulators. Present company excepted, of course.

SAGE. [lights a cigar, laughs] Perhaps I can convince you to work for us, Simeone. If you find ways the bank can reduce capital requirements I think you will get very large bonuses and you can stop driving around that old Toyota. [winks]

SIMEONE. [folding arms] Toyota is a great car, it never let me down in fifteen years. Unlike your top cars, I hear.

SALVA. Ok, leave toys alone. Let us talk about funding costs modeling, which is another very controversial topic in the industry.

SIMEONE. That was the subject of today's "useless meeting", as you say.

SALVA. [raising a calming hand] Yes, that is a quite popular subject. If you attend a practitioner conference, a lot of talks will be on consistent inclusion of funding costs, mostly through very patched solutions. Only a few works try to build a consistent picture where funding costs are included together with the aspects we discussed earlier, like CVA, DVA, collateral, closeout, etc. For example, the working paper Crépey (2011), then published in Crépey (2012a) and Crépey (2012b), is a comprehensive treatment. The only limitation is that it does not allow for underlying credit instruments in the portfolio and has possible issues with FX, but this has now been fixed in Crépey and Song (2014). Crépey (2011,2012a, 2012b, 2014) are technical papers based on a mathematical instrument called a backward stochastic differential equation (BSDE). A related framework that includes most recent literature as a special case is in Pallavicini, Perini, and Brigo (2011, 2012), who however does not resort to BSDEs explicitly and starts from different assumptions on the risk-free rate and the risk neutral measure. Another good example of related work studying the funding issue is Bielecki and Rutkowski (2013), where the issue is studied in the martingale pricing and hedging framework, also referring to BSDEs when needed. Earlier works are partial but still quite important.

SAGE. [puffs] Yaaawn, prof skip intro. Let's roll, like yesterday. [sips wine]

SIMEONE. [annoyed, unfolding arms, palms on the table] I am quite interested in the literature review, if you do not mind.

SALVA. [waving hand dismissively towards Sage] The industry paper Piterbarg (2010) considers the problem of replication of derivative transactions under collateralization but without default risk, and in a purely classical Black-Scholes framework, and then considers two relatively basic special cases. When you derive the first basic results that way, you have to be very careful about the way you formulate the self-financing condition.

That paper is very influential, but the self-financing condition was not well stated there. A similar problem regarding the self-financing condition happens in some other industry papers. Luckily, this does not seem to ruin the final result but when you write papers on funding you have to pay particular attention to the self-financing condition. This is just to give you an idea of the difficulties in this area. Even top quants cannot have everything 100% right.

SAGE. [puffs] Fascinating, but let us leave aside the sociology of the practitioner's industry awards, technical communications and interaction with academia. What is the point of collateral without default risk, as you mentioned in the reference above? What is collateral used for then?

SALVA. [smiling wryly, looks at Sage sideways] If you let me develop a discourse I can try to be more comprehensive. There is not only default risk in collateral, there is also liquidity risk, transaction costs... and you might still be modeling credit spread risk but not jump to default risk, a distinction that is natural in intensity models. This involves filtrations, by the way, but I won't scare you with such technicalities.

SAGE. [puffs] Oh please, scare me. I am at your mercy, my wonderful femme fatale. [drinks]

SALVA. [ignoring Sage] However, it is true that the main reason to have collateral is default risk. Otherwise one would not have collateral in the first place. And it is not easy to account for all such features together, which may explain the hand-waving of default risk in papers like Piterbarg (2010) and the claim that including default risk would be easy. The fact that it is not can be easily grasped by looking at Crépey (2011), then published in Crépey (2012a, 2012b), for example. Following Piterbarg (2010), the fundamental funding implications in presence of default risk have been considered, in simple settings, first in Morini and Prampolini (2011); see also Castagna (2011). These initial works focus on zero coupon bonds or loans. One important point in Morini and Prampolini (2011) is that in simple payoffs such as bonds or loans, DVA has to be interpreted as funding, in order to avoid double counting. However, this result is not general and does not extend to more general payoffs. In the general case different aspects interact in a more complex way and the general approach of Crépey (2011,2012a, 2012b) or Pallavicini, Perini, and Brigo (2011, 2012) is needed. To complete a first basic literature review, the paper by Fujii and Takahashi (2011a) analyzes implications of currency risk for collateral modeling, while the above-mentioned, Burgard and Kjaer (2011b) resorts to a PDE approach to funding costs. As I mentioned above, Crépey (2011,2012a, 2012b) or Pallavicini, Perini, and Brigo (2011, 2012) and Bielecki and Rutkowski (2013) remain the most general treatments of funding costs to date. Then...

SAGE. [resting cigar on ashtray rim, raising his glass] Ok, I don't care who is interested in this, I have had enough of references and bla bla bla. Can you summarize the funding problem difficulties? To me, funding seems to reduce to discounting, with different curves depending on the party we are dealing with and I don't get why it should be such a formidable problem. Explain please, but in plain English. Or I will simply look at you, which is a pleasant activity, but without listening to what you are saying.

SALVA. [crossing legs, adjusting skirt, concealing irritation behind a wry smile] All right, I''ll humor you and pretend to believe you are not interested in modeling details. In a nutshell, when one needs to manage a trading position, one has to obtain funds in order to hedge the position, post collateral, pay coupons and swaps resets and so on. These funds may be obtained from one's treasury department or in the market. One may also receive funds, not from the treasury, but as a consequence of being in the position: a coupon, a notional reimbursement, a positive mark-to-market move, getting some collateral back, a closeout payment, etc. All flows need to be remunerated: if one is borrowing funds, this will have a cost and if one is lending funds, this will provide revenues in form of interest. Including funding costs into valuation means properly accounting for such features.

SAGE. [sipping wine] Which, as I said, can be done through discounting with different curves. So what is difficult here?

SALVA. [sighing] The point is doing this consistent with all other aspects, especially with counterparty risk. I think you heard of this "funding valuation adjustment", or FVA, that would be additive, so that the total price of the netting set portfolio would be something like

MtM - DVA - CVA - FVA = MtM - TVA.

However, this is too much to expect. Proper inclusion of funding costs leads to an implicit (or recursive if you want, as in a fixed-point equation) pricing problem. This may be formulated as a backward stochastic differential equation (BSDE, as in Crépey (2011,2012a, 2012b) or Bielecki and Rutkowski (2013)), or to a discrete time backward induction equation or nonlinear PDE (as in Pallavicini, Perini, and Brigo (2011, 2012)). Moreover, there is some overlap (or double counting) between DVA and FVA. One possible way out of double counting, used in Crépey (2012b), is to replace the final formula FVA term above by LVA, a term for the funding liquidity valuation adjustment (and there is also a last term, dubbed RC, for replacement cost, reflecting the mismatch between the value of the contract right before the default and its valuation by the liquidator in the closeout procedure). This is a possible choice but is not the only one. For example, Pallavicini, Perini, and Brigo (2012) renounce separating the adjustment's different quantities beacuse they are deemed to be intrinsically non-separable. [sees Sage stirring, raises a flat hand]

I am getting there, be patient. I saw you flash your eyes when I mentioned recursion. I'm lucky you don't have heat vision. Why is the problem inherently recursive? The value of the cash and collateral processes depends on the price and on the future cash flows of the derivative which, in turn, depends on the choices for such processes, transforming the pricing equation into a recursive (and nonlinear) equation! Thus, funding and investing costs cannot be considered as a simple additive term to a price obtained by disregarding them. Importantly, identifying DVA with funding is wrong in general, except that in the simple cases in Morini and Prampolini (2011), but see also the discussion in Castagna (2011).

SAGE. [taking cigar] And you said above that this is done well with BSDEs?

SIMEONE. Indeed. I was wondering about that too. Salva, you know we are not as advanced as you are mathematically. Could you give us a summary on what BSDEs are?

SALVA. [smiling, looks at Sage sideways, then looks at Simeone] Since you are asking so nicely. [sips wine] You know that a pricing problem is usually represented either as an expected value or as an (integro) partial differential equation for the price...

SAGE. The two representations being connected by the Feynman-Kac theorem, we know.

SALVA. ...for simple pricing problems, the pricing can be expressed like a straight expectation, in which case we could roughly say that the pricing problem is linear. This is not the case for American put options, for example, where early exercise introduces a nonlinearity.

SIMEONE. Indeed, I remember this from my master's days.

SALVA. However, if you have a kind of recursive relationship similar to what we get with the funding problem, or if you have both early exercise and path dependency, you need to combine backward induction with forward simulation. If the size of the portfolio is very large, as is typical with CVA, you will be suffering with the curse of dimensionality if you use PDEs. The alternative is to use BSDEs. These equations need a terminal condition and then allow one to implement a sort of backward induction similar to what you see in least squared Monte Carlo for early exercise options. Actually, Brigo and Pallavicini (2007) first applied this idea to CVA, but in a case where the pricing problem is still linear. This idea, now called American Monte Carlo, is at the centre of the CVA system built by UBS and promoted in the book by Cesari, Aquilina, Charpillon, Filipovic, Lee, and Manda (2010). The general idea, implemented more generally through BSDEs, goes like this: basically, at every backwards induction step, on the forward simulated trajectories of the underlying assets, one uses a multilinear regression to estimate the value of the deal one time step earlier based on the underlying at the next time and on the price scenarios at the next time, which is known since we are going back in time. Because this is based on simulation and regression, it is also suited to high-dimensional situations, at least more so than the PDE approach. And if the problem is nonlinear one cannot trivially translate the PDE formulation into an expectation using the classical Feynman-Kac theorem, so that the use of BSDEs becomes relevant. On these technical issues, in relation to CVA, I recommend the recent book by Crépey (2013).

SAGE. This may sound convincing to a quant audience, but I doubt the banks will be willing to implement BSDEs and I also doubt the regulators will prescribe that. We need something simple coming out of this.

SIMEONE. As a regulator I think this last comment of yours is very important. It raises the question: if we realize that a project requires mathematical or methodological tools that are too advanced for the times, will we give up the related project and activities or should we keep it going being content with using completely inadequate methodologies?

SAGE. [puffs] I hope you are not suggesting that I should stop my business because I don't want to implement BSDEs. Do you realize how ridiculous that is?

SIMEONE. It's annoying for you, maybe, but not ridiculous. What if the pharmaceutical companies reasoned like you?

SAGE. Who says they don't? [laughs]

SALVA. Look, I am not saying you necessarily need to implement a BSDE, but definitely you need to start thinking along those lines and seriously consider the above recursion.

SIMEONE. I still think that, with uncertainty on the recoveries and occasionally on probabilities of default, volatilities and correlations that are very large, insisting on a fully rigorous and precise mathematical treatment of second order effects could be pointless. Is this right?

SALVA. I don't think so. The fact that you can't estimate recovery precisely does not mean you should give up properly understanding the remaining aspects. And in fact a mistake in the implementation of funding can cause quite large differences. Of course, if you are risk measuring CVA, then maybe you can accept being less precise than in pricing. However I think that Basel III is really simplifying too much.

SIMEONE. CVA VaR seems to mix everything together. Could you summarize CVA VaR for me?

SALVA. Let us say that Credit VaR measures the risk of losses you face due to the possible default of some counterparties you are doing business with. CVA measures the pricing component of this risk, i.e. the adjustment to the price of a product due to this risk. Fine. But now suppose that you revalue and mark to market CVA in time. Suppose that CVA moves in time and moves against you, so that you have to book negative losses, NOT because the counterparty actually defaults, but because the pricing of this risk has changed for the worse for you. So in this sense you are being affected by CVA volatility and you are booking CVA mark to market losses. To quote Basel III, even if I still haven't found out how they computed this 2/3 number:

"Under Basel II, the risk of counterparty default and credit migration risk were addressed but mark-to-market losses due to credit valuation adjustments (CVA) were not. During the financial crisis, however, roughly two-thirds of losses attributed to counterparty risk were due to CVA losses and only about one-third were due to actual defaults."

In other words, the variability of the price of this risk over time has done more damage than the risk itself. This is why Basel is considering setting up quite severe capital charges against CVA.

SIMEONE. That is a good summary. [looks at Salva approvingly] To measure the risk in CVA mark to market, we have to compute VaR for CVA. Can you summarize how this would be computed?

SALVA. You simulate basic market variables and risk factors under $\mathbb{P}$, up to the risk horizon. Then, in each scenario at the horizon, you price the residual CVA until the final maturity, using a $\mathbb{Q}$ expectation (but in a full approach this may be a recursion, as we commented above). You put all the prices at the horizon time together in a histogram and obtain a profit and loss distribution for CVA at the risk horizon. On this $\mathbb{P}$ distribution you select a quantile at the chosen confidence level. Now you will have computed VaR of CVA. If you take the tail expectation of this histogram, you have expected shortfall of CVA. I stress again that this does not measure default risk directly. It measures the risk to have a mark to market loss due either to default or to adverse CVA change in value over time.

SAGE. [puffs] Let's get to the juicy part. What is Basel III prescribing for this CVA VaR?

SALVA. Well the framework has changed several times: bond equivalent formula, multipliers, tables... One of the main issues has to do with (WWR). In some part of the Basel regulation it had been argued that you could calculate CVA as if there were no wrong-way risk and then use a standard multiplier/table/adjuster to account for wrong-way risk. In other words, you should assume independence between default of the counterparty and of the underlying portfolio, compute CVA and then multiply by a certain number to account for correlation risk, or something along those lines. However, this does not work. Depending on the specific dynamics of the underlying financial variables, on volatilities and correlations and on the chosen models, the hypothetical multipliers are very volatile. See again Brigo, Morini, and Tarenghi (2011), Brigo and Pallavicini (2007), Brigo, Chourdakis, and Bakkar (2008) and Brigo and Chourdakis (2008) for examples from several asset classes. Even if one were to use this idea only for setting capital requirements of well diversified portfolios, this could lead to bitter surprises in situations of systemic risk. [sits back, uncrosses and stretches legs, finishes glass of wine, orders a soft drink]

SAGE. Looks like a very intensive numerical procedure. Pricing CVA in each risk horizon scenario would require a simulation, so you have subpaths. And, if the portfolio PV is not known in closed form, even the residual NPV at default inside the CVA term requires a simulation, so you would need three levels of simulations, impossible even for current technology. Suppose you need $10000 = 10^4$ scenarios at every step. Then the total number of paths would be $10000^3 = 10^{12}$, one trillion scenarios. So it is clear you can't do that by brute force (yet).

SALVA. Very perceptive. [smiles]

SIMEONE. So the only way out of capital depleting CVA VaR charges is collateral?

SALVA. [straightening skirt] Well, as I said earlier collateral does not completely kill CVA, this is folklore. Gap risk is real. Gap risk shows up when default happens between margining dates, and a relevant negative mark to market move happens very quickly. Brigo, Capponi, and Pallavicini (2014) show a case of an underlying CDS with strong default contagion where even frequent margining in collateralization is quite ineffective. But it is recognized that, in cases without strong default contagion, collateral can be quite effective.

SAGE. [drops cigar on ashtray] You know, this has been very educational but more tiring than I thought. Could we stop here for today and perhaps conclude tomorrow?

SALVA. Fine by me. Today I am the one who did most of the talking so I would stop gladly. [smiles]

SIMEONE. I am tired as well.
[usual exchange of handshakes and kisses]

1.5 The Fourth Day

1.5.1 Counterparty Risk Restructuring. CCDS, Papillon, Floating Rate CVA and Margin Lending. Global Calibration. Global Valuation. Available CVA Books and Forthcoming CVA Books.

SIMEONE. My first question today is this. From what we said yesterday, Basel III may end up imposing heavy capital requirements for CVA. Collateralization is a possible way out, but it may become expensive for some firms and it may lead to a liquidity strain, while firms that are not organized for posting collateral will not be able to resort to collateral in the first place. The article by Watt (2011) reports the case of Lufthansa:

> "The airline's Cologne-based head of finance, Roland Kern, expects its earnings to become more volatile "not because of unpredictable passenger numbers, interest rates or jet fuel prices, but because it does not post collateral in its derivatives transactions."

For banks, the choice is whether to collateralize everything or to be subject to heavy CVA capital requirements. Is there a third way? I read about some attempts to restructure counterparty risk differently.

SAGE. I can tell you... There have been proposals for instruments that can hedge CVA, or reduce its capital requirements in principle. Contingent credit default swap (CCDS) anyone?

SIMEONE. CCDS, yes, recently standardized by ISDA after a long time.

SALVA. Yes I remember CCDS... a CCDS is similar to a CDS, but when the reference credit defaults, the protection seller pays protection on a notional that is not fixed but is rather given by the factor (1 - recovery) fraction of the residual value of the Portfolio netting set at that time, if positive. This matches exactly the CVA on the given portfolio if the refence credit of the CCDS is the counterparty of a deal. Back in 2008 the Financial Times commented:

> "[...] Rudimentary and idiosyncratic versions of these so-called CCDS have existed for five years, but they have been rarely traded due to high costs, low liquidity and limited scope. [...] Counterparty risk has become a particular concern in the markets for interest rate, currency and commodity swaps - because these trades are not always backed by collateral. [...] Many of these institutions - such as hedge funds and companies that do not issue debt - are beyond the scope of cheaper and more liquid hedging tools such as normal CDS. The new CCDS was developed to target these institutions (Financial Times, April 10, 2008)."

Interest in CCDS came back in 2011-2012 as CVA capital charges were becoming punitive. ISDA has launched CCDS on index portfolios, i.e. standardized portfolios that are thought to be representative of the typical netting sets components banks are facing. This is an attempt to increase standardization. However, CCDS do not fully solve the problem of CVA capital requirements. First of all, there is no default-free bank from which the CCDS protection can be bought, so that CCDS are themselves subject to counterparty risk. Second, I doubt the ISDA standard portfolios will match any realistic bank portfolio. So there will be only a partial hedge that ignores cross-asset classes correlation and especially the structure of realistic netting sets. Better than nothing, but very partial.

SAGE. Yeah, and this is prompting the industry to look for other solutions that may also be effective across several counterparties at the same time and on complex netting sets.

SIMEONE. CVA securitization could be considered, although the word "securitization" is not very popular these days.

SALVA. Is there any proposed form of CVA restructuring, or securitization?

SAGE. [concentrating, looking tired] I need a drink, but probably a coffee. There are a few structures. I am familiar with a few deals that I cannot disclose. However, there have also been deals discussed in the press and in the Financial Times Alphaville blog, in particular Pollack (2012a):

"In short, Barclays has taken a pool of loans and securitised TO: securitized them, but retained all but the riskiest piece. On that riskiest Euro 300m, Barclays has bought protection from an outside investor, e.g. hedge fund. That investor will get paid coupons over time for their trouble, but will also be hit with any losses on the loans, up to the total amount of their investment. To ensure that the investor can actually absorb these losses, collateral is posted with Barclays."

SALVA. Looks like a CDO?

SAGE. Yes, and collateral is key here. The blog continues:

"This point about collateral means that, at least in theory, Barclays is not exposed to the counterparty risk of the hedge fund. This is especially important because the hedge fund is outside the normal sphere of regulation, i.e. they aren't required to hold capital against risk-weighted assets in the way banks are. [...] And then there is the over-engineering element whereby some deals were and maybe still are, done where the premiums paid over time to the hedge fund are actually equal to or above the expected loss of the transaction. That the Fed and Basel Committee were concerned enough to issue guidance on this is noteworthy. It'll be down to individual national regulators to prevent "over-engineering" and some regulators are more hands-on than others."

SIMEONE. Wasn't another deal, called "Score", discussed as well?

SAGE. Yes, but it didn't work well, Pollack (2012c):

"RBS had a good go at securitising these exposures, but the deal didn't quite make it over the line. However, Euroweek reports that banks are still looking into it:

"Royal Bank of Scotland's securitisation of counterparty risk, dubbed Score 2011, was pulled earlier this year, but other banks are said to be undeterred by the difficulties of the asset class and are still looking at the market. However, other hedging options for counterparty risk may have dulled the economics of securitising this risk since the end of last year". "

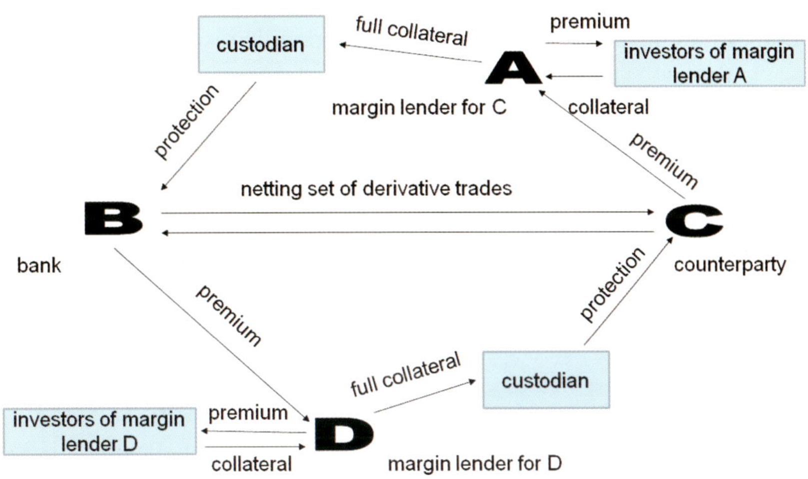

FIGURE 1.1: General counterparty scheme, including quadri-partite structure.

SIMEONE. The most effective one was probably the Credit Suisse one, Pollack (2012b):

"Last week Credit Suisse announced it had bought protection on the senior slice of its unusual employee compensation plan. The Swiss bank pays some of its senior bankers using a bond referencing counterparty risk, which also involves shifting some counterparty risk from the bank to its workers".

That is like buying protection from your own employees. Interesting concept if you think about it from a regulator's point of view. That way the employee, in theory, is incentivized in improving the risk profile of the company.

SAGE. [getting mad] You are kidding, right? I wouldn't be too happy if I were paid that kind of bonus. My decision doesn't have the capacity to affect the large scale performances of the group. That is more for the CEO and company. So let the CEO bonus be like that, but not my own.

SIMEONE. For your benefit, there are actually more innovative ideas. On CVA securitization see for example Albanese, Bellaj, Gimonet, and Pietronero (2011), that advocates a global valuation model. The more model–agnostic Albanese, Brigo, and Oertel (2013) explains how margin lending through quadri-partite or penta-partite structures involving clearing houses would be effective in establishing a third way. Let me borrow from Albanese, Bellaj, Gimonet, and Pietronero (2011) and Albanese, Brigo, and Oertel (2013), to which I refer for the full details. If I understood correctly, the structure is like [Fig. 1.1].

Traditionally, CVA is charged by the structuring bank B either on an upfront basis or it is built into the structure as a fixed coupon stream. This other Score deal we discussed above (there is also Papillon) is probably of this type too. Margin lending instead is based on the notion of floating rate CVA payments with periodic resets.

SALVA. Let me guess what Floating rate CVA is: assume for simplicity a bi-partite transaction between the default-free bank B and the defaultable counterparty C. Instead of charging CVA upfront at time 0 for the whole maturity of the portfolio, the bank may require a CVA payment at time 0 for protection on the exposure up to 6 months. Then in 6 months the bank will require a CVA payment for protection for further six months on what will be the exposure up to one year and on and on, up to the final maturity of the deal. Such a CVA would be an example of Floating Rate CVA.

SAGE. Yeah, makes sense.

SIMEONE. Good. I was saying that margin lending is based on the notion of floating rate CVA payments with periodic resets, and is designed in such a way so as to transfer the conditional credit spread volatility risk and the mark-to-market volatility risk from the bank to the counterparties. We may explain this more in detail by following the arrows in Fig. 1.1. The counterparty C has problems with posting collateral periodically in order to trade derivatives with bank B. To avoid posting collateral, C enters into a margin lending transaction. C pays periodically a floating CVA to the margin lender A, which the margin lender A pays to investors. This latest payment can have a seniority structure similar to that of a cash CDO. [raises hand] Wait. In exchange for this premium, for six months the investors provide the margin lender A with daily collateral posting and A passes the collateral to a custodian. This way, if C defaults within the semi-annual period, the collateral is paid to B to provide protection and the loss in taken by the investors who provided the collateral. At the end of the six months period, the margin lender may decide whether to continue with the deal or to back off. With this mechanism C is bearing the CVA volatility risk, whereas B is not exposed to CVA volatility risk, which is the opposite of what happens with traditional upfront CVA charges.

SAGE. The idea is interesting, but I can see a number of problems. First, proper valuation and hedging of this to the investors who are providing collateral to the lender is going to be tough. I recall there is no satisfactory standard for even simple synthetic CDO tranches. One would need an improved methodology.

SALVA. Actually, I heard that recently a paper on dynamic copulas and simultaneous defaults has been proposed that solves the CDO bottom-up / top-down dilemma by displaying both a realistic aggregated loss dynamics (good calibration of CDO tranches) and single names information. See Bielecki, Cousin, Crépey, and Herbertsson (2013a), who generalize earlier works such as Brigo, Pallavicini, and Torresetti (2007) (see also the summary in Brigo, Pallavicini, and Torresetti (2010)).

SAGE. But still…I am sure it remains a formidable problem. And the other problem that, I am sure, has not escaped Simeone is: what if all margin lenders pull out at some point due to a systemic crisis?

SIMEONE. That would be a problem. Albanese, Brigo, and Oertel (2013) submit that the market is less likely to arrive in such a situation in the first place if the wrong incentives to defaulting firms are stopped and an opposite structure, such as the one in Fig. 1.1, is implemented. I would also point out that this structure probably also helps with the funding implications.

SALVA. We haven't talked about CCPs and initial margins. You know, often people object that all these CVA and funding issues will disappear once a CCP is intermediating the deal.

SAGE. Well, as long as you are happy with the CCP knowing the details of your transactions, and you're also happy to pay the possibly daily initial margins, besides variation margins.

SIMEONE. Initial margins are supposed to protect you from wrong-way risk and gap risk. So they should be a good thing.

SAGE. Except that they are not passed to the client but are kept by the CCP itself. So what would become of your initial margin if the CCP defaulted? It happened in the past.

SALVA. You see, to analyze whether initial margins are fair or appropriate, you need exactly the type of credit and funding analytics that have been developed recently in papers like Crépey (2011,2012a, 2012b) and Pallavicini, Perini, and Brigo (2011, 2012).

SAGE. I think we have reached the border of this new land. We need to celebrate. Hey maestro, Prosecco please!

SIMEONE. I would ask Salva for suggestions for further reading while we wait for the bottle.

SAGE. You are really a spoilsport. Who cares about books now?

SALVA. [laughs] Why don't you drop the recital? You read more books than I do, the only difference is that you do it during the night (and I don't know how you manage such an energy level).

SAGE. Oh, by now I mostly meet people, shake hands and send email. I am a big boss now. [laughs]

SIMEONE. All right, all right. It should only take five minutes?

SALVA. Ok, no worries. So let us list the counterparty risk / CVA books.

- The first book we can mention is Pykhtin (2005). This is a collection of papers on CVA and is quite interesting, although it is now a little outdated.
- The book by Cesari, Aquilina, Charpillon, Filipovic, Lee, and Manda (2010) is more recent. From a modeling point of view it is rather basic but it also looks at the IT implications of building a CVA system. Relying on the so called American (or least square) Monte Carlo technique introduced for CVA computations in Brigo and Pallavicini (2007, 2008), it addresses a number of practical problems that deal with CVA for realistically large portfolios.
- The book by Gregory (2009, 2012) is technically quite simple but explains basic CVA concepts in a clear way and is good especially for managers and finance people who need to get a general grasp of CVA fundamentals with some elements about funding/discounting, without going too technical. It is quite popular and has been succesful.
- The book by Kenyon and Stamm (2012), although technically basic, is original in that it tackles current and relevant problems such as multi-curve modeling and credit valuation adjustments, with closeout and especially goodwill, which cannot be found anywhere else to the best of my knowledge. Funding costs, hints at systemic risk, regulation and Basel III are also considered.
- The book by Brigo, Morini, and Pallavicini (2013) is mostly based on Brigo's work with several co-authors during the period 2002-2012 and has been written to be widely accessible while being technically advanced. It has a very readable introductory part written in the form of a Platonic dialogue. The book deals with complex issues such as CVA and DVA, with several examples of advanced wrong-way risk calculations across asset classes. It also shows how to model gap risk, collateral in general, closeout, rehypothecation and funding costs. The funding part is based on Pallavicini, Perini, and Brigo (2011, 2012). Several numerical studies illustrate the fine structure of wrong-way risk. The final part of the book provides cutting edge research on CVA restructuring through CCDS, CDO tranches type structures, floating rate CVA and margin lending.
- The book by Crépey, Bielecki, and Brigo (2014) is probably the most advanced. It also has a very readable introductory part in the form of a long and entertaining Galileian dialogue... wait, I feel a strange sensation of being sucked into a different reality... into the book actually ... strange ... OK, I was saying that this is an advanced book dealing with CVA, DVA and funding, collateral and wrong-way risk. The funding part is based on Crépey (2011,2012a, 2012b) and is at the forefront of current research, similar to Pallavicini, Perini, and Brigo (2011, 2012) in terms of generality. By comparison with the book by Brigo, Morini, and Pallavicini (2013),

the emphasis is on the mathematical dependence structure of the problem, using mainstream stochastic analysis: BSDEs in particular, in line with Crépey (2013), to address the nonlinear recursive nature of the funding issue in a more systematic way, and dynamic copulas to reconcile bottom up and top down perspectives in credit models. These are the two puzzles referred to in the title, which make the book quite unique. By the way there is a possible connection between the two: dynamic copula models can also be analyzed in terms of BSDEs, as the final chapter illustrates (see also Crépey and Song (2014)). The book also contains a number of numerical studies.

These are the books I heard of, but there are probably others. But did you feel that strange pull I felt while speaking about Crépey, Bielecki, and Brigo (2014)?

SAGE. Actually yes, I felt a strange situation of being watched and then being embedded in a recursion, I cannot explain it...

SIMEONE. Oh you explained it even too well, I felt the same!!

SAGE. Oh, here comes the Prosecco. Great, let's drink so we can forget these strange feelings.

SIMEONE. But the label on the bottle says Taylor and Francis?? [thunder]

SAGE. What? [sky becomes suddenly dark]

SALVA. [eyes wide open, long hair in the wind] Oh my, you are right. Don't give me that. I am not part of a meta-narrative experiment. I have had enough of Grant Morrison, Jostein Gaarder and Jasper Fforde, and even of the final part of Brigo, Morini, and Pallavicini (2013). I am real, leave me alone.

SAGE. What? This can't be true...

Chapter 2

The Whys of the LOIS

In this chapter, we provide an economic analysis of the post-crisis multi-curve reality of financial markets, an important feature underlying several aspects of the studies carried out in this book.

The 2007 subprime crisis induced a persistent disharmony between the Libor derivative markets of different tenors and the OIS market. Commonly proposed explanations for the corresponding spreads refer to a combined effect of credit risk and liquidity risk. However, in the literature, the meaning of liquidity is often not stated precisely, or it is simply defined as a residual spread after removal of a credit component. In this chapter we develop a stylized indifference valuation model in which a Libor-OIS spread (named LOIS in Bloomberg) emerges as a consequence of:

- on the one hand, a credit component determined by the skew of the CDS curve of a representative Libor panelist (playing the role of the "borrower" in an interbank loan),
- on the other hand, a liquidity component corresponding to the volatility of the spread between the funding rate of a representative Libor panelist (playing the role of the "lender") and the overnight interbank rate.

The credit component is thus, in fact, a credit skew component, while the relevant notion of liquidity appears as the optionality of dynamically adjusting through time the amount of a rolling overnight loan, as opposed to lending a fixed amount up to the tenor horizon on Libor (as will be seen, this optionality is valued by the aforementioned volatility). In the "at-the-money" case, i.e. when the funding rate of the lender and the overnight interbank rate match on average, this results, under diffusive features, in a square root term structure of the LOIS, with a square root coefficient given by the above-mentioned volatility. Empirical observations reveal a square root term structure of the LOIS consistent with this theoretical analysis. Specifically, on the euro market considered in our study in the period from mid-2007 to mid-2012, LOIS is explained in an equal way by the credit and liquidity components through the beginning of 2009, and then dominantly explained by the liquidity component.

2.1 Financial Setup

The main reference rate for a variety of fixed income derivatives is the Libor in the USD market and the Euribor in the euro market. Libor (respectively Euribor) is computed daily as an average of the rates at which a designated panel of banks (respectively a prime bank) believe they can obtain unsecured funding, for periods of length up to one year. From now on we use the term Libor and the letter L to denote either of these two rates. Following the subprime triggered credit crunch, which severely impacted trust between financial institutions, overnight interest rate swaps (OIS) became more and more popular. In these financial instruments, the floating rate is obtained by compounding an overnight interbank rate (O/N) r_t, i.e. a rate at which overnight unsecured loans can be obtained in the interbank market (the Federal Funds rate in the USD fixed income market and the EONIA rate in the euro market). As a result, an OIS rate[1] R can be interpreted as a suitable average of r.

In theory, arbitrage relations imply that $L = R$. However, the interbank loan market has been severely impacted since the 2007 subprime crisis and the ensuing liquidity squeeze. The reference interbank rate remains the Libor, though, which still underlies most vanilla interest-rate derivatives like swaps, forward rate agreements (FRA), interest rate swaps (IRS), cap/floor and swaptions. The resulting situation where an underlying to financial derivatives has become in a sense arbitrarily fixed by a panel of key players in the derivatives market poses insider issues, as illustrated by the recent Libor manipulation affairs (Wheatley 2012). But foremost, it poses a crucial funding issue as, on the one hand, in parallel to the drying up of the interbank loan market, Libor got disconnected from OIS rates (see Fig. 2.1); while on the other hand, as more and more trades are collateralized, their effective funding rate is the corresponding collateral rate, which is typically indexed to the O/N r_t. This creates a situation where the price of an interest-rate product, even the simplest flow instrument such as a FRA, involves (at least) two curves, a Libor fixing curve and an OIS discount curve, as well as the related convexity adjustment, which in the case of optional products can be significant (cf. (Mercurio 2010b)). In view of the relations between counterparty risk and funding this also has some important CVA implications (see Part III).

Commonly advanced explanations for the Libor-OIS spread $(L - R)$, often called LOIS in the market, refer to combined effect of credit risk and/or liquidity risk. See (Bean 2007; Brunnermeier and Pedersen 2009; Smith 2010; Crépey, Grbac, and Nguyen 2012; Eisenschmidt and Tapking 2009; Filipović and Trolle 2013; Morini 2009). Nevertheless, in these explanations the meaning of liquidity is either not precisely stated or it is simply defined as a residue after removal of a credit component. In this chapter we propose a stylized economical model for evaluating the rate at which a bank would find it interesting to lend for a given tenor horizon, as opposed to rolling an overnight loan, which the bank

[1] See e.g. the formula (62) in Subsection 4.3 of (Crépey, Grbac, and Nguyen 2012) for an exact definition of an "OIS rate".

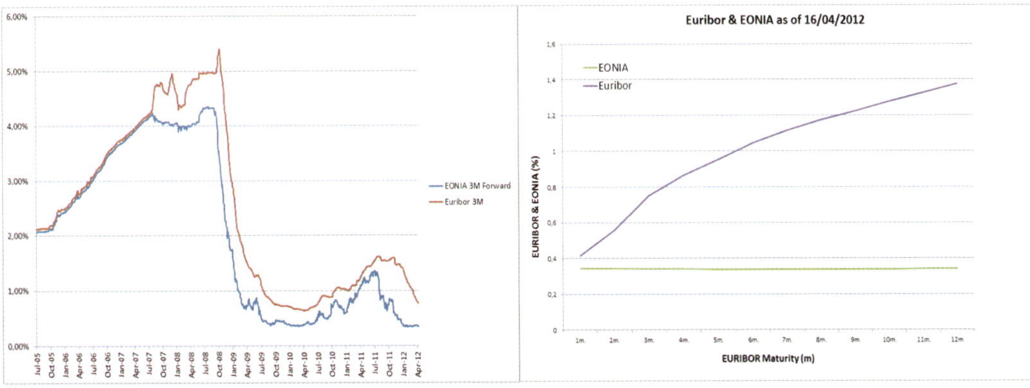

FIGURE 2.1: Divergence Euribor ("L") / EONIA-swap ("R") rates. Left: Sudden divergence between the 3m Euribor and the 3m EONIA-swap rate that occurred on Aug 6 2007. Right: Term structure of Euribor vs EONIA-swap rates, 16 April 2012.

can cancel at any moment. In this setup LOIS emerges as a consequence of the skew (as measured by λ_t below) of the credit curve of a representative Libor panelist (playing the role of the borrower in an interbank loan), and of the volatility of the spread $c_t = \alpha_t - r_t$ between the refinancing rate α_t of a representative Libor panelist (playing the role of the lender) and the O/N r_t.

We illustrate our study using 07/07/2005 to 16/04/2012 euro market Euribor/Eonia-swap Bloomberg data, covering both out-of-crisis and in-crisis data (2007-09 credit crisis and ongoing Eurozone sovereign debt crisis). Note that the euro market is even larger than the USD market; moreover, the Euribor and Eonia bank panels are the same, whereas the FF rate panel is larger than the USD Libor one. On related topics (Filipović and Trolle 2013), conducted empirical studies on both euro and USD markets and obtained very similar results in both cases.

2.2 Indifference Valuation Model

We assume that the funding rate (or refinancing rate, cost-of-capital, cost-of-liquidity...) of a lending bank with a short term debt D ("immediately callable debt" such as short term certificate deposits, say more broadly $\leq 1y$-liabilities) is given by a random function $\rho_t(D)$. In particular, denoting by D_t the time-t short term debt of the bank, $\alpha_t = \rho_t(D_t)$ is the annualized interest rate charged to the bank on the last euro that is borrowed for funding its loan (current refinancing or funding rate of the bank). Note that this time-t funding rate α_t is a complex output possibly impacted by, depending on the treasury management of the bank beyond its level of the time-t short term debt D_t (as compared to its immediately repositable capital C_t), various factors including the O/N r_t, the bank's CDS spread as well as other macro-economic global variables. More broadly, $\rho_t(D_t + x)$ is the annualized interest rate charged to the bank on the last euro of a total of €$(D_t + x)$ borrowed by the bank for funding its loan. Here the variable x represents the additional amount of capital borrowed by the bank to refund a loan (rolled overnight or on the Libor market). For liquidity reasons, $\rho_t(D)$ is increasing in the debt D (or $\rho_t(D_t + x)$ is increasing in x; see (2.4) for a linear specification), so that "the next euro borrowed costs more than the previous one". Since the marginal cost $\rho_t(D_t + x)$ needs to be integrated over x from 0 to N in order to give the global funding cost for the bank of lending a given global amount N (note that the bank needs to borrow money in the first place in order to fund its lending strategy), it follows that this global refinancing cost is convex in N (as the integral of an increasing function). This convexity reflects the optionality feature of the funding liquidity issue.

We denote by $\mathbb{P}$ and $\widehat{\mathbb{E}}$ the actuarial probability measure and the related expectation. Since we are dealing with short term debt with $T \leq 1$ yr, we do not introduce any discount factor. Discount factors would make no significant difference at this time horizon and would only obscure the analysis in the qualitative financial perspective of this chapter. Extending the model to longer term debt might require a correction in this regard.

Let n_t represent the amount of notional that the bank is willing to lend at the O/N rate r_t between t and $t + dt$. The problem of the bank lending overnight is modeled as follows: we need to maximize, over the whole stochastic process n, the expected profit of the bank, which we put in mathematical form as

$$\mathcal{U}(r; n) = \frac{1}{T}\widehat{\mathbb{E}}\left(\int_0^T n_t r_t dt - \int_0^T \int_0^{n_t} \rho_t(D_t + x)dxdt\right) \longleftarrow \max n. \qquad (2.1)$$

By contrast to this situation, when lending at Libor over a whole period of length T, a bank cannot modify a notional amount N which is locked between 0 and T. Since the composition of the Libor panel is updated at regular time intervals, during the life of a Libor loan, there is potential increase in credit risk embedded in the Libor loan as compared with credit risk embedded in an overnight loan rolled over the same period. Indeed, as time goes on, the refreshment mechanism of the panel guarantees a sustained credit quality of the names underlying the rolling overnight loan, whereas the Libor loan is contracted once for all with preservation of all the initial panelists; see (Filipović and Trolle 2013) for a detailed analysis, and see further developments following (2.6) below. Accordingly, let a stylized default time τ of the borrower reflect the deterioration of the average credit quality of the Libor contributors during the length of the tenor. This deterioration only affects the lender when lending at Libor or, in other words, the stylized default time τ only corresponds to those default events that could have been avoided had the loan been made as rolling overnight. It represents the "survivorship bias" that overnight loans benefit from. Let a constant N represent the amount of notional that the bank is willing to lend at Libor rate L over the period $[0, T]$. The related optimization problem of the bank can then be modeled as follows: to maximize with respect to the constant amount N the expected profit of the bank, which we put in mathematical form as

$$\mathcal{V}(L;N)=\frac{1}{T}\widehat{\mathbb{E}}\left(NL(T\wedge\tau)-\int_0^{T\wedge\tau}\int_0^N\rho_t(D_t+x)dxdt-\mathbb{1}_{\tau<T}N\right)\longleftarrow\max N \quad (2.2)$$

As we are dealing with short-term debt, we assume no recovery in case of default.

We stress that r and n represent stochastic processes in (2.1), whereas L and N are constants in (2.2). We stick to the stylized formulation (2.1)-(2.2) for tractability issues, and also in order to emphasize that the volatility terms which appear in the end-formula (2.12) are convexity adjustments reflecting an inherent optionality of LOIS, even without any risk premia.

Letting $U(r)=\sup_n\mathcal{U}(r;n)$ and $V(L)=\sup_N\mathcal{V}(L;N)$ represent the best utilities a bank can achieve by lending OIS or Libor, respectively, our approach for explaining the LOIS consists, given the O/N process r, in solving the following equation for L:

$$V(L)=U(r). \quad (2.3)$$

This equation expresses an indifference between the utility of lending rolling overnight versus Libor for a bank involved in both markets (indifference at the optimal amounts prescribed by the solution to the corresponding optimization problems). Note that r is supposed to be an endogenously given process here. A second problem, that we leave aside, would be in the determination of the O/N process r, depending on the supply/demand of liquidities and on base rates from the central bank.

To summarize: the key features differentiating the two lending strategies are, on the one hand, the deterioration in credit worthiness of a representative Libor "borrower", and on the other hand, the funding liquidity of a representative Libor "lender". Note that other issues, such as central banks' policies or possible manipulations of the rates, are not explicitly stated in the analysis. However, to some extent, these can be reflected in the model parametrization that we will specify now (see, in particular, the last paragraph of Subs. 2.2.1 and the discussion of Fig. 2.2). Also note that in our indifference utility analysis we only consider a rolling overnight loan and a Libor loan of a given maturity (tenor) T, equating in (2.3) the expected profits from optimal lending in each case, in order to eventually derive our LOIS formula (2.12). It appears that this formula does show a (square root) dependence in the tenor T. This relation doesn't need to consider the possibility of lending at Libor over different tenors T, but only rolled overnight versus at one fixed Libor of given maturity T.

2.2.1 Credit and Funding Costs Specification

For tractability we assume henceforth that the funding rate ρ is linear in x, i.e.

$$\rho_t(D_t+x)=\alpha_t+\beta_t x, \quad (2.4)$$

where $\alpha_t = \rho_t(D_t)$ is the time-t cost-of-capital of the lending bank and the coefficient β_t (positive in principle) represents the marginal cost of borrowing one more unit of notional for the bank already indebted at the level D_t. For instance, $\alpha_t = 2\%$ and $\beta_t = 50$bps means that the last euro borrowed by the bank was at an annualized interest charge of 2 cents, whereas if the bank would be indebted by €100 more, the next euro to be borrowed by the bank would be at an annualized interest charge of 2.5 cents.

In view of (2.4), we have

$$\mathcal{U}(r;n) = \frac{1}{T}\widehat{\mathbb{E}} \int_0^T \left((r_t - \alpha_t)n_t - \frac{1}{2}\beta_t {n_t}^2 \right) dt. \tag{2.5}$$

Denoting by λ_t the intensity of τ and letting $\gamma_t = \alpha_t + \lambda_t$ and $\ell_t = e^{-\int_0^t \lambda_s ds}$, then, by Lemma 13.7.5 we have:

$$\mathcal{V}(L;N) = \frac{1}{T}\widehat{\mathbb{E}} \int_0^T \left((L - \gamma_t)N - \frac{1}{2}\beta_t N^2 \right) \ell_t dt. \tag{2.6}$$

As explained after (2.2), the stylized default time τ reflects the deterioration of the average credit quality of a Libor representative borrower during the length of the tenor. Consistent with this interpretation, the intensity λ_t of τ can be proxyed by the slope of the credit curve of the Libor representative borrower (differential between the borrower's 1y CDS spread and the spread of its short term certificate deposits, currently 10 to a few tens of bp for major banks; see also the statistical estimates for λ_t in Filipović and Trolle (2013)). Accordingly we call λ_t the credit skew of a Libor representative borrower.

Note that central banks' liquidity policies can be reflected in the α_t and β_t components of the cost-of-liquidity ρ in (2.4). A possible manipulation effect, or incentive for a Libor contributor to bias its borrowing rate estimate in order to appear in a better condition than it is in reality (Wheatley 2012), can be included as a spread in the borrower's credit risk skew component λ.

2.3 LOIS Formula

Problems (2.1), (2.5) and (2.2), (2.6) are respectively solved, for given r and L, as follows. Writing $c_t := \alpha_t - r_t$, the OIS problem (2.1), (2.5) is solved independently at each date t according to

$$u_t(r_t;n_t) = c_t n_t - \frac{1}{2}\beta_t n_t^2 \longleftarrow \max n_t,$$

hence

$$n_t^\star = \frac{c_t}{\beta_t} \quad \text{and} \quad u_t(r_t;n_t^\star) = \frac{c_t^2}{2\beta_t}.$$

The expected profit of the bank over the period $[0, T]$ is

$$U(r) = \mathcal{U}(r; n^\star) = \widehat{\mathbb{E}}\left(\frac{1}{T}\int_0^T \frac{c_t^2}{2\beta_t} dt\right).$$

In the Libor problem (2.2), (2.6), we must solve

$$T\mathcal{V}(L; N) = N\widehat{\mathbb{E}}\int_0^T (L - \gamma_t)\ell_t dt - \frac{1}{2}N^2\widehat{\mathbb{E}}\int_0^T \beta_t \ell_t dt \longleftarrow \max N,$$

hence

$$N^\star = \frac{\widehat{\mathbb{E}}\frac{1}{T}\int_0^T (L - \gamma_t)\ell_t dt}{\widehat{\mathbb{E}}\frac{1}{T}\int_0^T \beta_t \ell_t dt} \quad \text{and} \quad V(L) = \mathcal{V}(L; N^\star) = \frac{\left(\widehat{\mathbb{E}}\frac{1}{T}\int_0^T (L - \gamma_t)\ell_t dt\right)^2}{2\widehat{\mathbb{E}}\frac{1}{T}\int_0^T \beta_t \ell_t dt} \tag{2.7}$$

We recall that the O/N process r is endogenously given. We are going to compute a stylized LOIS defined as LOIS $:= L^\star - R$, where $R = \widehat{\mathbb{E}}\frac{1}{T}\int_0^T r_t dt$ and where $L^\star$ is the solution to (2.3), assumed to exist. Note that the function V is continuous and increasing in L, so that if a solution $L^\star$ to (2.3) exists, then it is unique.

Now we will argue that LOIS ≥ 0, i.e. $L^\star \geq R$. First note that in case $\lambda = 0$, i.e. no risk of default (specifically, $\mathbb{P}(\tau = \infty) = 1$), one necessarily has $U(r) \geq V(R)$, since the constant $N^\star$ solving the Libor maximization problem (2.2) is a particular strategy (constant process $n_t = N^\star$) in the OIS maximization problem (2.1). Since V is an increasing function, the indifference pricing equation (2.3) in turn yields that $L^\star \geq R$.

Next, we consider the case $\lambda > 0$. In order to proceed, we denote by $\mathcal{V}_0(\cdot; N)$ the utility of lending Libor in case $\lambda = 0$. For each given amount N, one has via λ which is present in γ in (2.6) that $\mathcal{V}(R; N) \leq \mathcal{V}_0(R; N)$ (up to the second order impact of the discount factor ℓ). Hence $V(R) \leq V_0(R) \leq U(r)$ follows, the latter inequality resulting from the inequality already proven in case $\lambda = 0$ joint to the fact that τ doesn't appear in $\mathcal{U}(r; n)$. We conclude as in the case $\lambda = 0$ that $L^\star \geq R$.

For notational convenience let us introduce the following time-space probability measures on $\Omega \times [0, T]$, as well as the corresponding time-space averages of a process $f = f_t(\omega)$:

$$\overline{\mathbb{P}} = \mathbb{P} \otimes \frac{dt}{T}, \ \bar{f} = \overline{\mathbb{E}} f = \widehat{\mathbb{E}}\frac{1}{T}\int_0^T f_t dt$$
$$d\widetilde{\mathbb{P}}(t, \omega) = \ell_t(\omega) d\overline{\mathbb{P}}(t, \omega)/\bar{l}, \ \widetilde{f} = \widetilde{\mathbb{E}} f = \overline{\mathbb{E}}\left[f\ell/\bar{\ell}\right]$$

(so, in this notation, $R = \bar{r}$). Similarly, for any processes f, g let

$$\overline{\mathbb{C}\mathrm{ov}}(f, g) = \overline{\mathbb{E}}(fg) - \overline{\mathbb{E}} f \overline{\mathbb{E}} g, \ \bar{\sigma}_f^2 = \overline{\mathbb{E}}(f - \bar{f})^2, \ \widetilde{\sigma}_f^2 = \widetilde{\mathbb{E}}(f - \widetilde{f})^2. \tag{2.8}$$

Since

$$U(r) = \overline{\mathbb{E}}\left[\frac{c^2}{2\beta}\right] \quad \text{and} \quad V(L) = \frac{\bar{\ell}^2 (L - \widetilde{\gamma})^2}{2\overline{\mathbb{E}}[\beta\ell]},$$

then, equating $V(L^\star) = U(r)$ yields

$$\bar{\ell}^2 \left(L^\star - \widetilde{\gamma}\right)^2 = \overline{\mathbb{E}}[\beta\ell]\overline{\mathbb{E}}\left[\frac{c^2}{\beta}\right], \tag{2.9}$$

in which

$$\overline{\mathbb{E}}[\beta\ell]\overline{\mathbb{E}}\left[\frac{c^2}{\beta}\right] = \overline{\mathbb{E}}\left[c^2\ell\right] - \overline{\mathbb{C}\mathrm{ov}}\left[\beta\ell, \frac{c^2}{\beta}\right].$$

So,

$$\bar{\ell}^2 (L^\star - \widetilde{\gamma})^2 = \bar{\ell}\widetilde{\mathbb{E}}\left[c^2\right] - \overline{\mathbb{C}\mathrm{ov}}\left[\beta\ell, \frac{c^2}{\beta}\right]. \tag{2.10}$$

A particularly interesting formula follows in the case that we call "at-the-money," when $R = \widetilde{\alpha} = \widetilde{\gamma} - \widetilde{\lambda}$, i.e. the funding rate of the lender and the overnight interbank rate match on average, so that (recall $R = \bar{r}$) $\widetilde{c} = \widetilde{\alpha} - \widetilde{r} = \bar{r} - \widetilde{r}$. Then (2.10) reads (cf. (2.8)):

$$\bar{\ell}(L^\star - R - \widetilde{\lambda})^2 = \widetilde{\sigma}_c^2 + \bar{\ell}^{-1}(\bar{r} - \widetilde{r})^2 - \overline{\mathbb{C}\mathrm{ov}}\left[\beta\ell/\bar{\ell}, c^2/\beta\right]. \tag{2.11}$$

A reasonable guess is that $\bar{\ell}^{-1}(\bar{r} - \widetilde{r})^2$ and the covariance are negligible in the right-hand-side (in particular, these terms vanish when the credit risk deterioration intensity λ is zero and the marginal cost-of-capital coefficient β is constant). More precisely, for the sake of the argument, let us postulate a diffusive behavior of the instantaneous funding spread process $c_t = \alpha_t - r_t$, i.e. $dc_t = \sigma^\star dW_t$ for some "reference volatility" $\sigma^\star$ and a Brownian motion W. Let us also assume a constant $\lambda_t = \lambda^\star$ ("reference credit skew" of the borrower) and a constant marginal cost of borrowing β. Neglecting the impact of the "discount factor" $\ell_t = e^{-\lambda^\star t} \approx 1 - \lambda^\star t$ for small t in (2.11) (so that $\widetilde{\mathbb{P}} \approx \overline{\mathbb{P}}$), it follows that

$$\widetilde{\sigma}_c^2 \approx \bar{\sigma}_c^2 = (\sigma^\star)^2 \widehat{\mathbb{E}} \frac{1}{T}\int_0^T W_t^2 dt = (\sigma^\star)^2 T/2,$$

and our "LOIS formula" follows from the above as[2]:

$$\text{LOIS} \approx \lambda^\star + \sigma^\star \sqrt{T/2}. \tag{2.12}$$

From a broader perspective, according to the formula (2.11), the two key drivers of the LOIS are:

- a suitable average $\lambda^\star$ of the borrower's credit skew λ, which can be seen as the "intrinsic value" component of the LOIS and is the borrower's credit component;
- a suitable volatility $\sigma^\star$ of the instantaneous funding spread process c_t; this second component can be seen as the "time-value" of the LOIS and interpreted as the lender's liquidity component.

---

[2]Admitting that $\text{LOIS} - \widetilde{\lambda} \geq 0$, a natural assumption as Libor lending should at least compensate for the credit risk over the tenor horizon period; cf. the comments following the equation (2.7).

A possible application of the LOIS formula (2.12) could be to imply the value $\sigma^\star$ "priced" by the market from an observed LOIS and a borrower's CDS slope taken as a proxy for $\lambda^\star$. The value $\sigma^\star$ thus implied through (2.12) can be compared by a bank to an internal estimate of its "realized" funding spread volatility, so that the bank can decide whether it should rather lend Libor or OIS, much like with going long or short an equity option depending on the relative position of the implied and realized volatilities of the underlying stock.

In adition, a potential implication of the LOIS formula regarding the modeling of funding costs will be seen in Chapter 6.

2.4 Numerical Study

The LOIS formula (2.12) indicates a square-root term structure of the LOIS against $T/2$, with slope and intercept coefficients respectively related to the volatility of the funding spreads and the slope of the credit curve (funding spreads and credit curve of a representative Libor panelist). In this section we confront these theoretical predictions with the results of square root regressions of market time series of LOIS term structures, with a focus on $T = $ 3m or 6m (we recall that 3m is the most liquid tenor on the Libor markets and 6m is the second most liquid one). Note that by the formula (2.12), we have, adding the maturity as an explicit argument of the LOIS:

$$\text{LOIS}(3m) \approx \lambda^\star + \sigma^\star\sqrt{3m/2},\ \ \text{LOIS}(6m) - R \approx \lambda^\star + \sigma^\star\sqrt{6m/2},$$

so that the liquidity components of the 3m- and 6m- LOIS are respectively given by $\sigma^\star\sqrt{3m/2}$ and $\sigma^\star\sqrt{6m/2}$ (and the credit component of the LOIS is $\lambda^\star$ in both cases). Moreover, in view of (2.12) again, $\sigma^\star\sqrt{3m/2}$ and $\sigma^\star\sqrt{6m/2}$, the liquidity components of the 3m- and 6m- LOIS as just stated, respectively correspond to the slope of the LOIS viewed as a function of $\sqrt{T/3\text{m}}$ and $\sqrt{T/6\text{m}}$.

Fig. 2.2 shows the euro market 15/08/2007 to 16/04/2012 time series of the intercept, slope and $R2$ coefficients of the linear regression of the LOIS term structure against $\sqrt{T/3\text{m}}$ or $\sqrt{T/6\text{m}}$, for T varying from 1m to 1yr (of course choosing $\sqrt{T/3\text{m}}$ or $\sqrt{T/6\text{m}}$ as a regressor only affects the slope coefficient of the regression, by a factor of $\sqrt{2}$). If we believe in our LOIS theoretical formula (2.12), the intercept in these regressions can therefore be matched with the credit component $\lambda^\star$ of the LOIS (independent of the tenor T), while the slope coefficients can be matched with $\sigma^\star\sqrt{3m/2}$ and $\sigma^\star\sqrt{6m/2}$, the liquidity components of the 3m- and 6m- LOIS. So, according to our model, the red and blue (respectively red and purple) curves on the figure are the market-implied values for the credit and liquidity components of the 3m (respectively 6m) Libor. Before Aug 2007 the LOIS is negligible so that the regression (not displayed on the figure) is insignificant. After the"advent" of the LOIS in mid-Aug 2007, we can distinguish three market regimes. In a first phase, until Q1 of 2009 (turmoil of the subprime crisis), the market seems to "try to understand" is happening, with R2 becoming significant together with very large and volatile credit ("red intercept") and liquidity ("blue or purple slope") LOIS components. Note in particular the spike of both components at the turn of the credit crisis following Lehman's default in Sept 2008, during which the interbank drop of trust created both a credit and a liquidity crisis. Between Q2 of 2009 and mid-2011, the situation seems "stabilized" with R2 close to 1, a liquidity LOIS component of the order of 30bps on the 3m or 45bps on the 6m and a much smaller credit LOIS component. The ongoing Eurozone crisis, prompted by the US downgrade mid-2011, reveals a third pattern, with a much higher liquidity LOIS component, of the order of 60bps on the 3m or 90bps on the 6m, revealing increased funding liquidity concern of banks, due to harder regulatory constraints (e.g. government bonds no longer repositable). To illustrate the three "market regimes" in this analysis, Fig. 2.3 shows the fit between a theoretical square root term structure and the empirical LOIS term structure corresponding to the Euribor/Eonia-swap data of 14 Aug 2008, 28 Apr 2010 and 16 Apr 2012 (data of the right panel in Fig. 2.1). The last two terms that we neglected in (2.11) to deduce (2.12) are a possible explanation for (minor) departures of the actual LOIS spread curve from the theoretical square root term structure implied by (2.12). In relation with the discussion of the economical determinants of λ_t and α_t in Subs. 2.2.1, note that intercepts of e.g. 10 bp appear reasonable for a "credit skew", i.e. the differential between the one year CDS spread and the short term certificate deposit credit spread of a major bank, while the coefficient of $\sqrt{T/2}$, ranging between 100 and 200 bp/yr (corresponding to magnifying on Fig. 2.2 by a factor 2 the purple curve, or slope coefficients of the regression against $\sqrt{T/6\text{m}}$), is quite in line the volatilities of major banks' CDS spreads – certainly a reasonable lower bound, as funding spreads are complex and may depend on other less volatile inputs.

Conclusion

Since the 2007 subprime crisis, OIS and Libor markets (Eonia and Euribor in the euro market) diverged. In this chapter we show how, by optimizing their lending between Libor and OIS markets, banks are led to apply a spread (LOIS) over the OIS rate when lending at Libor. Theory implies that the LOIS has two components: one corresponding to the credit skew λ_t of a representative Libor borrower in an interbank loan, and one corresponding to the liquidity funding spread $c_t = \alpha_t - r_t$ of a representative Libor lender, where α_t and r_t respectively denote the instantaneous refinancing rate of the lender and the overnight interbank rate. Assuming a diffusive evolution of the instantaneous funding spread c_t, the above-mentioned optimization results in a square root term structure of the LOIS given by the formula (2.12), where the intercept $\lambda^\star$ can be proxied by the slope of a representative Libor credit curve and the coefficient $\sigma^\star$ is a volatility of c_t. These theoretical developments are corroborated by empirical evidence on the euro market studied in this chapter over the period from mid-2007 till mid-2012, with LOIS explained in a balanced way by credit and liquidity until the beginning of 2009 and predominantly explained by liquidity since then. The methodology that we develop here is relevant for any market in which credit and funding liquidity are the main drivers of interbank risk. Residual discrepancies between the theory and the data can be explained by the existence of other features such as Libor manipulations[3].

With respect to a multi-curve pricing approach where market OIS and Libor curves are simply fitted, the indifference utility approach of this chapter could allow a bank to arbitrage the LOIS by preferably lending Libor (respectively OIS) whenever its internally estimated funding spread is statistically found less (respectively more) volatile than $\sigma^\star$ implied from the market through the LOIS formula (2.12). Moreover, a potential application to the modeling of funding costs will be seen in Chapter 6.

[3]As explained at the end of Subs. 2.2.1, Libor manipulations can also be modeled within the present framework by a spread in the borrower's credit risk skew component λ_t.

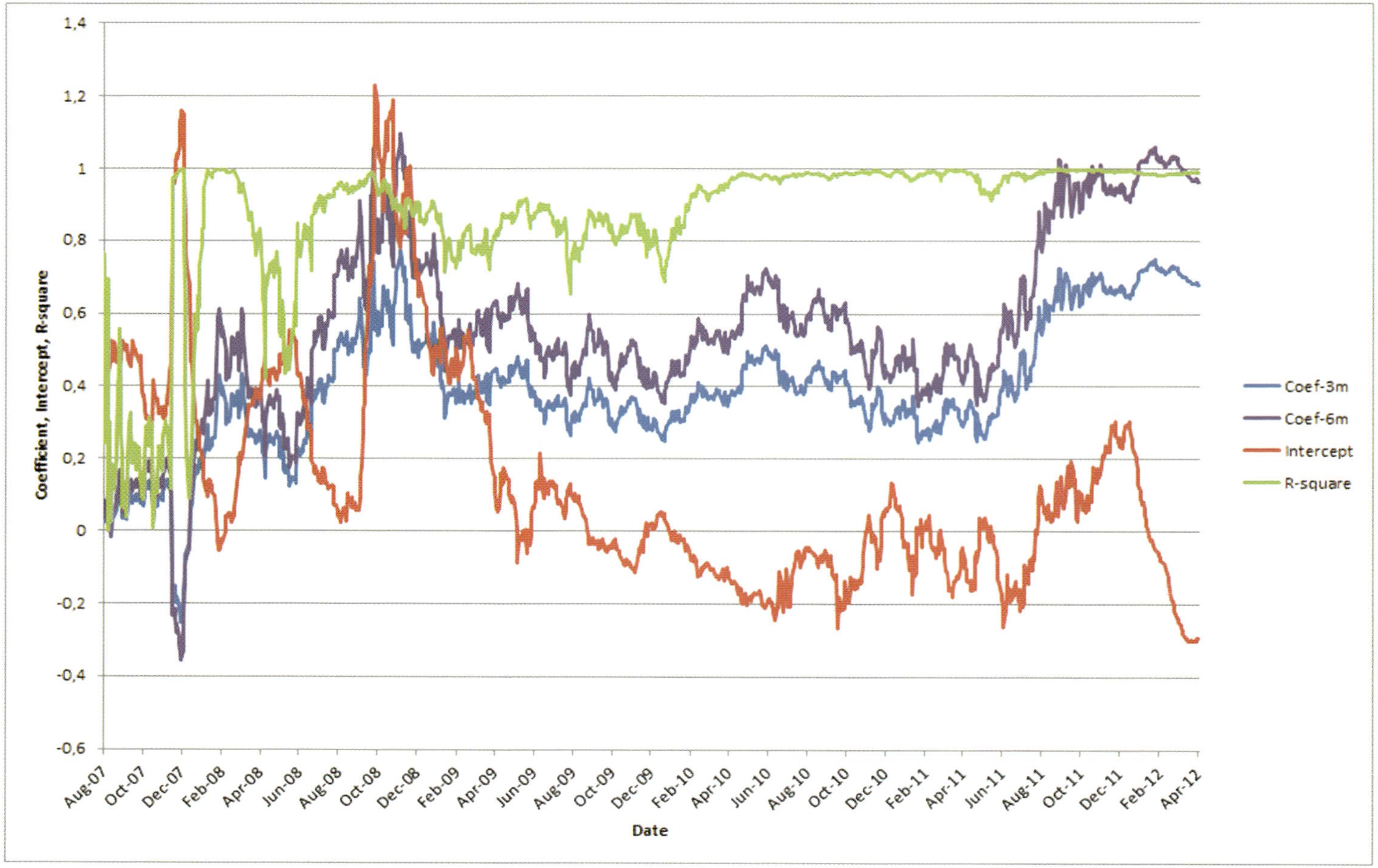

FIGURE 2.2: Time series of the "red" intercepts (in %; credit component of the LOIS), "blue" and "purple" slopes (in %; liquidity component of the 3m- and 6m- LOIS) and "green" R2 coefficients of the regressions of the 1m to 1yr LOIS against $\sqrt{T/3\text{m}}$ or $\sqrt{T/6\text{m}}$, over the period from 15/08/2007 to 16/04/2012.

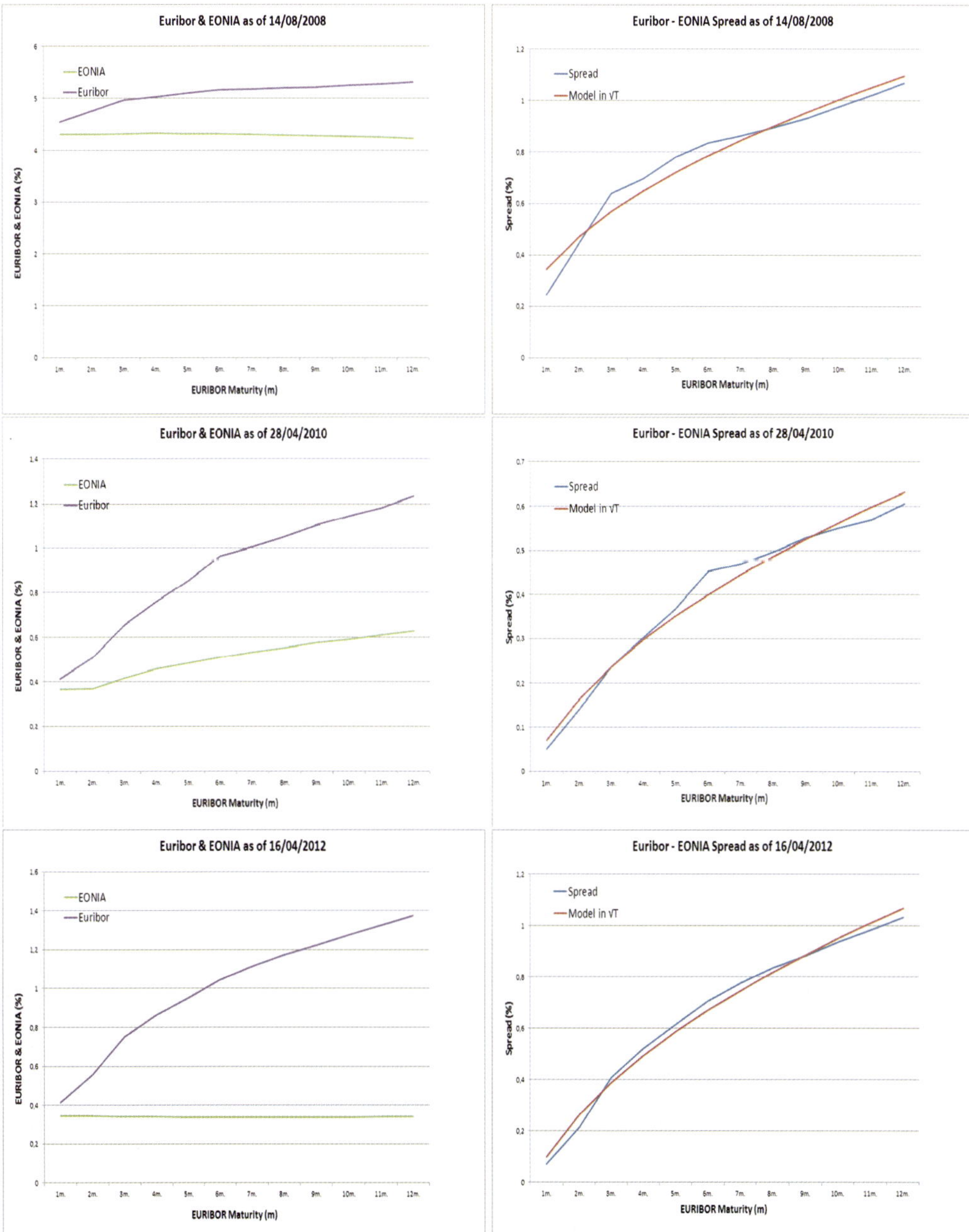

FIGURE 2.3: Euribor / EONIA-swap rates (left) and square root fit of the LOIS (right), $T = 1\text{m}$ to 12m. Top to bottom: 14 Aug 2008, 28 Apr 2010 and 16 Apr 2012 (data of the right panel in Fig. 2.1).

Part II

Model-Free Developments

A recurring theme of the book is the decomposition of the counterparty risky price Π_t of an OTC derivative contract (or portfolio of contracts) between two parties. The generic decomposition is of the form: (essentially, but see (3.12) for the precise statement)

$$\Pi_t = P_t - \Theta_t,$$

where P_t is the counterparty clean price of the contract and Θ_t is the total valuation adjustment. Throughout the text we will derive various, more specific, decompositions of the form:

$$\Pi_t = P_t - \sum_i A_t^i,$$

where A^i represent various adjustments, such as the CVA, DVA, LVA, RVA and RC mentioned in the preface of the book. Thus, essentially, we will provide various decompositions of a TVA:

$$\Theta_t = \sum_i A_t^i.$$

From the practical point of view, the key issue is to provide useful representations for these adjustments, as well as to compute their sensitivities with respect to relevant quantities, in order to be able to dynamically price and hedge them.

In this part of the book we describe in mathematical, but model-free terms all the basic elements of the above framework. Specific, model based discussion will be given in later parts.

Chapter 3

Pure Counterparty Risk

We describe here in mathematical, but model-free terms all the basic elements of the pricing and hedging framework that are relevant with regard to counterparty risk. We will represent a TVA process in terms of CVA and DVA components, as well as in terms of specific measures of counterparty exposures such as the exposure at default, the expected positive exposure and the expected negative exposure. In this chapter we focus on a classical setting of a single funding curve (single-curve set up); that is, we assume that all positions are funded (and discounted) using the same (risk-free) interest rate. We call this setting the "classical" setup, as it was prevalent prior to the financial crisis of the year 2007–2009 in the sense that bases between Libor rates and rates such as OIS or Eonia were negligible. Consequently, the funding issue, i.e. the issue of funding positions using various rates, was not really consequential. The funding issue becomes relevant in case of multiple and significantly different funding rates. This issue (in the case of bilateral counterparty risk in particular) will be added to the discussion in the next chapter.

3.1 Cash Flows

Let $(\Omega, \mathcal{G}_T, \mathbb{G})$ stand for a filtered space with a finite horizon T, which is used throughout this book as the space containing random events that underlie modeling the stochastic evolution of a financial market. The filtration $\mathbb{G}$, as well as any other filtration in the book, are assumed to satisfy the usual conditions. All our random variables are $\mathcal{G}_T$-measurable; all our random times are $[0,T] \cup \{\infty\}$-valued $\mathbb{G}$-stopping times; all our processes are defined over $[0,T]$ and are $\mathbb{G}$-adapted.

The horizon T represents the maturity in a Credit Support Annex (CSA) regarding a "contract", in the sense of a generic netted portfolio of OTC derivatives between two parties. By "netted portfolio" we mean that in case of default of a party at a time $\tau \leq T$, the debt of either party to the other will be valued by a liquidator on the basis of the cumulative value of all the ingredients of the portfolio, rather than as the cumulative debt over the different ingredients in the nonnetted case. Note that this setup can be considered as general since, for any partition of a portfolio into netted sub-portfolios, the results may be applied separately to every sub-portfolio. The results at the portfolio level are then simply derived as the sum of the results of the sub-portfolios.

We call the two parties the bank and the counterparty. The parties are assumed to be default prone. We denote by τ_b and τ_c the default times of the bank and the counterparty, i.e. they are the times at which promised dividends and margin calls cease to be paid by a distressed party. We also define the "first-party-default" time $\tau = \tau_b \wedge \tau_c$, with related survival indicator process denoted by $J_t = \mathbb{1}_{t<\tau}$. It needs to be emphasized that we do not exclude simultaneous defaults of the bank and its counterparty[1]. We do assume, however, that the default times cannot occur at fixed times, an assumption that is for instance satisfied in all the intensity models of credit risk; consequently, in particular, the scenario $\{\tau = T\}$ has zero probability and is immaterial in any price and hedge, so that typically we simply ignore it for simplicity (writing e.g. $\{\tau < T\}$ interchangeably with $\{\tau \leq T\}$). We will study unilateral counterparty risk from the perspective of the bank (i.e. $\tau_b = \infty$ and $\tau_c < \infty$, so $\tau = \tau_c$), as well as the bilateral counterparty risk (i.e. $\tau_b < \infty$ and $\tau_c < \infty$). We denote

$$\bar{\tau} = \tau \wedge T,$$

where $\bar{\tau}$ represents the effective time horizon of our problem, since there will be no cash flows after it.

Remark 3.1.1 (Unilateral or Bilateral Counterparty Risk?) As will be seen, in a classical risk-neutral derivative pricing perspective, the possibility of own default should be accounted for by a suitable correction, actually standing as a benefit, to the value of the contract. This "benefit" (if any), the so-called debt valuation adjustment (DVA), corresponds to the value of the part of own debt which is not reimbursed at own default time. However, there is a debate, both at a theoretical level and among practitioners, regarding the relevance of accounting for own credit risk as a benefit through bilateral counterparty risk valuation. The point is that selling protection on oneself is hardly doable in practice (as nobody would buy it unless it is in a fully funded form, in which case it is practically a bond; in CDS form it is frequently even illegal). To hedge one's own jump to default risk one may try to repurchase one's own bond (see Sect. 4.6 for a study in a toy model), but this is not practical in general. Therefore it is not really possible to hedge one's own jump-to-default risk. In 2011 it was announced that Goldman Sachs was proxy-hedging its DVA risk through peers (so called proxy hedging). This may partially hedge the DVA spread risk (one may also consider that DVA, like CVA, is a natural hedge to other things, being contra-cyclical), but not the DVA jump to default risk. Given that hedging and therefore monetization of bilateral counterparty risk is difficult, the principle of risk-neutral valuation of bilateral counterparty risk (own jump-to-default risk in particular) is thus questionable. But the practical justification for using a model of bilateral counterparty risk is that unilateral valuation of counterparty risk induces a significant gap between the values computed by the two parties. This implies that in case the two parties are risky it will be difficult for them to agree on a deal on the basis of unilateral counterparty risk valuations alone.

[1]In case of a credit derivative we allow for simultaneous defaults between the parties to the contract and the names in the underlying portfolio.

If in the end one does not want to account for bilateral counterparty risk (or the counterparty risk is indeed purely one-sided), one simply considers a model of unilateral counterparty risk, which corresponds in our formalism to letting $\tau_b = \infty$ everywhere below (for unilateral counterparty risk from the perspective of the bank).

3.1.1 Promised Dividend

We let a finite variation process D represent the promised (or clean) cumulative dividend process of the contract from the bank to the counterparty, with jump process denoted by Δ, i.e. $\Delta_t = D_t - D_{t-}$. All cash flows are considered from the bank's perspective in the sense that $\Delta_t = 1$ means $+1$ to the bank. A promised dividend is only effectively paid at time t if none of the parties defaulted by time t, resulting in the effective dividend process C given as

$$dC_t = \mathbb{1}_{t<\tau} dD_t = J_t dD_t.$$

Moreover, the two parties are tied by the CSA, a legal agreement which prescribes the collateralization scheme and the closeout cash flow in case of default of either party, meant to mitigate counterparty risk. Collateral consists of cash or various possible eligible securities posted through margin calls as default guarantee by the two parties. The closeout cash flow is the last (terminal) cash flow $\mathfrak{R}$, based on the value Q_τ of the contract assessed by the liquidator (such as PricewaterhouseCoopers in the case of bankruptcy of Lehman Brothers International) and including the accumulated collateral at time τ, which closes the bank's position at that time (if $\tau < T$). Cash flows related to the defaultability of either party (margin calls and CSA closeout cash flow) will be called CSA cash flows.

3.1.2 Collateral

Collateral consists of cash or various possible eligible securities posted through CSA regulated margin calls as default guarantee by the two parties. Noncash collateral is valued at some haircut to compensate for the fact that the corresponding assets may lose some value during the period, called the margin period of risk, between the last margin call and the time the contract and/or these assets can be liquidated. But in order to simplify the presentation, we ignore such practicalities, assuming throughout the book that margins are posted in cash to the other party and the cumulative margin amount is remunerated at some rate (a risk-free OIS rate r_t in this chapter and a possibly different rate in the next one). We model the value of the collateral by means of an algebraic margin amount Γ_τ passed from the counterparty to the bank at time $\tau < T$. So, before τ, a positive Γ_t represents an amount "lent" by the counterparty to the bank (and remunerated as such by the bank at some rate), but intended to become the property of the bank in case of default of either party at time τ (if $\tau < T$). It is worth stressing that, according to industry standards, in case of default of the bank at time τ, this (algebraic) collateral will also become property of the bank, unless a special segregation procedure is in force (see Sect. 3.3.4). Symmetric remarks apply to negative Γ_t (swap the roles of the counterparty and bank in the above description). So, before τ, a positive $(-\Gamma_t)$ represents an amount "lent" by the bank to the counterparty (and remunerated as such by the counterparty), but intended to become the property of the counterparty in case of default of either party at time $\tau < T$. We emphasize that we consider just one collateral account, which may take either positive or negative value. Of course, a negative value from the perspective of one of the two counterparties means a positive value from the perspective of the other counterparty and vice versa. Three prototypal collateralization schemes addressed in this book are the naked scheme ($\Gamma = 0$), the so-called continuous scheme ($\Gamma = Q$) and the ISDA scheme introduced (in two variants) in Sect. 3.3.2.

Remark 3.1.2 In the case of centrally cleared transactions, the ownership convention regarding the collateral is different, i.e. the collateral becomes the ownership of the receiving party as soon as it has been posted to it. In particular, the collateral is not remunerated. This induces some subtle differences in the pricing, analyzed in Cont, Mondescu, and Yu (2011).

3.1.3 Closeout Cash Flow

We define a $\mathcal{G}_\tau$-measurable random variable χ, for $\tau < \infty$, as

$$\chi = Q_\tau + \Delta_\tau - \Gamma_\tau, \tag{3.1}$$

where Q and Γ are the CSA closeout valuation process (see Sect. 3.3.1) and the collateral process (see Sect. 3.3.2), respectively, and where $\Delta_\tau = D_\tau - D_{\tau-}$ denotes the jump of D at τ, representing a promised bullet dividend at τ. From the point of view of financial interpretation, χ represents the (algebraic) debt of the counterparty to the bank at the first-party-default time τ, accounting for the legal value of the portfolio at that time, plus any bullet dividend which should be paid at time τ by the counterparty to the bank, less the margin amount Γ_τ, the ownership of which has already "instantaneously" been transferred at time τ (from the counterparty to the bank if positive and vice versa if negative).

In the single-curve setup of this chapter, the closeout cash flow $\mathfrak{R}$ is given as

$$\mathfrak{R} = \Gamma_\tau + \mathbb{1}_{\tau=\tau_c}\left(R_c\chi^+ - \chi^-\right) - \mathbb{1}_{\tau=\tau_b}\left(R_b\chi^- - \chi^+\right) - \mathbb{1}_{\tau_b=\tau_c}\chi, \tag{3.2}$$

where the $[0,1]$-valued $\mathcal{G}_{\tau_c}$- and $\mathcal{G}_{\tau_b}$- measurable random variables R_b and R_c, respectively, denote the recovery rates of the bank and of its counterparty upon default. So:

- if the bank defaults first at time $\tau_b < \tau_c \wedge T$, then, at time $\tau = \tau_b$, the closeout cash flow takes place in the amount of $\Gamma_\tau - \left(R_b\chi^- - \chi^+\right)$;
- if the bank's counterparty defaults first at time $\tau_c < \tau_b \wedge T$, then, at time $\tau = \tau_c$, the closeout cash flow takes place in the amount of $\Gamma_\tau + R_c\chi^+ - \chi^-$;
- if the bank and the counterparty default simultaneously at time $\tau_c = \tau_b < T$, then, at time $\tau = \tau_b = \tau_c$, the closeout cash flow takes place in the amount of $\Gamma_\tau + R_c\chi^+ - R_b\chi^-$.

Note that the formulation (3.2) for the CSA closeout cash flow $\mathfrak{R}$ offers a good theoretical paradigm, but it is not the only possible one; it is still a stylized payoff and it does not cover the totality of cases encountered in practice. For instance, in the case of a collateral segregated by a third-party, the collateral which is lent can in certain cases be recovered at a better rate than the rest of the debt. Accordingly, Capponi (2013) and Pallavicini, Perini, and Brigo (2011, 2012) consider a CSA closeout cash flow generalizing (3.2) to different levels of recoveries applicable to different components of the exposure.

Remark 3.1.3 At a more fundamental level, one may object that in case of the default of a counterparty the closeout cash flow $\mathfrak{R}$ always has the same functional form (3.2) regardless of what happens with the reference dividend process. For example, in case of modeling the counterparty risk relative to an underlying CDS contract, we postulate that the recovery structure in case of a default of a counterparty is the same regardless whether or not the counterparty defaults at the same time as the obligor referencing the CDS contract. It might seem at first sight that this tacit assumption excludes wrong-way risk from the model, namely the risk that the value of the contract be particularly high from the perspective of the other party at the moment of default of the counterparty, which is a major issue regarding counterparty risk. In fact, the recovery structure in case of a default of a counterparty is the same, but the ingredients of the structure (χ, specifically) may support wrong-way risk, with for instance a significant chance that Δ_τ in χ be more important than a "typical" Δ_t, as will be the case with the common-shock counterparty credit risk model in Part IV.

3.2 Valuation and Hedging

We will focus on the bank "buying" (simply in the sense of our sign convention that $\Delta_t = 1$ means $+1$ to the bank) a contract from the counterparty and left with the task of setting-up a hedging, collateralization and funding portfolio. Of course symmetrical considerations apply to the bank's counterparty, but with potentially nonsymmetrical hedging positions and funding conditions. In the classical single-curve setup of the present chapter this is immaterial, but in the multi-curve setup of the next chapter this asymmetry of data is effective and results in a "buyer price" of the bank typically being different from the "seller price" of the counterparty. This is why we need to focus on a given party, called "the bank" in this book. Again, symmetrical considerations and methodology apply to the counterparty, but with nonsymmetrical data and an ensuing price discrepancy in a multi-curve setup (see the remark 4.2.2 for comments on the related "violation of money conservation" issue).

Remark 3.2.1 In this book, we stick to the paradigm of only one risk-neutral measure $\mathbb{Q}$ chosen by the market. Consequently, this terminology of "buyer price" versus "seller price" has nothing to do here with different pricing measures that can coexist in an incomplete market (see e.g. El Karoui and Quenez (1995) for a seminal reference or, more recently, Eberlein, Madan, Pistorius, and Yor (2013)), it simply reflects different hedging and funding policies.

We endow the measurable space $(\Omega, \mathcal{G}_T)$ with a fixed probability measure $\mathbb{Q}$, which will later be interpreted as a martingale pricing measure in some sense (and is supposed to be chosen by the market). We assume in particular that $\mathbb{Q}$ is equivalent to the historical probability measure $\mathbb{P}$ over $(\Omega, \mathcal{G}_T)$. We denote by $\mathbb{E}_t$ the conditional expectation under $\mathbb{Q}$ given $\mathcal{G}_t$.

By funding assets we mean in this book riskless, finite variation assets which are used for funding a position. In the classical single-curve setup of this chapter with a locally risk-free rate r_t, there is only one funding asset, the so-called savings account, growing at rate r_t. The savings account is thus the inverse of the risk-free discount factor $\beta_t = e^{-\int_0^t r_s ds}$. In the ensuing classical risk neutral valuation framework we can proceed with presenting the following definitions, which are consistent with the standard theory of arbitrage (cf. Delbaen and Schachermayer (2006)).

3.2.1 Valuation of the Contract

Recall that D_t stands for the cumulative promised (or clean) cash flows (dividend process) of the contract on $[0, T]$, ignoring counterparty risk.

Definition 3.2.2 **(i)** For $t \in [0, T]$, the (counterparty-)clean price process[2] P_t of the contract is given by:

$$\beta_t P_t = \mathbb{E}_t \int_t^T \beta_s dD_s. \tag{3.3}$$

The clean cumulative value process of the portfolio is given by:

$$\widehat{P}_t = P_t + p_t, \tag{3.4}$$

whereπ_t p_t represents the discounted cumulative clean dividend process up to time t, so

$$\beta_t p_t = \int_0^t \beta_s dD_s. \tag{3.5}$$

(ii) The (counterparty-)risky price Π_t of the contract is given, for $t \in [0, \bar{\tau}]$, by:

$$\beta_t \Pi_t = \mathbb{E}_t \left(\int_t^{\bar{\tau}} \beta_s dC_s + \beta_{\bar{\tau}} \mathbb{1}_{\tau < T} \mathfrak{R} \right). \tag{3.6}$$

The risky cumulative value process of the portfolio is given, for $t \leq \bar{\tau}$, by:

$$\widehat{\Pi}_t = \Pi_t + \pi_t, \tag{3.7}$$

[2]In the market terminology, it is frequently called the mark-to-market (MtM).

where

$$\beta_t \pi_t = \int_0^t \beta_s dC_s. \tag{3.8}$$

In the context of this book, where the focus is on counterparty risk, it is convenient to distinguish two categories of cash flows involving the bank:

- dividends, in the sense of all pre-default cash flows (or we specify "promised dividends" if we want to refer to D);
- closeout cash flows, i.e. cash flows at the default time τ (if $< T$).

Note that, consistent with this convention, the closeout cash flow $\mathfrak{R}$ is excluded from dividends and accounted for separately as a terminal condition in (3.6). In particular, at time $t = \bar{\tau}$, we obtain from (3.6) that

$$\Pi_{\bar{\tau}} = \mathbb{1}_{\tau<T}\mathfrak{R}.$$

Note that with respect to the usual mathematical finance ex-dividend and cum-dividend prices convention, our choice in this book (for risky prices) corresponds to a hybrid convention: "cum-" regarding $\mathfrak{R}$ and "ex-" regarding D (or C). So, defining the counterparty risky cumulative cash flow process as$\mathcal{D}_t$

$$\mathcal{D}_t = C_t + (1 - J_t)\mathfrak{R} \tag{3.9}$$

(dividends, in our restricted sense of C, along with the close out cash flow $\mathfrak{R}$), we have that (cf. (3.7))

$$\Pi_t^{cum} = \widehat{\Pi}_t - \pi_t^{cum}, \ \Pi_t^{ex} = \widehat{\Pi}_t - \pi_t^{ex}, \tag{3.10}$$

where π_t^{cum} and π_t^{ex} are given by

$$\beta_t \pi_t^{cum} = \int_{(0,t)} \beta_s d\mathcal{D}_s, \quad \beta_t \pi_t^{ex} = \int_0^t \beta_s d\mathcal{D}_s = \int_{(0,t]} \beta_s d\mathcal{D}_s \tag{3.11}$$

(in accordance, regarding the second integral, with the notational convention stipulated in the preface of the book). The classical decompositions given in (3.10)-(3.11) are less convenient for our purposes, and that is why we do not use them, but instead use (3.7).

Remark 3.2.3 In principle, when dealing with counterparty risk valuation, one could consider not one but two filtered risk-neutral pricing models: $(\Omega, \mathbb{G}, \mathbb{Q})$ and $(\Omega, \mathbb{F}, \mathbb{Q})$. Here the filtration $\mathbb{F} = (\mathcal{F}_t)_{t\in[0,T]}$ would represent the counterparty risk-free filtration, not carrying any direct information about the default times τ_b and τ_c or about any factors that might be specific to evolution of the credit standards ("ratings") of the counterparties. This is the proper filtration that would normally be used for pricing the counterparty risk-free contracts, which serves as a reference so to assess the counterparty riskiness of actual contracts being priced and hedged. Mathematically speaking, writing $\mathbb{H}^b = (\mathcal{H}_t^b)_{t\in[0,T]}$ and $\mathbb{H}^c = (\mathcal{H}_t^c)_{t\in[0,T]}$ for the natural filtrations of the indicator processes of τ_b and τ_c, we have that $\mathcal{H}_t^b \not\subseteq \mathcal{F}_t$ and $\mathcal{H}_t^c \not\subseteq \mathcal{F}_t$. The filtration $\mathbb{G} = (\mathcal{G}_t)_{t\in[0,T]}$ would represent the counterparty risky filtration and it is a filtration such that $\mathcal{F}_t \vee \mathcal{H}_t^b \vee \mathcal{H}_t^c \subseteq \mathcal{G}_t$.

The discount factor β, the clean cumulative dividend process D and the collateral process Γ would thus typically be assumed to be $\mathbb{F}$-adapted, and the counterparty clean price process of the portfolio would be given by[3]

$$\beta_t P_t = \widetilde{\mathbb{E}}_t \int_t^T \beta_s dD_s,$$

where we denote by $\widetilde{\mathbb{E}}_t$ the conditional expectation under $\mathbb{Q}$ given $\mathcal{F}_t$. See the following parts of this book and Crépey and Song (2014) for more developments in this regard.

Remark 3.2.4 The practical assessment and evaluation of the counterparty risk is of course quite sensitive to the choice of closeout formulation. It is implicitly understood above that the CSA closeout price process Q is an exogenous process, as in the standard clean CSA closeout pricing scheme $Q = P$. An a priori unusual situation from this point of view is the so-called pre-default CSA closeout pricing scheme $Q = \Pi$. The impact of alternative closeout cash flows will be discussed in Sect. 3.3.1.

3.2.2 Valuation of Counterparty Risk

We will begin with defining TVA as a counterparty risk (total) valuation adjustment process. Then, in Proposition 3.2.8, we will demonstrate that a TVA is a price process for a so-called contingent credit default swap (CCDS), which pays exposure at default. Since all we will need for practical purposes is in fact a "cumulative" TVA process on the time interval $[0, \bar{\tau}]$ (see the remark 3.2.7), we now define the TVA process as follows.

Definition 3.2.5 *The TVA process, denoted as* Θ, *is given, for* $t \in [0, \bar{\tau}]$, *as:*

$$\Theta_t = \widehat{P}_t - \widehat{\Pi}_t = P_t - \Pi_t + \mathbb{1}_{t=\tau<T}\Delta_\tau. \tag{3.12}$$

Remark 3.2.6 In the single-curve setup of this chapter, the left equality appears as a natural definition for what a (cumulative) counterparty risk valuation adjustment should be. The right equality follows since, by the definition 3.2.2, for $t \le \bar{\tau}$, we have:

$$p_t - \pi_t = \mathbb{1}_{t=\tau<T}\Delta_\tau.$$

Note that in the multi-curve setup of the next chapter, the TVA will still be given by the right-hand-side in (3.12), whereas the corresponding notions of cumulative prices will be less straightforward.

[3] We stress again that the clean price process P above is the process of the "clean contract", that is the contract in which any counterparty risk (and also "excess funding costs", coming as a spread with r_t, in the next chapter) is disregarded.

Remark 3.2.7 The term $\mathbb{1}_{t=\tau<T}\Delta_\tau$ is needed in the right hand side of (3.12) so that we get a cumulative TVA. If there are no promised cash flows at τ, so that $\Delta_\tau = 0$, then of course this term vanishes. The reader may ask at this point why is it so that we are interested in a cumulative TVA. The reason is that a cumulative TVA is a (local) martingale (under $\mathbb{Q}$), which is an important property from the point of view of dynamic hedging of the counterparty risk.

In the rest of this section we will discuss various representations of the TVA.

3.2.3 Exposure at Default

We define the exposure at default (ED) by the $\mathcal{G}_\tau$-measurable random variable ξ such that

$$\begin{aligned}
\xi &:= P_\tau + \Delta_\tau - \mathfrak{R} \\
&= P_\tau - Q_\tau + \chi - \mathbb{1}_{\tau=\tau_c}\left(R_c\chi^+ - \chi^-\right) + \mathbb{1}_{\tau=\tau_b}\left(R_b\chi^- - \chi^+\right) + \mathbb{1}_{\tau_b=\tau_c}\chi \qquad (3.13)\\
&= P_\tau - Q_\tau + \mathbb{1}_{\tau=\tau_c}(1-R_c)\chi^+ - \mathbb{1}_{\tau=\tau_b}(1-R_b)\chi^-,
\end{aligned}$$

where the second equality is by definition (3.2) of $\mathfrak{R}$ and the third one follows by the definition (3.1) of χ.

Proposition 3.2.8 *For $t \in [0, \bar{\tau}]$, we have:*

$$\beta_t\Theta_t = \mathbb{E}_t\left[\beta_\tau \mathbb{1}_{\tau<T}\xi\right]. \qquad (3.14)$$

Proof. This follows from the martingale property of $\beta_t\Theta_t$, which is apparent on the left-hand-side in (3.12), whereas we have by the right-hand side:

$$\Theta_{\bar{\tau}} = P_{\bar{\tau}} - \Pi_{\bar{\tau}} + \mathbb{1}_{\tau<T}\Delta_\tau = \mathbb{1}_{\tau<T}(P_\tau + \Delta_\tau - \mathfrak{R}) = \mathbb{1}_{\tau<T}\xi.$$

□

As first observed in Brigo and Pallavicini (2008) (and Brigo and Capponi (2008a) for the bilateral case), we get an interpretation of the TVA as the price of a contingent credit default swap (CCDS), which is an option on the debt χ (that is present in ξ) of the counterparty toward the bank at time τ.

Remark 3.2.9 **(i)** Since we consider the general case of the bilateral counterparty risk, the TVA process may take negative values. In case of the unilateral counterparty risk (from the bank's perspective), this process always takes nonnegative values (see Sect. 3.2.4.2).
(ii) In general, a representation of the form (3.14) does not uniquely define ξ. It is an open question under what assumptions, on the model filtration $\mathbb{G}$ in particular, a representation (3.14) uniquely defines ξ.

A major issue in regard to counterparty credit risk is the so-called wrong-way risk. From the perspective of the bank this occurs when the exposure to the counterparty is adversely correlated with the credit quality of the latter (risk that the value of the contract is particularly high at the counterparty's default). As will be the case in the credit portfolio common-shock model of chapters 8 through 10, an extreme form of wrong-way risk ("instantaneous default contagion") can also be present via the term Δ_τ in χ (see (3.1)).

3.2.4 TVA and CVA/DVA/RC

From (3.14), we see that the TVA process Θ decomposes, for $t \in [0, \bar{\tau}]$, as follows:

$$\Theta_t = \mathrm{CVA}_t + \mathrm{DVA}_t + \mathrm{RC}_t, \tag{3.15}$$

where

$$\mathrm{CVA}_t = \beta_t^{-1}\mathbb{E}_t\left[\beta_\tau \mathbb{1}_{\tau<T}(1 - R_c)\mathbb{1}_{\tau=\tau_c}\chi^+\right] \tag{3.16}$$

represents a so called credit valuation adjustment (cost to the bank of the default of the counterparty),

$$\mathrm{DVA}_t = -\beta_t^{-1}\mathbb{E}_t\left[\beta_\tau \mathbb{1}_{\tau<T}(1 - R_b)\mathbb{1}_{\tau=\tau_b}\chi^-\right] \tag{3.17}$$

represents a so called debt valuation adjustment (windfall benefit of the bank at its own default from not reimbursing in totality the counterparty) and

$$\mathrm{RC}_t = \beta_t^{-1}\mathbb{E}_t\left[\beta_\tau \mathbb{1}_{\tau<T}(P_\tau - Q_\tau)\right] \tag{3.18}$$

represents a so called replacement cost (mismatch, unless $Q = P$, between the clean price P and the CSA value Q of the contract assessed by the liquidator at default time).

It is implicit in (3.15) that we conventionally choose to work with a positive CVA and a negative DVA adding up, together with RC, to the TVA. Often people[4] use a positive DVA, which is then substracted from the CVA. In Part III of the book the TVA will also include any liquidity funding costs in excess over the risk-free rate. The sign of the resulting additional LVA (liquidity valuation adjustment) is not determined in general (this also applies to replacement costs), so we prefer to write all the adjustments with a "+".

Remark 3.2.10 Although banks report on DVA in their earnings reports, it is not included in determining the capital levels. This is also stated in (Basel Committee on Banking Supervision 2011a, Paragraph 75) as

> "Derecognise in the calculation of Common Equity Tier 1, all unrealised gains and losses that have resulted from changes in the fair value of liabilities that are due to changes in the bank's own credit risk."

[4] As in the introductory dialogue of Chapter 1!

Therefore, Basel III framework does not allow banks to account for DVA in their regulatory capital calculations (see also Basel Committee on Banking Supervision (2012) for a detailed discussion). The main reason of this treatment of DVA in Basel III is not to allow banks to have the value of their liabilities decrease while their credit risk is increasing.

Note that the above decomposition is only one among many possible variants that can be found in the literature. In fact, the most comprehensive decomposition in this book will come later, as (6.16) (under an immersion assumption, but this can be relaxed as demonstrated in Crépey and Song (2014)).

3.2.4.1 Expected Positive/Negative Exposures

In case β is deterministic, the following representation of the CVA follows from (3.14):

$$\begin{aligned}\mathrm{CVA}_0 &= \mathbb{E}\left(\beta_\tau(1-R_c)\mathbb{1}_{\tau=\tau_c<T}\chi^+\right)\\ &= \int_0^T \beta_s \mathbb{E}\left((1-R_c)\mathbb{1}_{\tau=\tau_c\in ds}\chi^+\right)\\ &= \int_0^T \beta_s \mathbb{E}\left((1-R_c)\chi^+ \,\middle|\, \tau_c = s \le \tau_b\right)\mathbb{Q}(\tau_c\in ds,\ \tau_b \ge s)\\ &= \int_0^T \beta_s \mathrm{EPE}(s)\mathbb{Q}(\tau_c\in ds,\ \tau_b\ge s),\end{aligned} \tag{3.19}$$

where the expected positive exposure EPE, also known as the asset charge, is the function of time defined by, for $t\in[0,T]$,

$$\mathrm{EPE}(t) = \mathbb{E}\left[(1-R_c)\chi^+|\tau_c = t\le\tau_b\right]. \tag{3.20}$$

Similarly, we have for the time-0 DVA the following representation:

$$\mathrm{DVA}_0 = -\int_0^T \beta_s \mathrm{ENE}(s)\mathbb{Q}(\tau_b\in ds,\ \tau_c\ge s), \tag{3.21}$$

where the expected negative exposure ENE, also known as the liability benefit, is the function of time defined, for $t\in[0,T]$, by:

$$\mathrm{ENE}(t) = \mathbb{E}\left[(1-R_b)\chi^-|\tau_b = t\le\tau_c\right]. \tag{3.22}$$

3.2.4.2 Unilateral Counterparty Risk

In the context of unilateral counterparty risk, i.e. $\tau_b=\infty$, and assuming $Q=P$, then, by application of Proposition 3.2.8, the TVA Θ reduces to a (unilateral) CVA given by:

$$\beta_t\Theta_t = \mathbb{E}_t\left[\beta_\tau \mathbb{1}_{\tau<T}\xi\right], \tag{3.23}$$

where $\tau=\tau_c$ and

$$\xi = (1-R_c)\chi^+,\ \ \chi = Q_\tau + \Delta_\tau - \Gamma_\tau. \tag{3.24}$$

Similarly, (3.19) reduces to:

$$\mathrm{CVA}_0 = \int_0^T \beta_s \mathrm{EPE}(s)\mathbb{Q}(\tau \in ds), \tag{3.25}$$

where the expected positive exposure (EPE) is the function of time defined, for $t \in [0, T]$, by:

$$\mathrm{EPE}(t) = \mathbb{E}\left[\xi|\tau = t\right] = \mathbb{E}\left[(1 - R_c)\chi^+|\tau_c = t\right]. \tag{3.26}$$

Remark 3.2.11 The terminology used in the counterparty risk literature is quite fluid. For example, what we call expected positive exposure, and what we denote as EPE(t), is frequently called expected (conditional) exposure and is denoted as EE(t). Then the expected positive exposure, over, say, the nominal lifetime of the portfolio, i.e. over the interval $[0, T]$, is defined as the average

$$\frac{1}{T}\int_0^T EE(t)dt.$$

In fact, frequently in the counterparty risk literature both the EE and the corresponding expected positive exposure are computed under the statistical probability, rather than under the pricing measure $\mathbb{Q}$.

Remark 3.2.12 Frequently, in the case of unilateral counterparty risk, the EPE is computed under the assumption that $R_c = 0$. A loss-given-default factor is then introduced at the stage of CVA computation based on this “zero recovery EPE”.

3.2.5 Dynamic Hedging of Counterparty Risk

Since a TVA is the price process of the corresponding CCDS, then in principle it can be hedged according to usual practices of hedging of financial derivatives. Here, we give some generalities regarding the issue of dynamic hedging of TVA in a model-free set-up. Specific, model based discussion will be given later in the book. The starting point for dynamic hedging of the TVA is the derivation of the TVA dynamics, based on the following generic formula.

Lemma 3.2.13 *For any $t \in [0, \bar{\tau}]$, we have:*

$$d\Theta_t \quad = \quad J_t(dP_t - d\Pi_t) - \xi dJ_t. \tag{3.27}$$

Proof. Before τ, we have from the right-hand side in (3.12) that $d\Theta_t = dP_t - d\Pi_t$. On $\tau < T$, (3.14) implies that $\Delta\Theta_\tau = \xi - \Theta_{\tau-}$. □

The dynamics of Θ splits into the “pre-first-party-default" component $J_t(dP_t - d\Pi_t)$ and the “at-first-party-default" component $-(\xi - \Theta_{t-})dJ_t$. This splitting will be important regarding hedging of counterparty risk.

3.2.5.1 Min-Variance Hedging

In the classical single-curve funding setup of this chapter, given a set of hedging instruments with vector process of cumulative prices denoted by $\widehat{\mathcal{P}}$, the time-t tracking error e_t is defined as

$$\beta_t e_t(\zeta) = \beta_t \Theta_t - \int_0^t \zeta_u d(\beta_u \widehat{\mathcal{P}}_u), \tag{3.28}$$

where ζ is the hedging strategy (in row-vector form). Given that there are enough hedging instruments available to span all the risks embedded in the TVA process, one can replicate the latter using a self-financing portfolio based on these instruments. We refer to Bielecki, Jeanblanc, and Rutkowski (2009) for a comprehensive study of replication of defaultable claims that is also relevant in the context of counterparty risk. But, typically, the TVA process can't be replicated and thus we need to revert to approximate hedging. We refer the reader to Schweizer (2001) for a survey about various quadratic hedging approaches that can be used in an incomplete market. From the practical perspective it is of course more desirable to hedge under the statistical measure $\mathbb{P}$. But minimizing the hedging error under $\mathbb{P}$ entails some difficult technicalities. Therefore, for the sake of tractability, we only consider in this book minimization under the martingale pricing measure $\mathbb{Q}$, which is a rather straightforward exercise using the conditional covariance sharp bracket formula in the second line in (13.28), along with a standard min-variance hedging regression formula (see e.g. Sect. 4.2.3.1 in (Crépey 2013)). To emphasize this difference we write in this book min-variance hedging, under $\mathbb{Q}$ (as opposed to mean-variance hedging, under $\mathbb{P}$).

Remark 3.2.14 The numerical results of De Franco, Tankov, and Warin (2013) indicate that, at least in some of the cases reported, minimizing under $\mathbb{P}$ or $\mathbb{Q}$ does not make such a difference in practice.

3.3 CSA Specifications

3.3.1 Close Out Valuation Schemes

The specification $Q_\tau = P_\tau$ (or, as will be discussed later in the book, $Q_\tau = P_{\tau-}$) appears to be considered as the current market standard. But this is not the only possibility. A different one is $Q_\tau = \Pi_{\tau-}$, which represents the case (admittedly a bit artificial, mainly for the sake of the argument) of a bank in a "dominant" position, able to enforce the value from its own perspective ("cost of its own hedge"), i.e. Π, for the CSA closeout valuation process Q. We observed in Crépey, Jeanblanc, and Zargari (2010) that, in the simple reduced-form set-up of this paper, adopting either convention made little difference in practice. However, this might not be the case in other (e.g. structural) set-ups.

Another alternative discussed in Brigo and Morini (2010a) is to set

$$Q_\tau = P_\tau + \text{UDVA}_\tau,$$

where UDVA represents a univariate debt valuation adjustment of the surviving party. The latter alternative seems to be supported by ISDA (2009) Closeout Amount Protocol, where it is stated that

> "In determining a Closeout Amount, the Determining Party may consider any relevant information, including, without limitation, one or more of the following types of information: (i) quotations (either firm or indicative) for replacement transactions supplied by one or more third parties that may take into account the creditworthiness of the Determining Party at the time the quotation is provided".

The rationale behind the above is that it makes valuation more continuous: upon default we still price including the DVA, as we were doing before default. See also Brigo, Buescu, and Morini (2012) for a discussion of the impact of alternative closeout conventions.

3.3.2 Collateralization Schemes

Modeling of the margin process Γ is a key issue in counterparty risk modeling. For instance, the bailout of the monoline AIG in 2008 was largely triggered by its inability to face increasing margin calls on its sell-protection CDS positions (in particular, on the distressed Lehman). The "continuous scheme" $\Gamma_\tau = Q_{\tau-}$ is, arguably, the extreme case of the collateral (collateral continuously updated to track the left-limit of the CSA value of the contract, the left-limit in $Q_{\tau-}$ reflecting an "infinitesimal" cure period, i.e. time needed for a liquidator to close all the positions of a defaulted party; see Sect. 3.3.3). In this case, we obtain from (3.1) and (3.13):

$$\begin{aligned}\xi = (P_\tau - Q_\tau) &+ (1 - R_c)\mathbb{1}_{\tau=\tau_c}\left((D+Q)_\tau - (D+Q)_{\tau-}\right)^+ \\ &- (1 - R_b)\mathbb{1}_{\tau=\tau_b}\left((D+Q)_\tau - (D+Q)_{\tau-}\right)^- .\end{aligned}$$

Remark 3.3.1 This corresponds to the case of bilateral counterparty risk. In the unilateral case, the continuous scheme reads $\Gamma_\tau = Q^+_{\tau-}$, with corresponding ED given by:

$$\xi = (P_\tau - Q_\tau) + (1 - R_c)\mathbb{1}_{\tau=\tau_c}\left((Q^+ + D)_\tau - (Q^+ + D)_{\tau-} - Q^-_\tau\right)^+ . \quad (3.29)$$

In addition to the obvious "naked" scheme $\Gamma = 0$ and the above continuous scheme $\Gamma = Q_-$, other reference collateralization schemes are the ISDA schemes to be described now. According to International Swaps and Derivatives Association (2010), page 57, the paradigm for the level of collateral amount is the following:

> "Collateral value = (i) the [collateral taker's] exposure plus (ii) the aggregate of all independent amounts applicable to the [collateral provider], if any, minus (iii) the aggregate of all independent amounts applicable to the collateral taker, if any, minus (iv) the [collateral provider]'s threshold."

The "exposure" in the above terminology refers to the CSA value process Q_t of the reference portfolio. Here, we propose an algorithm that is meant to generate the collateral process that, right after every margin call time, conforms to the above paradigm. That is to say, since there are no "independent amounts" as of items (ii) and (iii) in our set-up, therefore "collateral value = CSA value minus threshold", where threshold refers to bounds that are set on the admissible values of the "debt" χ between the parties (so the parties need to adjust the collateral Γ in case χ in (3.1) leaves these bounds).

Remark 3.3.2 In the case of centrally cleared transactions, in addition to the above "variation margin", an independent amount (also called initial margin, even though it is also readjusted on a possibly daily basis) is posted by both parties in order to mitigate gap risk. This is likely to be soon imposed in bilateral transactions as well (see Sect. 3.3.3).

Towards this end, we denote by $t_0 = 0 < t_1 < \cdots < t_n < T$ the margin call dates. Thus, we assume, as it is done in practice, that margin calls are executed according to a discrete tenor of dates.

Remark 3.3.3 The time interval between the default time τ and the last margin call date preceding it constitutes the first part of the so called margin period of risk, the second part being the cure period δ which will be dealt with in Sect. 3.3.3.

In order to model a realistic collateral process, we need to introduce:

- the nominal thresholds ("free credit lines") of the bank and the counterparty, $\eta_b \leq 0$ and $\eta_c \geq 0$,
- the minimum transfer amounts of the bank and the counterparty, $\epsilon_b \leq 0$ and $\epsilon_c \geq 0$.

Then we define the effective thresholds for the bank and the counterparty as $\widehat{\eta}_b = \eta_b + \epsilon_b$ and $\widehat{\eta}_c = \eta_c + \epsilon_c$. We define

$$\widetilde{\chi}_t = Q_t + \Delta_t - \Gamma_t, \tag{3.30}$$

so that $\chi = \widetilde{\chi}_\tau$ (see (3.1)). In the ISDA collateralization scheme we construct the right continuous, piecewise-constant collateral process Γ by setting $\Gamma_{0-} = 0$ and by postulating that, at every $t_i < \tau$,

$$\begin{aligned}\Delta\Gamma_{t_i} = \Gamma_{t_i} - \Gamma_{t_i-} &= \mathbb{1}_{\widetilde{\chi}_{t_i-} > \widehat{\eta}_c}(\widetilde{\chi}_{t_i-} - \eta_c)^+ - \mathbb{1}_{\widetilde{\chi}_{t_i-} < \widehat{\eta}_b}(\widetilde{\chi}_{t_i-} - \eta_b)^- \\ &= \mathbb{1}_{\widetilde{\chi}_{t_i-} > \widehat{\eta}_c}(\widetilde{\chi}_{t_i-} - \eta_c) + \mathbb{1}_{\widetilde{\chi}_{t_i-} < \widehat{\eta}_b}(\widetilde{\chi}_{t_i-} - \eta_b).\end{aligned} \tag{3.31}$$

Then Γ stays constant on every interval (t_i, t_{i+1}). Note that the amount (3.31) of collateral, which is transferred at the call times according to the above scheme, satisfies the following natural properties:

- $\Delta\Gamma_{t_i} > 0$ if the debt $\widetilde{\chi}$ right before t_i exceeds the counterparty's threshold $\widehat{\eta}_c$; this means that at time t_i the bank makes a margin call and the counterparty delivers $\Delta\Gamma_{t_i}$ worth of (cash) collateral; intuitively, the counterparty thus brings $\widetilde{\chi}$ down to η_c at t_i if it exceeded $\widehat{\eta}_c$ right before t_i,

- $\Delta\Gamma_{t_i} < 0$ if $\widetilde{\chi}_{t_i-}$ is less than the bank's threshold $\widehat{\eta}_b$; this means that at time t_i the counterparty makes a margin call and the bank delivers $(-\Delta\Gamma_{t_i})$ worth of (cash) collateral; intuitively, the bank brings it up to η_b at t_i if it was lower than $\widehat{\eta}_b$ right before t_i,,
- $\Delta\Gamma_{t_i} = 0$ if $\widetilde{\chi}_{t_i-}$ is within the bounds $[\widehat{\eta}_b, \widehat{\eta}_c]$; this means that at time t_i no margin call is made and no collateral is transferred by either of the two parties; debt remains unadjusted.

The identity (3.32) below shows that, at margin call times, the ISDA collateralization scheme conforms to the requirements of ISDA.

Proposition 3.3.4 *We have, at every t_i,*

$$\Gamma_{t_i} = Q_{t_i-} - \left(\mathbb{1}_{\widetilde{\chi}_{t_i-}>\widehat{\eta}_c}\eta_c + \mathbb{1}_{\widetilde{\chi}_{t_i-}<\widehat{\eta}_b}\eta_b + \mathbb{1}_{\widehat{\eta}_b\le\widetilde{\chi}_{t_i-}\le\widehat{\eta}_c}\widetilde{\chi}_{t_i-}\right), \tag{3.32}$$

so that $Q_{t_i-} - \Gamma_{t_i} \in [\widehat{\eta}_b, \widehat{\eta}_c]$, with $Q_{t_i-} - \Gamma_{t_i} = \widehat{\eta}_c$ if $\widetilde{\chi}_{t_i-} > \widehat{\eta}_c$ and $Q_{t_i-} - \Gamma_{t_i} = \widehat{\eta}_b$ if $\widetilde{\chi}_{t_i-} < \widehat{\eta}_b$.

Proof. From (3.31), at every t_i, we have:

$$\Gamma_{t_i} = \Gamma_{t_i-} - \left(\mathbb{1}_{\widetilde{\chi}_{t_i-}>\widehat{\eta}_c}\eta_c + \mathbb{1}_{\widetilde{\chi}_{t_i-}<\widehat{\eta}_b}\eta_b + \mathbb{1}_{\widehat{\eta}_b\le\widetilde{\chi}_{t_i-}\le\widehat{\eta}_c}\widetilde{\chi}_{t_i-}\right) + \widetilde{\chi}_{t_i-},$$

which is (3.32) in view of (3.30), where it should be noted that the left-limit process of Δ_t (itself the jump process of the càdlàg process D) is 0. □

The above ISDA scheme is not the only possibility. Defining the thresholded exposure

$$\widehat{Q}_t = (Q_t - \eta_b)^+ - (Q_t - \eta_c)^-, \tag{3.33}$$

a closely related scheme goes as follows: $\Gamma_{0-} = 0$ and, at every $t_i < \tau$:

$$\Gamma_{t_i} - \Gamma_{t_i-} = \mathbb{1}_{\widehat{Q}_{t_i-}-\Gamma_{t_i-}\notin[\epsilon_b,\epsilon_c]}(\widehat{Q}_{t_i-} - \Gamma_{t_i-}). \tag{3.34}$$

Proposition 3.3.5 *Under the alternative ISDA scheme* (3.33)-(3.34)*, we have, at every t_i, that $Q_{t_i-} - \Gamma_{t_i} \in [\widehat{\eta}_b, \widehat{\eta}_c]$ and $Q_{t_i-} - \Gamma_{t_i} \in [\eta_b, \eta_c]$ provided $Q_{t_i-} - \Gamma_{t_i-} \notin [\epsilon_b, \epsilon_c]$.*

Proof. If $\widehat{Q}_{t_i-} - \Gamma_{t_i-} \in [\epsilon_b, \epsilon_c]$, then $\Gamma_{t_i} = \Gamma_{t_i-}$, so that

$$Q_{t_i-} - \Gamma_{t_i} = Q_{t_i-} - \widehat{Q}_{t_i-} + \widehat{Q}_{t_i-}\Gamma_{t_i} \in [\eta_b + \epsilon_b, \widehat{\eta}_c + \epsilon_c] = [\widehat{\eta}_b, \widehat{\eta}_c].$$

Otherwise $\Gamma_{t_i} = \widehat{Q}_{t_i-}$, so that

$$Q_{t_i-} - \Gamma_{t_i} = Q_{t_i-} - \widehat{Q}_{t_i-} + \widehat{Q}_{t_i-}\Gamma_{t_i} = Q_{t_i-} - \widehat{Q}_{t_i-} \in [\eta_b, \eta_c].$$

□

Remark 3.3.6 Translation from cash based collateralization in this book to collateralization through assets would need to be done via haircuts. That is, if the collateral transferred at time t_i is posted in some asset different from cash, then the total value of that asset that needs to be posted is $(1 + h_{t_i})\Delta\Gamma_{t_i}$, where h_{t_i} is the appropriate haircut to be applied at time t_i. In case of a portfolio of collateral assets, one distributes $\Delta\Gamma_{t_i}$ among the assets, and applies appropriate haircut to each portion. Moreover, in that case the value Γ_t is no longer constant between t_i but fluctuates as the one of the collateral assets.

3.3.3 Cure Period

In practice there is a time lag $\delta > 0$, called the cure period and typically taken to be (at least) $\delta =$ two weeks, between the default time τ and the closeout cash flow, which thus occurs at time $\tau + \delta$. One calls margin period of risk the time lag between the last margin call preceding τ and the time $\tau + \delta$ of the closeout cash flow. The cure period thus constitutes the second part of the margin period of risk, the first part being the time lag between the default time τ and the last margin call preceding it. These two components of the margin period of risk play rather distinct roles in the modeling. The role of the first component has been analyzed in Sect. 3.3. The second component (cure period) represents a time period δ between the "effective" default time τ, in the sense of the time at which promised dividends and margin calls cease to be paid by the distressed party, and the "legal" default time $\tau + \delta$ of the closeout cash flow. Accounting for a positive cure period δ, the closeout cash flow $\mathfrak{R}$ is still given by (3.2), but for χ in (3.2) now given, instead of (3.1), by:

$$\beta_{\tau+\delta}\chi^{\delta} = \beta_{\tau+\delta}Q_{\tau+\delta} + \int_{[\tau,\tau+\delta]} \beta_t dD_t - \beta_{\tau+\delta}\Gamma_{\tau}, \tag{3.35}$$

for an $\mathcal{G}_{\tau+\delta}$-measurable CSA value $Q_{\tau+\delta}$. Also, the recoveries R_c and R_b are now given as $\mathcal{G}_{\tau_c+\delta}$- and $\mathcal{G}_{\tau_b+\delta}$- (instead of $\mathcal{G}_{\tau_c}$- and $\mathcal{G}_{\tau_b}$- previously) measurable random variables. For example, if the counterparty stops payments at time $\tau = \tau_c < \tau_b$, then, at time $\tau + \delta$ the closeout cash flow takes place in the amount of $\Gamma_\tau + \left(R_c\chi^{\delta,+} - \chi^{\delta,-}\right)$. Of course, in case $\delta = 0$, this reduces to the above no-cure-period case. In case of a positive cure period δ, it follows by a straightforward adaptation of the proof of Proposition 3.2.8 that

$$\beta_t\Theta_t = \mathbb{E}_t\left[\beta_{\tau+\delta}\mathbb{1}_{\tau<T}\xi^{\delta}\right], \tag{3.36}$$

for

$$\xi^{\delta} = P_{\tau+\delta} - Q_{\tau+\delta} + (1 - R_c)\mathbb{1}_{\tau=\tau_c}\chi^{\delta,+} - (1 - R_b)\mathbb{1}_{\tau=\tau_b}\chi^{\delta,-}.$$

The margin period of risk induces gap risk, which is the risk of increased mismatch between the (closeout CSA) value of a position and of the collateral, due to the variation of value of the contract during the margin period of risk. Gap risk motivates an additional layer of counterparty risk mitigation in the form of so called initial margins, maintained as a reserve by each of the two parties and segregated by a third one (see the remark

3.3.2). The determination of the initial margin has recently been the topic of intensive discussions and debate between banks and regulators (see Basel Committee on Banking Supervision and Board of the International Organization of Securities Commissions (2012, 2013)).

For simplicity, henceforth in this book we assume that $\delta = 0$; see Crépey and Song (2014) for more in this regard.

3.3.4 Rehypothecation Risk and Segregation

When discussing cash flows of an OTC contract in the presence of collateralization, it is important to point out an important risk factor, rehypothecation, sometimes mitigated by segregation. Rehypothecation refers to the possibility for the counterparty (the symmetric issue arises for the latter with respect to the bank) to post as collateral, in the context of another transaction with a third party, assets that it had previously received from the bank as collateral (and therefore still belong to the bank, at least until a default happens). To mitigate this risk, collateral is sometimes kept in a segregated, third-party account. In case of a default, the segregated collateral has a higher seniority than other components of the debt, which can be accomodated in our formalism by suitable amendments to the process $\mathfrak{R}$ (see the comments following (3.2)).

3.3.5 Haircuts

According to Basel Committee on Banking Supervision (2004), page 31, different forms of collateral may be used, such as:

- gold,
- debt securities rated by a recognized external credit assessment institution, where these are either:
  - at least BB- when issued by sovereigns or PSEs that are treated as sovereigns by the national supervisor;
  - or at least BBB- when issued by other entities (including banks and securities firms);
  - or at least A-3/P-3 for short-term debt instruments;
- debt securities not rated by a recognised external credit assessment institution, where these are:
  - issued by a bank;
  - and listed on a recognised exchange;
  - and classified as senior debt;
  - and all rated issues of the same seniority by the issuing bank are rated at least BBB- or A-3/P-3;

  - and the bank holding the securities as collateral has no information to suggest that the issue justifies a rating below BBB- or A-3/P-3 (as applicable);
  - and the supervisor is sufficiently confident about the market liquidity of the security;

- equities (including convertible bonds) that are included in a main index,
- undertakings for collective investments in transferable securities (UCITS) and mutual funds.

However, in practice, it is mostly cash that is used for collateral, with some instances of sovereign bonds. It is not clear whether credit derivatives have ever been used as collateral, although thee European Central Bank indicates that CDS contracts can be used for this purpose (see European Central Bank (2009)).

3.3.6 Centrally Cleared Trading

The current trend of the regulation is to push dealers to negotiate centrally, via clearing houses (or CCPs, i.e. central counterparties) interfacing them and asking for intraday variation margin and daily initial margin (see the remark 3.3.2). Moreover, by default in principle, starting January 2015, an OTC derivative transaction initiated between two dealers (as opposed to a transaction involving end-clients, for which this would put an excessive liquidity pressure) must be collateralized, in terms of variation margins as in current CSAs but also of initial margins (see Basel Committee on Banking Supervision and Board of the International Organization of Securities Commissions (2012, 2013). Exceptions would include foreign-exchange derivatives, for which the operational cost of collateralization is not be justified with regard to their relatively low counterparty risk. Also, this will only regard newly initiated transactions.

Even though we don't directly deal with centrally cleared trading in this book, most of the counterparty risk and funding analysis that we develop can be adapted to this case by introduction of the initial margin as done in Pallavicini and Brigo (2013). The generalization of CCPs and collateralization also poses a severe liquidity constraint on the market, as well as serious systemic and concentration risk issues (see Duffie (2010), Cont and Kokholm (2012), Cont, Mondescu, and Yu (2011), Cont, Santos, and Moussa (2013), Singh and Aitken (2009), Singh (2010), Levels and Capel (2012)).

Chapter 4

Bilateral Counterparty Risk under Funding Constraints

4.1 Introduction

Under current market conditions, banks are no longer default-free and one should take a bilateral counterparty risk perspective. Since it makes no sense to assume that a default-risky bank can lend and borrow cash at a common and risk-free rate, a companion issue is a proper assessment of the cost for the bank of funding its position. This question has become a major concern for the industry, reflected for instance in Piterbarg (2010), Morini and Prampolini (2011), Burgard and Kjaer (2011a,2011b, 2012), Hull and White (2013b, 2013a, 2013c, 2013d) or Castagna (2011).

In the previous chapter we worked under the classical assumption that there exists a locally risk-free asset, accruing at a single rate r_t, that can be used for funding purposes. The object of this chapter is to extend the previous developments to the case of multiple funding costs. Towards this goal, we develop in this chapter, still in a model-free fashion, an "additive, multi-curve" extension of the classical "multiplicative (i.e. discounted), single-curve" risk-neutral pricing approach.

4.1.1 Outline

In Sect. 4.2 we review all cash flows involved in a multi-curve setup. In Sect. 4.3 we characterize the hedging error arising from a given pricing and hedging scheme for the contract. Given potential nonlinearities in the funding cash flows, we cannot proceed through the usual discounting approach (at the risk-free rate or even with a credit spread). In Sect. 4.4 the cash flows are priced under a "multi-curve, additive, flat" extension of the classical "single-curve, multiplicative, discounted" risk-neutral assumption. We also derive the dynamic hedging interpretation of this "additive risk-neutral" price. Sect. 4.5 is a first model-free attempt toward a counterparty risk and funding TVA. Sect. 4.6 provides an illustrative example, discussing in the additive martingale pricing perspective of this chapter the situation in a complete Black-Scholes market, considered in Burgard and Kjaer (2011a, 2011b).

4.2 Market Model

This section specifies a "multi-curve" market model for a contract (understood here as a generic CSA portfolio of OTC derivative transactions with time horizon T between the bank and the counterparty), corresponding hedging assets and corresponding funding assets.

We call external funder (or funder for short) a generic third-party, possibly composed of several entities and/or devices, ensuring funding of a position taken by the bank. "External" (with respect to the contract) stands here in contrast to the "internal" sources of funding of a portfolio, which are provided to the bank via remuneration of the hedge and of the collateral, and which will be discussed in Sect. 4.2.1 and 4.2.2. In this book we only consider a counterparty risk of the bank with regard to a single CSA, so we really are only concerned with analysis regarding a single counterparty, in particular, just a single contract (i.e. single netted portfolio). Consequently, "external with respect to the contract" may be interpreted as external with respect to the bank. Thus, with this understanding, the external funding is (typically) channeled to the contract by the bank's treasury desk, but it originates outside of the bank. In a sense, the external funder plays the role of "lender/borrower of last resort" that is used after exhaustion of the internal sources of funding provided to the bank by the dividend and funding gains on its hedge and/or via the remuneration of the margin amount. For simplicity we assume the funder to be default-free (accounting for the default risk of the funder would be important for the study of systemic risk).

Consistent with our convention throughout the book, we distinguish two categories of cash flows, where in each bullet below the first item was already present and discussed in the previous chapter and the remaining ones are newly introduced for further discussion related to funding:

- dividends, in the sense of all pre-default cash flows involving the bank, decomposing into:
  - counterparty clean (or promised) contract dividends;
  - gains/losses on the hedging instruments before time τ;
  - the cost/benefit of funding-the-position/investing-in-it prior to time τ; this includes in particular the remuneration of the collateral;
- closeout cash flows, that is cash flows occurring at the default time τ (if $< T$) and consisting of:
  - the CSA closeout cash flow, i.e. the recovery on the contract paid by the counterparty to the bank upon default of a party; this includes in particular the delivery of the collateral;
  - a closeout funding cash flow from the funder to the bank, in case of a default of the bank.

Remark 4.2.1 Recall that throughout the book we consider all cash flows from the bank's perspective, e.g. $\Delta_t = 1$ means a promised cash flow of $+1$ to the bank. In Crépey (2012a, 2012b), Crépey, Gerboud, Grbac, and Ngor (2013) and Crépey, Grbac, Ngor, and Skovmand (2013), the sign convention is the opposite one, which eases the mathematical exposition a little bit, but makes it less intuitive from a financial point of view.

Note that, apart from the promised dividends of the contract and the remuneration of the collateral, which are exchanged between the two parties (and are symmetric[1]), in a multi-curve setup all the other cash flows typically are not symmetric relative to the two parties. This cash flow asymmetry between the parties results in the valuation asymmetry: the values of the contract as seen by the two parties will not be the same (except for the sign difference). And this even without considering different pricing measures which could be jointly considered in the valuation (see the remark 3.2.1). This is why we need to focus on a given party, the bank in this book. Of course analogous considerations will apply to the bank's counterparty, but with nonsymmetrical hedging positions and funding conditions.

Remark 4.2.2 The fact that the economic value (in a sense of cost of hedging, as we will see below) of a contract is different to the two parties poses the practical problem of agreeing on a price between them. Asymmetry of the economic value of the contract is as factual as, for instance, market incompleteness is in the real world. One just has to live with it, and to find trading agreements in this context (still better knowing it than ignoring it). One might claim that asymmetry only regards the funding side, but not the (bilateral) counterparty risk side, so that "maybe at first order" the asymmetry is not such an important issue. However, the level and the impact of asymmetry on the contract valuation depends on the CSA underlying the contract[2]. Moreover, as it will be seen in the next chapter, counterparty risk and funding are tied together, so that one cannot really disentangle them. Thus, asymmetry in valuation needs to be taken seriously. In this context, we refer to discussion in Castagna (2011), Pallavicini, Perini, and Brigo (2011, 2012) or Hull and White (2013a, 2013c, 2013d) regarding different views about the law of one price or the notions of price versus value.

4.2.1 Hedging Assets

Let $\mathcal{P}$ denote the $\mathbb{R}^d$-valued semimartingale price process of a family of hedging assets, and let a finite variation process $\mathcal{C}$ represent the corresponding $\mathbb{R}^d$-valued cumulative promised dividend process (for a unit buy-and-hold position in each of the hedging assets). Note that we ignore the counterparty risk of the third parties with which the bank trades its hedging assets.

We distinguish hedging assets traded in swapped form, at no upfront payment, from hedging assets traded upfront on a primary market. Hedging assets traded in swapped

[1]That is, they do differ just by the sign – positive for one party and negative for the other party.
[2]See Sect. 3.3 and 6.3.

form include, in particular, (counterparty risk clean) CDS contracts on the two parties, which are typically used for hedging the counterparty jump-to-default exposure of the contract.

Remark 4.2.3 A fixed CDS contract, with a given (fixed) contractual spread in particular, cannot be traded dynamically in the market the way that, say, equity is traded. Only newly emitted CDS contracts can be entered into, at no cost and at the related fair market spread, at a given time. What is used in practice for hedging corresponds to the concept of a rolling CDS, formally introduced in Bielecki, Jeanblanc, and Rutkowski (2008)[3], which is essentially a self-financing trading strategy in market CDS. So, much like with futures contracts, the value of a rolling CDS is null at any point in time, but due to the trading gains of the strategy the cumulative value process is not zero.

Note that the case of hedging assets traded in swapped form also covers, in a formalism developed below, the situation of a physical[4] hedging asset traded via a repo market. Therefore we can safely assume that every hedging asset can be traded in swapped form, either as a natively swapped instrument rolled over time or, for a physical asset, via a corresponding repo market. In mathematical terms, trading the hedging asset with price $\mathcal{P}_t^i$ in swapped form effectively means than one uses, instead of the original (physical or fixed swap) asset, a synthetic asset with price process $\mathcal{S}_t^i = 0$ (see below) and gain process of the form q_t^i

$$d\mathcal{P}_t^i - \left(r_t\mathcal{P}_t^i + q_t^i\right)dt + d\mathcal{C}_t^i, \tag{4.1}$$

where:

- in case of an asset traded via a repo market, q_t^i corresponds to a repo basis; the meaning of all other terms in (4.1) is clear, with the provision that $\mathcal{P}_t^i \neq \mathcal{S}_t^i = 0$ (i.e. the value of a "repo swap" referencing the i-th asset with market price $\mathcal{P}_t^i$);
- in case of a natively swapped asset rolled over time[5], the different terms in (4.1) are to be understood as[6]

$$d\mathcal{P}_t^i = d\bar{\mathcal{P}}_t^{i,t_0}\mid_{t_0=t}, \;\; \mathcal{P}_t^i = \bar{\mathcal{P}}_t^{i,t} = \mathcal{S}_t^i = 0, \;\; q_t^i = 0, \;\; d\mathcal{C}_t^i = d\bar{\mathcal{C}}_t^{i,t_0}\mid_{t_0=t}, \tag{4.2}$$

  where $(\bar{\mathcal{P}}_t^{i,t_0})_{t\geq t_0}$ is the price process at time t of the corresponding fixed (as opposed to rolled) swap emitted at time $t_0 \leq t$, with dividend process $(\bar{\mathcal{C}}_t^{i,t_0})_{t\geq t_0}$, and where the notation $\mathcal{S}_t^i$ is introduced here for future use (cf. Sect. 4.3.1).

[3] See also the related concept of floating rate CDS in Brigo (2005).
[4] As opposed to natively swapped.
[5] E.g. a rolling CDS (see the remark 4.2.3).
[6] See Bielecki, Jeanblanc, and Rutkowski (2008) and Bielecki, Crépey, Jeanblanc, and Rutkowski (2011) for more details.

4.2.2 Funding Assets

This subsection provides a possible specification of the funding assets. The corresponding notion of a self-financing trading strategy will be derived in Sect. 4.3.1. A general formulation of the pricing and hedging problem under abstract funding conditions will then be given in Sect. 4.3.2. All the funding costs, before occurrence of the first default, are expressed as a spread relative to the risk-free cost at rate r_t such as, in the case of $\mathcal{P}^i$ above, the repo basis q^i in (4.1).[7] We assume that the bank can lend money to, respectively borrow money from, its external funder at an excess cost over the risk-free rate r_t determined by a funding (credit and liquidity) spread λ, respectively $\bar{\lambda}$. In case the bank is indebted to its funder at time $\tau = \tau_b < T$, the bank will in principle not be in a position to reimburse the totality of its external debt, but only a fraction $\bar{R}_b$ of it, where a $[0, 1]$-valued $\mathcal{G}_{\tau_b}$-measurable random variable represents the recovery rate of the bank towards its external funder. As we show below, this results in a closeout funding cash flow, proportional to $(1 - \bar{R}_b)$, from the external funder to the bank in case $\tau = \tau_b < T$. This cash flow corresponds to the funding side of "the bank benefiting from its own default".

In case $\bar{R}_b < 1$, the bank defaults at time τ_b not only on its commitments in the contract with regard to the counterparty, but also on its related funding debt. The case $\bar{R}_b = 1$ can be seen as a partial default in which at time τ_b the bank only defaults on its contractual commitments with regard to the counterparty, but not on its funding debt with respect to its funder. This can be used for modeling the (rather unusual) situation of a bank in a short-term lender position, so that it actually does not need any external lender (in case cash is needed for funding a position, the bank simply uses its own cash reserves). Also note that, in case of unilateral counterparty risk (from the perspective of the bank, i.e. τ_b set equal to ∞), the value of $\bar{R}_b$ is immaterial.

Remark 4.2.4 In this chapter we no longer postulate the existence of the savings account. The risk-free rate r_t simply corresponds to the time-value of money and we can only think of β_t^{-1} as a "fictitious" savings account. What we have instead is coexistence of various funding assets with different growth rates in the economy. This raises the question of arbitrage that might result from trading between these rates. These can simply reflect different levels of credit-riskiness, so that a related arbitrage opportunity is only a pre-default view, disregarding losses-upon-defaults. Even without credit risk, different funding rates may consistently coexist in an economy, reflecting trading constraints or, in other words, liquidity funding costs. The rationale here is that a given funding rate may be only accessible for a definite notional and for a specific purpose, so that "funding arbitrage strategies" are either not possible or not sought by the parties. An example that arises in the context of counterparty risk is that of the collateral, in which the two parties must have contractually prescribed amounts (denoted below as $\Gamma_t^{\pm}$) at any point in time.

Regarding the collateral (cash-only collateral for simplicity in this book), we follow

[7] There is some abuse of terminology here: typically a "basis" means the differential between two interest rates. For convenience, as dictated by our presentation, we take the liberty of calling q^i the basis.

the most common CSA covenant under which the party getting the collateral can use it in its trading, as opposed to a covenant where the collateral is segregated by a third party in order to avoid the related rehypothecation risk (see Sect. 3.3.4). Specific CSA rates $(r_t + b_t)$ and $(r_t + \bar{b}_t)$ are typically used to remunerate the collateral owned by either party. Recalling that, under our sign convention, $\Gamma_t > 0$ means collateral posted by the counterparty, this results in a remuneration of the margin amount (prior to the first default) which is worth

$$-(r_t + \bar{b}_t)\Gamma_t^+ dt + (r_t + b_t)\Gamma_t^- dt = -r_t\Gamma_t dt + \big(b_t\Gamma_t^- - \bar{b}_t\Gamma_t^+\big)dt$$

to the bank (and the opposite to the counterparty).

Regarding funding of the hedging portfolio, we suppose that the hedge in a given instrument is either entirely swapped (e.g. traded in a secured way on repo market), or funded in totality by the external lender as an unsecured trade; we assume that this choice is made and fixed for the entire life of the contract for every hedging asset. We let superscript ^{s} refer to the subset of the hedging instruments traded in swapped form, and we let $^{\bar{s}}$ refer to the subset, complement of ^{s}, of (physical) hedging instruments which are traded directly on a primary market (and are therefore funded together with the contract by the external funder).

Remark 4.2.5 In the above specification, it is assumed for simplicity that the funding cost is not transaction specific but always the same across transactions within the given contract, distinguishing only the direction of the flow (funding or investment). A different model of treasury, more on the micro scale, would allocate different funding to deals with different characteristics (see Pallavicini, Perini, and Brigo (2011, 2012)).

In order to account for the above funding specification in classical formalism of self-financing trading strategies, we introduce the following funding assets on $[0, \bar{\tau}]$:

- two collateral funding assets, B^0 and $\bar{B}^0$, evolving as

$$dB_t^0 = (r_t + b_t)B_t^0 dt, \ \ d\bar{B}_t^0 = (r_t + \bar{b}_t)\bar{B}_t^0 dt, \tag{4.3}$$

  dedicated to the funding of the negative and the positive part of the margin account (i.e. the bank remunerates the collateral posted by the counterparty at the rate $r + \bar{b}$ when $\Gamma > 0$; conversely, when $\Gamma < 0$, the bank is remunerated for the collateral it posted to the counterparty at the rate $r + b$);

- two external funding assets, B^f and $\bar{B}^f$, evolving as

$$dB_t^f = (r_t + \lambda_t)B_t^f dt, \ \ d\bar{B}_t^f = (r_t + \bar{\lambda}_t)\bar{B}_t^f dt - (1 - \bar{R}_b)\bar{B}_{t-}^f \boldsymbol{\delta}_{\tau_b}(dt), \tag{4.4}$$

  where $\boldsymbol{\delta}_{\tau_b}$ represents a Dirac measure at time τ_b; these are the assets used by the bank for its external investing and funding purposes, respectively.

4.3 Trading Strategies

In this section we characterize the hedging error arising from a given pricing, collateralization, hedging and funding scheme for the bank, detailing in particular the funding cash flows.

A hedge is defined as a left-continuous[8] and locally bounded, $\mathbb{R}^d$-valued row-vector process ζ over $[0, \bar{\tau}]$, where $(-\zeta)$ represents the numbers of units of the hedging assets which are held in the hedging portfolio. This "short position" notation (coming with a "$-$") for hedging positions is chosen for consistency with the idea (just to fix the mindset) that the contract is "bought" by the bank (which, again, simply reflects the sign convention that $\Delta_t = 1$ means $+1$ to the bank).

By price-and-hedge of the contract for the bank we mean any pair-process $(\bar{\Pi}, \zeta)$ over $[0, \bar{\tau}]$, where ζ is a hedge and $\bar{\Pi}$ is a real semimartingale such that $\bar{\Pi}_{\bar{\tau}} = \mathbb{1}_{\tau<T}\mathfrak{R}^c$, where a CSA closeout cash flow $\mathfrak{R}^c$ will be specified later. By hedging error process of the price-and-hedge $(\bar{\Pi}, \zeta)$ we mean $\varrho = \bar{\Pi} + \bar{\mathcal{W}}$, where the process $\bar{\mathcal{W}}$ is the value process of the collateralization, hedging and funding portfolio. Consistent with the idea that the contract is "bought" by the bank, the initial wealth of the bank is modeled as

$$w = \bar{\mathcal{W}}_0 = -\bar{\Pi}_0,$$

and the bank must find a strategy (price-and-hedge) such that $\bar{\mathcal{W}}$ "tracks" $(-\bar{\Pi})$ "as well as possible" so that, in particular, $\varrho_{\bar{\tau}}$ is "as small as possible". Replication corresponds to $\varrho_{\bar{\tau}} = 0$ almost surely (as will be evident from Sect. 4.4.3).

4.3.1 Self-Financing Condition

We begin with a notational remark: for any vector $V \in \mathbb{R}^d$, we denote by V^s and $V^{\bar{s}}$ sub-vectors given as $V^s = (V^i, i \in s)$ and $V^{\bar{s}} = (V^i, i \in \bar{s})$. Also, we set $V^{\emptyset} = 0$.

Given the position of the bank being funded as described in Sect. 4.2.2 we have, for $t \in [0, \bar{\tau}]$,

$$\bar{\mathcal{W}}_t = -\Gamma_t - \zeta_t \mathcal{P}_t + (\bar{\mathcal{W}}_t + \Gamma_t + \zeta_t^{\bar{s}} \mathcal{P}_t^{\bar{s}}), \tag{4.5}$$

where the terms in the parentheses, in which Γ_t is the value of the collateral received by the bank and $\zeta_t^{\bar{s}} \mathcal{P}_t^{\bar{s}} = \zeta_t \mathcal{P}_t$ comes from the proceeds of the shortening of the hedging assets, represent the amounts invested by the bank in the funding assets.

Remark 4.3.1 More precisely let $(-\zeta^s)$ and $(-\zeta^{\bar{s}})$ represent the swapped and non swapped components of the hedging positions. In the notation introduced in Sect. 4.2.1, we thus have:

$$\zeta_t \mathcal{P}_t = \zeta_t^s S_t^s + \zeta_t^{\bar{s}} \mathcal{P}_t^{\bar{s}},$$

[8] This will be explained in the remark 4.3.2.

where $\mathcal{S}^s_t = 0$. In particular, hedging instruments traded in swapped form do not directly contribute to the value $\bar{\mathcal{W}}_t$. But they contribute to the dynamics of $\bar{\mathcal{W}}_t$, via their gain process (4.1).

Equivalent to (4.5), we can write

$$\bar{\mathcal{W}}_t = \eta^0_t B^0_t + \bar{\eta}^0_t \bar{B}^0_t - \zeta^s_t \mathcal{S}^s_t - \zeta^{\bar{s}}_t \mathcal{P}^{\bar{s}}_t + \eta^f_t B^f_t + \bar{\eta}^f_t \bar{B}^f_t, \tag{4.6}$$

with

$$\eta^0_t = \frac{\Gamma^-_t}{B^0_t},\ \bar{\eta}^0_t = -\frac{\Gamma^+_t}{\bar{B}^0_t},\ \eta^f_t = \frac{(\bar{\mathcal{W}}_t + \Gamma_t + \zeta^{\bar{s}}_t \mathcal{P}^{\bar{s}}_t)^+}{B^f_t},\ \bar{\eta}^f_t = -\frac{(\bar{\mathcal{W}}_t + \Gamma_t + \zeta^{\bar{s}}_t \mathcal{P}^{\bar{s}}_t)^-}{\bar{B}^f_t}.$$

We then say in view of (4.6) that a price-and-hedge $(\bar{\Pi}, \zeta)$ of the bank is self-financing if and only if $\bar{\mathcal{W}}_0 = -\bar{\Pi}_0$ and it holds, that for $t \in [0, \bar{\tau}]$:

$$\begin{aligned} d\bar{\mathcal{W}}_t &= dC_t + \eta^0_t dB^0_t + \bar{\eta}^0_t d\bar{B}^0_t \\ &-\zeta^s_t \left(d\mathcal{P}^s_t - (r_t \mathcal{P}^s_t + q^s_t)dt + dC^s_t \right) - \zeta^{\bar{s}}_t (d\mathcal{P}^{\bar{s}}_t + dC^{\bar{s}}_t) + \eta^f_t dB^f_t + \bar{\eta}^f_{t-} d\bar{B}^f_t, \end{aligned} \tag{4.7}$$

where the left-limit ("minus" in the subscript t) in $\bar{\eta}^f_{t-}$ is needed[9] because $\bar{B}^f_t$ jumps at time τ_b and process $\bar{\eta}^f$ is not predictable.

Remark 4.3.2 The possibility taking a left-limit in $\bar{\eta}^f$ is the reason why we restrict ourselves to left-continuous hedges ζ, as opposed to (more general) predictable hedges, typically used in the mathematical finance literature. This restriction does no harm in practice; see for example the hedges found in Sects. 4.6, 5.3, 7.3.3 or 8.3.2.

Fig. 4.1 provides a graphical representation of all the cash flows on $[0, \bar{\tau}]$.

For every real number w and $\mathbb{R}^d$-valued row-vector ς, we define:

$$\begin{aligned} \bar{\mathfrak{X}}_t(w, \varsigma) &= -\left(w + \Gamma_t + \varsigma^{\bar{s}} \mathcal{P}^{\bar{s}}_t\right), \\ \bar{g}_t(w, \varsigma) &= b_t \Gamma^-_t - \bar{b}_t \Gamma^+_t + \lambda_t \left(w + \Gamma_t + \varsigma^{\bar{s}} \mathcal{P}^{\bar{s}}_t\right)^+ - \bar{\lambda}_t \left(w + \Gamma_t + \varsigma^{\bar{s}} \mathcal{P}^{\bar{s}}_t\right)^- + \varsigma^s q^s_t, \end{aligned} \tag{4.8}$$

where $\bar{\mathfrak{X}}_{t-}(\bar{\mathcal{W}}_{t-}, \zeta_{t-})$ will be interpreted as the (algebraic) debt of the bank towards its external funder at time t, and $\bar{g}_t(\bar{\mathcal{W}}_t, \zeta_t)$ will be interpreted as the excess-funding-benefit of the bank prior to the first default. Finally, in order to avoid unnecessary appearance of Dirac measures in several equations, for $t \in [0, \bar{\tau}]$, we introduce:

$$\mathcal{W}_t = \bar{\mathcal{W}}_t - \mathbb{1}_{t \geq \tau_b} \bar{\mathfrak{R}}^f,\ \Pi^\star_t = \bar{\Pi}_t + \mathbb{1}_{t \geq \tau_b} \bar{\mathfrak{R}}^f, \tag{4.9}$$

where $\bar{\mathfrak{R}}^f = (1 - \bar{R}_b) \bar{\mathfrak{X}}^+_{\tau_b -}(\mathcal{W}_{\tau_b -}, \zeta_{\tau_b -})$ will appear in the last line of (4.11) as the closeout cash flow from the external funder to the bank at time $\tau = \tau_b < T$.

[9] We thank Marek Rutkowski for pointing this out as well as for a significant clarification of this section.

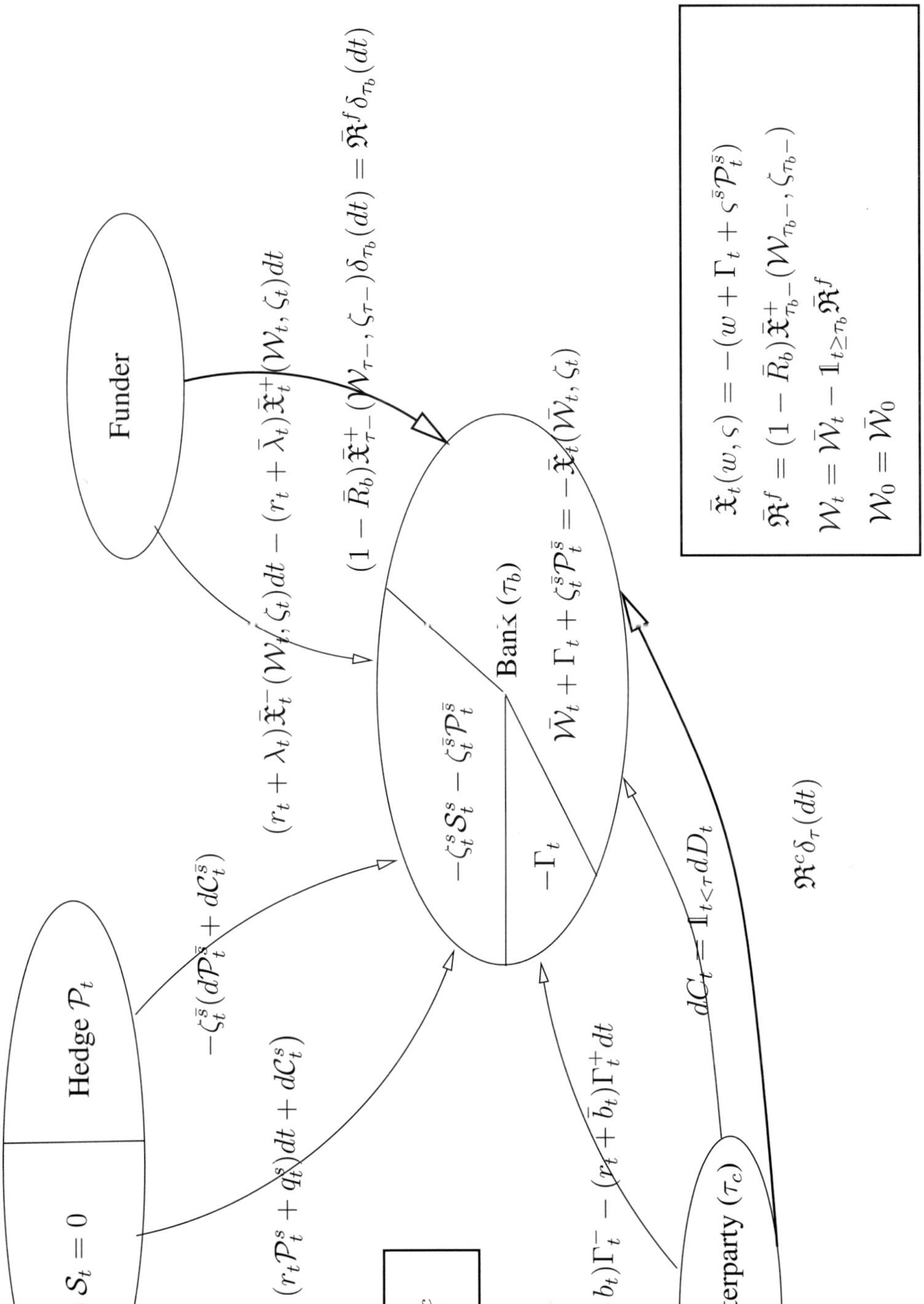

FIGURE 4.1: Cash flows of the bank over $[0, \bar{\tau}]$.

Proposition 4.3.3 *Under the funding specification of Sect. 4.2.2, a price-and-hedge* $(\bar{\Pi}, \zeta)$ *is self-financing if and only if* $\mathcal{W}_0 = -\Pi_0^\star$ *and it holds, that for* $t \in [0, \bar{\tau}]$*:*

$$d\mathcal{W}_t = dC_t + \big(r_t\mathcal{W}_t + \bar{g}_t(\mathcal{W}_t, \zeta_t)\big)dt - \zeta_t(d\mathcal{P}_t - r_t\mathcal{P}_t dt + d\mathcal{C}_t). \tag{4.10}$$

Proof. Substituting (4.3)-(4.4) into (4.7) and also using the current specification of the funding policy regarding hedging assets, we obtain that the strategy is self-financing if and only if $\bar{\mathcal{W}}_0 = -\bar{\Pi}_0$ and it holds, that for $t \in [0, \bar{\tau}]$:

$$\begin{aligned} d\bar{\mathcal{W}}_t =& dC_t + (r_t + b_t)\Gamma_t^- dt - (r_t + \bar{b}_t)\Gamma_t^+ dt - \zeta_t(d\mathcal{P}_t + d\mathcal{C}_t) + \zeta_t^s\,(r_t\mathcal{P}_t^s + q_t^s)\,dt \\ &+ (r_t + \lambda_t)\big(\bar{\mathcal{W}}_t + \Gamma_t + \zeta_t^{\bar{s}}\mathcal{P}_t^{\bar{s}}\big)^+ dt - (r_t + \bar{\lambda}_t)\big(\bar{\mathcal{W}}_t + \Gamma_t + \zeta_t^{\bar{s}}\mathcal{P}_t^{\bar{s}}\big)^- dt \\ &- \bar{\eta}_{\tau-}^f(1 - \bar{R}_b)\bar{B}_{\tau-}^f\boldsymbol{\delta}_{\tau_b}(dt) \\ =& dC_t + r_t(\bar{\mathcal{W}}_t + \zeta_t\mathcal{P}_t)dt - \zeta_t(d\mathcal{P}_t + d\mathcal{C}_t) + b_t\Gamma_t^- dt - \bar{b}_t\Gamma_t^+ dt + \zeta_t^s q_t^s dt \\ &+ \lambda_t\big(\bar{\mathcal{W}}_t + \Gamma_t + \zeta_t^{\bar{s}}\mathcal{P}_t^{\bar{s}}\big)^+ dt - \bar{\lambda}_t\big(\bar{\mathcal{W}}_t + \Gamma_t + \zeta_t^{\bar{s}}\mathcal{P}_t^{\bar{s}}\big)^- dt \\ &+ (1 - \bar{R}_b)\,\big(\bar{\mathcal{W}}_{\tau-} + \Gamma_{\tau-} + \zeta_{\tau-}^{\bar{s}}\mathcal{P}_{\tau-}^{\bar{s}}\big)^-\,\boldsymbol{\delta}_{\tau_b}(dt) \\ =& dC_t + r_t\bar{\mathcal{W}}_t dt - \zeta_t(d\mathcal{P}_t - r_t\mathcal{P}_t dt + d\mathcal{C}_t) + \bar{g}_t(\bar{\mathcal{W}}_t, \zeta_t)dt + \mathfrak{R}^f\boldsymbol{\delta}_{\tau_b}(dt). \end{aligned} \tag{4.11}$$

In view of (4.9), this proves the result. □

4.3.2 General Price-and-Hedge

As illustrated in Sect. 4.2.2, the exact nature of the funding cash flows depends on the specification of a funding policy defined in terms of related funding assets (once the contract and its the hedging strategy are specified). For the sake of clarity and generality we will henceforth work with the following abstract definition of a (self-financing) price-and-hedge, in which the funding shows up through an abstract excess-benefit-funding coefficient $\bar{g}_t(w, \varsigma)$, for $t < \bar{\tau}$, and a funding closeout cash flow $(1 - \bar{R}_b)\bar{\mathfrak{X}}_t^+(w, \varsigma)$, without explicit reference to specific funding assets[10]. Here, a $\mathcal{G}_{\tau_b}$-measurable random variable $\bar{R}_b$ represents, as before, a recovery rate of the bank towards an external funder (assumed risk-free), the case $\bar{R}_b = 1$ also covering by convention the situation of unilateral counterparty risk where τ_b is set equal to ∞ everywhere; $\bar{\mathfrak{X}}$ represents an abstract debt function of the bank to its external funder.

The following definition of a general price-and-hedge is put in the form of a forward-backward stochastic differential equation (FBSDE; see Ma and Yong (2007)) in the variables $(\mathcal{W}, \Pi^\star, \zeta, \varrho)$, where the hedging error ϱ equals $\Pi^\star + \mathcal{W}$ in view of (4.12). Solving this FBSDE would mean solving the related control problem, i.e. finding a general price-and-hedge $(\bar{\Pi}, \zeta)$ such that the related process ϱ has "nice" properties in terms of arbitrage ("ϱ being a martingale under some equivalent probability measure") and replication ("ϱ being small in some appropriate norm"). But in view of the coupled ("forward-backward") initial condition $\mathcal{W}_0 = -\Pi_0^\star$, this is a rather nonstandard FBSDE; see Horst,

[10]There are also the internal sources of funding, specified by the data $b, \bar{b}$ and c, provided to the bank by the remuneration of the collateral and by the swapped component of its hedge.

Hu, Imkeller, Reveillac, and Zhang (2011) for related technical issues. In fact we will not try to solve this FBSDE, but a more tractable backward stochastic differential equation (BSDE) introduced in the next section.

Definition 4.3.4 (General Price-and-Hedge) Let real semimartingales $\mathcal{W}, \Pi^\star$, ϱ and a hedge ζ satisfy the initial condition $\mathcal{W}_0 = -\Pi^\star_0$, $\varrho_0 = 0$, and for $t \in [0, \bar{\tau}]$:

$$\begin{aligned} d\mathcal{W}_t &= dC_t + \big(r_t\mathcal{W}_t + \bar{g}_t(\mathcal{W}_t, \zeta_t)\big)dt - \zeta_t(d\mathcal{P}_t - r_t\mathcal{P}_t dt + d\mathcal{C}_t), \\ d\Pi^\star_t &= -dC_t - \big(r_t\mathcal{W}_t + \bar{g}_t(\mathcal{W}_t, \zeta_t)\big)dt + \zeta_t(d\mathcal{P}_t - r_t\mathcal{P}_t dt + d\mathcal{C}_t) + d\varrho_t, \end{aligned} \tag{4.12}$$

along with a terminal condition $\Pi^\star_{\bar{\tau}} = \mathbb{1}_{\tau<T}\mathfrak{R}$, where

$$\mathfrak{R} = \mathfrak{R}^c + \mathbb{1}_{\tau=\tau_b}\bar{\mathfrak{R}}^f, \tag{4.13}$$

in which $\bar{\mathfrak{R}}^f = (1 - \bar{R}_b)\bar{\mathfrak{X}}^+_{\tau-}(\mathcal{W}_{\tau-}, \zeta_{\tau-})$.

We then call general price-and-hedge with hedging error ϱ the pair-process $(\bar{\Pi}, \zeta)$ where, for $t \in [0, \bar{\tau}]$,

$$\bar{\Pi}_t = \Pi^\star_t - \mathbb{1}_{t\geq\tau_b}\bar{\mathfrak{R}}^f.$$

We say that $(\bar{\Pi}, \zeta)$ is a replicating strategy if $\varrho_{\bar{\tau}} = 0$ almost surely.

Observe that $\mathfrak{R}$ represents the total closeout cash flow received by the bank at time τ if $\tau < T$: CSA closeout cash flow $\mathfrak{R}^c$ from the counterparty, plus closeout funding cash flow $\mathbb{1}_{\tau=\tau_b}\bar{\mathfrak{R}}^f$ from the external funder. Also note that under the funding specification of Sect. 4.2.2, the definition 4.3.4 is consistent with the developments made in Sect. 4.3.1. In the abstract definition 4.3.4, we focus on the processes $\Pi^\star$ and $\mathcal{W}$ rather on the processes $\bar{\Pi}$ and $\bar{\mathcal{W}}$ that showed up in the specific setup of Sect. 4.2.2 and 4.3.1, because $\Pi^\star$ (and ultimately Π introduced in the definition 4.4.5 below) and $\mathcal{W}$ lead to nicer expressions (regarding the treatment of Dirac measures). By abuse of terminology we call $\mathcal{W}$ the value of the hedging portfolio (to be precise, it is process $\bar{\mathcal{W}}$ which corresponds to what should be called exactly the value of the collateralization, hedging and funding portfolio).

Also observe that in case $\bar{R}_b = 1$ and $\bar{g} = c = 0$ (classical single-curve funding setup of the previous chapter), one recovers the usual notion of a self-financing hedging strategy with related wealth process $\mathcal{W}$. The funding closeout cash flow $\bar{\mathfrak{R}}^f$ and the funding bases $\bar{g}$ and c can thus be interpreted as the corrections to a classical single-curve setup.

4.4 Martingale Pricing Approach

In this section we deal with pricing of the contract bought by the bank from the counterparty, under the funding conditions of the bank given in terms of the external recovery rate $\bar{R}_b$ and the funding bases. Given nonlinearities in the funding (unless $\bar{R}_b =$ 1, and unless the funding coefficient $\bar{g}$ is linear), a special analysis of the funding costs has to be conducted beyond discounting [11] (compare with the previous chapter). Cash flows

[11]Unless one resorts to an implicit discount factor depending on the contract's value, which we seek.

will be priced instead under an "additive, flat" extension of the classical "multiplicative, discounted" risk-neutral assumption. We also derive the dynamic hedging interpretation of such an additive risk-neutral price, including consistency with pricing by replication in the case of a complete market.

4.4.1 Primary Market

We start by considering a market in which all hedging assets are such that their gains form a process, denoted as $\mathcal{M}$, satisfying: $\mathcal{M}_0 = 0$ and, for $t \in [0, \bar{\tau}]$,

$$d\mathcal{M}_t = d\mathcal{P}_t - \big(r_t\mathcal{P}_t + q_t\big)dt + d\mathcal{C}_t. \tag{4.14}$$

In view of (4.1) this market would need to be considered as the primary market if all hedging instruments were traded in swapped form. Clearly, such a market should not admit arbitrage opportunities (see the discussion following the assumption 4.4.1). Accordingly, our standing probability measure $\mathbb{Q}$ is henceforth interpreted as a risk-neutral pricing measure in such a primary market of hedging instruments, in the sense that:

Assumption 4.4.1 The primary gain process $\mathcal{M}$ is an $\mathbb{R}^d$-valued $\mathbb{Q}$-martingale.

Note that, even though some hedging instruments are not traded in the swapped form, the above assumption qualifies the probabilistic nature of their corresponding gains under measure $\mathbb{Q}$.

An arbitrage is a self-financing strategy with almost surely nonnegative gain and positive gain with a positive probability, under the historical or any equivalent probability measure. Since our standing probability measure $\mathbb{Q}$ is equivalent to the historical probability measure $\mathbb{P}$, Assumption 4.4.1 bans arbitrage opportunities in the market of the hedging assets traded in swapped form.

Remark 4.4.2 The expression (4.14) for the primary gain process $\mathcal{M}$ (and similar expressions related to valuation of the contract below) involves no discount factors. In this sense, this approach is an "additive" version of the "multiplicative" risk-neutral assumption more commonly used throughout the language of discounting at the risk-free rate r_t in the single-curve literature. For instance, in a Black-Scholes model on a stock S (with constant risk-free rate r, nil repo basis on S and zero dividends, for notational simplicity), our "additive" martingale $\mathcal{M}$ and the usual single-curve "multiplicative" martingale are respectively written as:

$$\mathcal{M}_t = \int_0^t dS_s - \int_0^t rS_s ds, \;\; \mathcal{N}_t = e^{-rt}S_t,$$

so that $d\mathcal{N}_t = e^{-rt}d\mathcal{M}_t$. More generally, in any classical "single-curve" setup, an additive and a multiplicative approach are equivalent and the multiplicative view is more convenient, since it allows one to get rid of the funding issue, which becomes "absorbed"in the discounting at the risk-free rate. But, in a bilateral counterparty risk and multi-curve setup, the funding issue has to be accounted for explicitly, i.e. "additively". Note that, unless $\bar{R}_b = 1$, the funding terms are nonlinear in a bilateral counterparty risk setup, via the positive part in $(1 - \bar{R}_b)\bar{\mathfrak{X}}^+$.

Under the assumption 4.4.1 it is convenient to rewrite (4.10) in martingale form as

$$d\mathcal{W}_t = \big(r_t\mathcal{W}_t - g_t(-\mathcal{W}_t, \zeta_t)\big)dt + dC_t - \zeta_t d\mathcal{M}_t, \qquad (4.15)$$

where, for any $\pi \in \mathbb{R}$ and $\varsigma \in \mathbb{R}^d$, respectively representing the contract's price[12] and hedge in the financial interpretation:

$$g_t(\pi, \varsigma) = -\bar{g}_t(-\pi, \varsigma) + \varsigma q_t. \qquad (4.16)$$

We likewise let $\mathfrak{X}_t(\pi, \varsigma) = \bar{\mathfrak{X}}_t(-\pi, \varsigma)$, so that the funding closeout cash flow of the bank is given in terms of $(1 - \bar{R}_b)\mathfrak{X}_t^+(\pi, \varsigma)$ (see Sect. 4.3.2).

Example 4.4.3 Under the funding specification of Sect. 4.2.2 we obtain, in view of (4.8):

$$\begin{aligned}
&\mathfrak{X}_t(\pi, \varsigma) = \pi - \Gamma_t - \varsigma^{\bar{s}}\mathcal{P}_t^{\bar{s}}, \\
&g_t(\pi, \varsigma) = \bar{b}_t\Gamma_t^+ - b_t\Gamma_t^- + \bar{\lambda}_t\big(\pi - \Gamma_t + \varsigma^{\bar{s}}\mathcal{P}_{t-}^{\bar{s}}\big)^+ \\
&\qquad - \lambda_t\big(\pi - \Gamma_t + \varsigma^{\bar{s}}\mathcal{P}_{t-}^{\bar{s}}\big)^- + \varsigma^{\bar{s}}q_t^{\bar{s}},
\end{aligned} \qquad (4.17)$$

which depend on ς through $\varsigma^{\bar{s}}$, the vector of the positions in the externally funded hedging assets. It is only in the fully swapped case, i.e. when $\bar{s} = \emptyset$, that $\mathfrak{X}_t(\pi, \varsigma)$ and $g_t(\pi, \varsigma)$ do not depend on ς (see (4.18) and the beginning of Sect. 4.3.1).

Remark 4.4.4 In the example 4.4.3, at first sight one might wonder why the coefficients q^i present in g are exactly those of the hedging instruments which are *not* traded in swapped form. But one should keep in mind that all the coefficients q^i are already present, "hidden" in the pricing measure $\mathbb{Q}$ through the assumption 4.4.1, which is stated in terms of the gain process $\mathcal{M}$ (where the latter involves all the q^i). Therefore, coefficients q^i are expected in g for the assets *not* traded in swap form, so that the impacts of the corresponding q^i through the measure $\mathbb{Q}$ and in g compensate each other (and the price of the contract does eventually *not* depend on these q^i), by application of a Girsanov theorem. On the contrary, the coefficients q^i of the hedging assets traded in swapped form, being present in $\mathbb{Q}$ but not in g, do impact the price, as they should. See the example 4.4.8 for concrete computations in the toy Black-Scholes model of the remark 4.4.2.

From the BSDE point of view, a particularly simple situation will be the one where $\mathfrak{X}_t$ and g_t are independent of ς, i.e.

$$\mathfrak{X}_t(\pi, \varsigma) = \mathfrak{X}_t(\pi, 0) =: \mathfrak{X}_t(\pi), \;\; g_t(\pi, \varsigma) = g_t(\pi, 0) =: g_t(\pi). \qquad (4.18)$$

We call this the fully swapped hedge case in reference to its financial interpretation under the funding specification of Sect. 4.2.2. Otherwise we will speak of an externally funded hedge.

[12] Or, more precisely, the opposite of the value of the collateralization, hedging and funding portfolio (cf. (4.15)).

4.4.2 $\mathbb{Q}$-Price-and-Hedge BSDE

The class of general price-and-hedges introduced in the definition 4.3.4 is too large for practical purposes. This leads us to introduce the following more restrictive definition. Given a hedge ζ and a semimartingale Π, we write $\mathfrak{R} = \mathfrak{R}^c + \mathbb{1}_{\tau=\tau_b}\mathfrak{R}^f$, where

$$\mathfrak{R}^f = (1 - \bar{R}_b)\mathfrak{X}^+_{\tau-}(\Pi_{\tau-}, \zeta_{\tau-}). \tag{4.19}$$

Let us stress that $\mathfrak{R}^f$ and $\mathfrak{R}$ implicitly depend on $(\Pi_{\tau-}, \zeta_{\tau-})$ in this notation.

Definition 4.4.5 ($\mathbb{Q}$-price-and-hedge) Let a pair (Π, ζ) composed of a semimartingale Π and a hedge ζ satisfy the following BSDE on $[0, \bar{\tau}]$:

$$\begin{aligned} &\Pi_{\bar{\tau}} = \mathbb{1}_{\tau<T}\mathfrak{R} \text{ and, for } t \in [0, \bar{\tau}] : \\ &d\Pi_t + dC_t - \big(r_t\Pi_t + g_t(\Pi_t, \zeta_t)\big)dt = d\nu_t \end{aligned} \tag{4.20}$$

for some martingale ν null at time 0. Letting, for $t \in [0, \bar{\tau}]$,

$$\bar{\Pi}_t = \Pi_t - \mathbb{1}_{t\geq\tau_b}\mathfrak{R}^f,$$

the process $(\bar{\Pi}, \zeta)$ is then said to be a $\mathbb{Q}$-price-and-hedge. The related cost process is the martingale ε defined by $\varepsilon_0 = 0$ and, for $t \in [0, \bar{\tau}]$,

$$d\varepsilon_t = d\nu_t - \zeta_t d\mathcal{M}_t. \tag{4.21}$$

Equivalent to the BSDE (4.20) in differential form, we can write in integral form, for $t \in [0, \bar{\tau}]$,

$$\beta_t\Pi_t = \mathbb{E}_t\Big(\int_t^{\bar{\tau}} \beta_s dC_s - \int_t^{\bar{\tau}} \beta_s g_s(\Pi_s, \zeta_s)ds + \beta_{\bar{\tau}}\mathbb{1}_{\tau<T}\mathfrak{R}\Big). \tag{4.22}$$

The reader is referred to Sect. 13.5 and Crépey (2013) about BSDEs in finance (see also El Karoui, Peng, and Quenez (1997) for a seminal reference, in particular see the example 1.1 there regarding different borrowing and lending rates).

The $\mathbb{Q}$-price-and-hedge BSDE (4.20) is made nonstandard by the random terminal time $\bar{\tau}$, the dependence of the terminal condition $\mathfrak{R}$ in $(\Pi_{\tau-}, \zeta_{\tau-})$, the contract effective dividend term dC_t and the fact that it is not driven by an explicit set of fundamental martingales like Brownian motions and/or compensated Poisson measures. In this last respect, the representation (4.21) rather suggests that this BSDE will be solved with respect to the "primary martingale" $\mathcal{M}$, up to a (typically orthogonal) martingale ε.

Remark 4.4.6 (BSDEs modeling) An analysis of counterparty risk under funding constraints led us to the $\mathbb{Q}$-price-and-hedge BSDE (4.20) as a simplification of the general price-and-hedge FBSDE of Sect. 4.3.2. BSDE modeling should not be really seen as a choice here, but rather as an output of our analysis. This is mainly due to the intrinsically nonlinear feature of the funding issue under bilateral counterparty risk, via the positive part in $(1 - \bar{R}_b)\mathfrak{X}^+$ (unless $\bar{R}_b = 1$; see the remark 4.4.2). As repeatedly observed in the mathematical finance literature, BSDEs and FBSDEs emerge naturally in the context of nonlinear pricing problems. In this regard, the reader is referred to the literature on large counterparty, insiders and pricing impact; see for instance Cvitanic and Ma (1996). Regarding counterparty risk, another advantage of BSDEs is that, since they are a "non-linear pricing tool", they are well-suited to deal with with recursive (i.e. implicit) features, which will appear in the next chapter.

Again, the $\mathbb{Q}$-price-and-hedge BSDE (4.20) is rather non standard. However, we will see in the follow-up chapter that it can be reduced to a more classical pre-default TVA BSDE. All this also raises interesting questions from the point of view of BSDEs per se. First, the solution to the general price-and-hedge FBSDE that we got, before simplification into the more tractable BSDE (4.20), seems to be an open problem. Second, via the large-size and high-dimensional features of the TVA computational pricing problem, this provides an important field of motivation and application for numerical BSDEs. Third, via path-dependent collateralization, this also points to the theory of time-delayed BSDEs (see the remark 5.2.7).

With regard to numerics, it's interesting to note that the BSDE modeling approach is consistent with the American Monte Carlo technology which is advocated for practical TVA computations in Cesari, Aquilina, Charpillon, Filipovic, Lee, and Manda (2010). The regression scheme that is used in Brigo and Pallavicini (2008) for computing the clean price P at all points of a simulated grid also has a flavor of numerical BSDEs.

As it will be shown in Lemma 4.4.7 below, a $\mathbb{Q}$-price-and-hedge $(\bar{\Pi}, \zeta)$ is a general price-and-hedge in the sense of the definition 4.3.4. In Sect. 4.4.3 we will comment upon a $\mathbb{Q}$-price-and-hedge from the points of view of arbitrage, hedging and computational tractability. Toward this end we now derive the equations of the wealth $\mathcal{W}$ and of the hedging error ϱ of a $\mathbb{Q}$-price-and-hedge.

Lemma 4.4.7 *Given a $\mathbb{Q}$-price-and-hedge $(\bar{\Pi}, \zeta)$ and the related process Π, let a process $\mathcal{W}$ be defined by the first line in* (4.12)*, starting from the initial condition $\mathcal{W}_0 = -\Pi_0$. We denote, for $t \in [0, \bar{\tau}]$,*

$$\Pi^\star_t = \bar{\Pi}_t + \mathbb{1}_{\tau=\tau_b}\bar{\mathfrak{R}}^f,$$

where $\bar{\mathfrak{R}}^f = (1 - \bar{R}_b)\bar{\mathfrak{X}}^+_{\tau-}(\mathcal{W}_{\tau-}, \zeta_{\tau-})$. Then $(\bar{\Pi}, \zeta)$ is a general price-and-hedge, with wealth $\mathcal{W}$ of the hedging portfolio such that, for $t \in [0, \bar{\tau}]$,

$$\Big(\beta_t\Pi_t - \int_0^t \beta_s g_s(\Pi_s, \zeta_s)ds\Big) + \Big(\beta_t\mathcal{W}_t - \int_0^t \beta_s \bar{g}_s(\mathcal{W}_s, \zeta_s)ds\Big) = \int_0^t \beta_s d\varepsilon_s \qquad (4.23)$$

and, with hedging error $\varrho = \Pi^\star + \mathcal{W}$ such that, for $t \in [0, \bar{\tau}]$,

$$\begin{aligned} d\varrho_t &= d\varepsilon_t + \big(r_t\varrho_t + g_t(\Pi_t, \zeta_t) + \bar{g}_t(\mathcal{W}_t, \zeta_t)\big)dt \;+\; \mathbb{1}_{\tau=\tau_b}(1 - \bar{R}_b)(\bar{\mathfrak{R}}^f - \mathfrak{R}^f)\boldsymbol{\delta}_\tau(dt) \\ &= d\varepsilon_t + \big(r_t\varrho_t + g_t(\Pi_t, \zeta_t) + \bar{g}_t(\mathcal{W}_t, \zeta_t)\big)dt \\ &\quad + \mathbb{1}_{\tau=\tau_b}(1 - \bar{R}_b)\Big(\bar{\mathfrak{X}}^+_{\tau-}(\mathcal{W}_{\tau-}, \zeta_{\tau-}) - \mathfrak{X}^+_{\tau-}(\Pi_{\tau-}, \zeta_{\tau-})\Big)\boldsymbol{\delta}_\tau(dt) \end{aligned} \qquad (4.24)$$

(and $\varrho_0 = \varepsilon_0 = 0$).

Proof. Identity (4.23) immediately follows from (4.15), (4.20), (4.21) and $\mathcal{W}_0 = -\Pi_0$. Rewritten in term of the hedging error $\varrho = \Pi^\star + \mathcal{W}$, the equation (4.15) for the value $\mathcal{W}$ of the hedging portfolio of $(\bar{\Pi}, \zeta)$ yields that, for $t \in [0, \bar{\tau}]$,

$$d\Pi^\star_t = -\big(r_t + \bar{g}_t(\mathcal{W}_t, \zeta_t)\big)dt - dC_t + \zeta_t d\mathcal{M}_t + d\varrho_t. \tag{4.25}$$

Moreover, the equation part (second line) in the $\mathbb{Q}$-price-and-hedge BSDE (4.20) can be written in terms of the cost $d\varepsilon_t = d\nu_t - \zeta_t d\mathcal{M}_t$ of (4.21) as

$$d\Pi_t = \big(r_t\Pi_t + g_t(\Pi_t, \zeta_t)\big)dt - dC_t + \zeta_t d\mathcal{M}_t + d\varepsilon_t. \tag{4.26}$$

Since $\Pi_t - \Pi^\star_t = \mathbb{1}_{t \geq \tau_b}(\mathfrak{R}^f - \bar{\mathfrak{R}}^f)$, subtracting (4.26) from (4.25) yields (4.24). □

4.4.3 Arbitrage, Replication and Computational Issues

Assume, first, that it is possible to find a $\mathbb{Q}$-price-and-hedge process $(\bar{\Pi}, \zeta)$ with a vanishing cost process $\varepsilon = 0$ and, second, that for this $(\bar{\Pi}, \zeta)$ and the related process Π, uniqueness holds for the following forward SDE in Y: $Y_0 = \Pi_0$ and, for $t \in [0, \bar{\tau}]$,

$$d(\beta_t Y_t) - \beta_t g_t(Y_t, \zeta_t)dt = d(\beta_t \Pi_t) - \beta_t g_t(\Pi_t, \zeta_t)dt.$$

Via the BSDE machinery (see for instance El Karoui, Peng, and Quenez (1997)), the first assumption is typically met by application of a martingale representation property (whenever available), whereas the second assumption is a technical requirement guaranteeing that Π and $\mathcal{W}$ coincide if they solve the same forward SDE. Under these assumptions we get, by (4.23) with $\varepsilon = 0$, that $\mathcal{W} = -\Pi$. It follows that $\mathfrak{R}^f = \bar{\mathfrak{R}}^f$ and therefore, by (4.24) that $\varrho = \varepsilon = 0$. In this case the $\mathbb{Q}$-price-and-hedge process $(\bar{\Pi}, \zeta)$ is a replicating strategy.

The following pure liquidity case (no counterparty risk) is a complement to the remarks 4.4.2 and 4.4.8.

Example 4.4.8 Introducing proportional repo basis ρS for some constant ρ in the Black-Scholes model of the remark 4.4.2, the $\mathbb{Q}$-martingale $\mathcal{M}$ there becomes

$$dS_t - (r + \rho)S_t dt. \tag{4.27}$$

Assuming an option with payoff $\phi(S_T)$ hedged by S_t and funded at the risk-free rate r_t, the $\mathbb{Q}$-price-and-hedge BSDE in (Π, ζ) for the option is written as (on $[0, T]$ since this is a pure liquidity funding case, without counterparty risk): $\Pi_T = 0$ and, for $t \in [0, T]$,

$$d\Pi_t + \boldsymbol{\delta}_T(dt)\phi(S_T) - \big(r_t\Pi_t + \rho\zeta_t S_t\big)dt = \zeta_t d\mathcal{M}_t + d\varepsilon_t, \tag{4.28}$$

respectively

$$d\Pi_t + \boldsymbol{\delta}_T(dt)\phi(S_T) - r_t\Pi_t dt = \zeta_t d\mathcal{M}_t + d\varepsilon_t, \tag{4.29}$$

depending on whether the hedge in S_t is funded primarily or via a repo market (for some cost martingale ε starting from 0). From standard Itô calculus, solutions are found in the form $(\Pi_t, \zeta_t, \varepsilon_t) = (u(t, S_t), \partial_S u(t, S_t), 0)$, where the pricing function u satisfies the following Black-Scholes equation, where $u(T, S) = \phi(S)$ and, for $t < T$:

$$\partial_t u + rS\partial_S u + \frac{\sigma^2 S^2}{2}\partial^2_{S^2} u - ru = 0 \tag{4.30}$$

(independent of ρ), respectively

$$\partial_t u + (r + \rho)S\partial_S u + \frac{\sigma^2 S^2}{2}\partial^2_{S^2} u - ru = 0 \tag{4.31}$$

(which depends on ρ). So, despite the fact that ρ appears in (4.28) and not in (4.29), the solution to (4.29) depends on ρ, since it is given in terms of u solving (4.31), and the solution to (4.28) does not depend on ρ, since it is given in terms of u solving (4.30). The explanation for this "paradox" is that S_t itself (or equivalently, the measure $\mathbb{Q}$ under which the martingale condition on (4.27) is postulated) already depends on ρ in both cases.

We refer the reader to Sect. 4.6, expanding upon Burgard and Kjaer (2011b), for a practical example of replication also involving some counterparty risk. Since replication ultimately relies on a martingale representation property, it typically holds (or not) not only for a particular contract, but for any financial derivative. Henceforth we will refer to this case as the "complete market" case.

In a more general "incomplete" market, the cost ε of a $\mathbb{Q}$-price-and-hedge $(\bar{\Pi}, \zeta)$, and in turn its hedging error ϱ, can only be reduced up to a level "proportional" to the "degree of incompleteness" of the primary market. The bank can only partially hedge its position, ending-up with a nonvanishing hedging error $\varrho_{\bar{\tau}}$.

Remark 4.4.9 (Arbitrage) In a complete market (where, in particular, one can replicate the jump to default) or if $\bar{R}_b = 1$ (in particular, under unilateral counterparty risk), the Dirac-driven term vanishes in (4.24). Then, up to suitable technical conditions, we can change the measure $\mathbb{Q}$ into an equivalent measure $\widetilde{\mathbb{Q}}$ such that the hedging error ϱ is a $\widetilde{\mathbb{Q}}$-martingale. Therefore, $\varrho_{\bar{\tau}}$ cannot be nonnegative almost surely and positive with positive probability. Consequently, a $\mathbb{Q}$-price-and-hedge $(\bar{\Pi}, \zeta)$ cannot be an arbitrage in this case. On the contrary, in an incomplete market, where additionally $\bar{R}_b < 1$, a $\mathbb{Q}$-price-and-hedge $(\bar{\Pi}, \zeta)$ is, in principle, arbitrable.

A nonarbitrable strategy would be a general price-and-hedge $(\bar{\Pi}, \zeta)$ such that the quadruplet $(\mathcal{W}, \Pi^*, \zeta, \varrho)$ in the definition 4.3.4 solves the related FBSDE, in the particular sense that, that the hedging error ϱ would be a martingale under some equivalent probability measure. However, in an incomplete market with also $\bar{R}_b < 1$, this FBSDE seems intractable (see Horst, Hu, Imkeller, Reveillac, and Zhang (2011) for related technical issues). The $\mathbb{Q}$-price-and-hedge BSDE can be viewed as a simplified version of this theoretical FBSDE. The price to pay for this simplification is that it opens the door to an arbitrage (unless the market is complete or $\bar{R}_b = 1$, in which cases the $\mathbb{Q}$-price-and-hedge BSDE and the above FBSDE are essentially equivalent). However we believe that this arbitrage is quite theoretical (the corresponding "free lunch" seems quite difficult to obtain).

In view of the above arbitrage, hedging and computational considerations, we restrict ourselves to $\mathbb{Q}$-price-and-hedges in the sequel. For brevity henceforth, we write "a price-and-hedge (Π, ζ)" when the related pair-process $(\bar{\Pi}, \zeta)$ is a $\mathbb{Q}$-price-and-hedge; by price related to a hedge ζ we mean any process Π such that (Π, ζ) is a price-and-hedge (solves the BSDE (4.20)); we call (4.20) the price BSDE. This shift of terminology is rather immaterial since nobody cares about the price of the contract at time $\bar{\tau}$ (we set $\bar{\Pi}_{\bar{\tau}} = \mathbb{1}_{\tau<T}\mathfrak{R}^c$ for consistency with our definition of the hedging error ϱ as $\bar{\Pi} + \bar{\mathcal{W}}$). What matters in practice is the price for $t < \bar{\tau}$, for which $\Pi_t = \bar{\Pi}_t = \Pi_t^\star$.

Remark 4.4.10 (Symmetries) Similar to the funding benefit coefficient $g^b = g$ and the external funding recovery rate $\bar{R}_b$ of the bank, we can introduce a funding cost coefficient g^c and an external funding recovery rate $\bar{R}_c$ for the counterparty. It's only in the case where $\bar{R}_b = \bar{R}_c = 1$ and $g^b = g^c = g(\pi)$ that all the cash flows are symmetric from the point of view of the two parties. Then the buyer price of the bank will agree with the seller price of the counterparty. An example of symmetric funding costs is the setup of Fujii and Takahashi (2011b), where the funding closeout cash flows are not represented (implicitly, $\bar{R}_b = \bar{R}_c = 1$). The only excess funding costs in their model are collateral bases, b and $\bar{b}$ in our notation. Since collateral remuneration cash flows are between the two parties of the contract (they do not involve external entities), collateral bases do not break the symmetry in our sense[13].

It is worth emphasizing that, as soon as $\bar{R}_b$ or $\bar{R}_c < 1$, or the g-coefficients depend on ς, or they don't depend on ς but $g^b \neq g^c$, funding induces an asymmetry between the two parties, resulting in a buyer price for the bank different from the seller price for the counterparty (and in turn different TVAs in the follow-up chapter). See the remark 4.2.2 for the practical "violation of money conservation" issue.

Another notable specification, corresponding to the setup in Piterbarg (2010), is the linear case where $\bar{R}_b = 1$ and $g = g(\pi)$ is linear. Then the bank has a common buyer and seller price. Under the funding specification of Sect. 4.2.2, the linear case corresponds to $\bar{R}_b = 1$, $b = \bar{b}$ and $\lambda = \bar{\lambda}$.

Finally, the single-curve setup corresponds to the case where $\bar{R}_b = 1$ and $g = 0$. The only funding rate in the economy[14] is then the risk-free interest rate r (there are also the repo bases q, but these do not compromise the linearity of the setup).

[13] Fujii and Takahashi (2011b) consider a different notion of symmetry, which may be broken even in their setup.

[14] Assuming the existence of a riskless asset with growth rate r_t.

4.5 TVA

This section extends Proposition 3.2.8 (representation of the TVA as the price of a CCDS) to a nonlinear funding setup. The resulting TVA accounts not only for counterparty risk, but also for excess funding costs. The CCDS is then a dividend-paying option, where the dividends correspond to these costs.

4.5.1 Clean Price

In order to define the TVA process Θ, one first needs to introduce the clean price process P of the contract. This is a fictitious value process, which would give the price of the contract (cost of its hedge) without counterparty risk or excess funding costs. Consistent with the definition 3.2.2(i), this clean price process P is formally defined, for $t \in [0, T]$, by:

$$\beta_t P_t = \mathbb{E}_t \int_t^T \beta_s dD_s, \tag{4.32}$$

so that by the tower law, for $t \in [0, \bar{\tau}]$,

$$\beta_t P_t = \mathbb{E}_t \Big[\int_t^{\bar{\tau}} \beta_s dD_s + \beta_{\bar{\tau}} P_{\bar{\tau}} \Big] \tag{4.33}$$

and the discounted cumulative clean price

$$\beta_t P_t + \int_{[0,t]} \beta_s dD_s, \tag{4.34}$$

is a martingale.

4.5.2 CSA Close-Out Cash-Flow

In the sequel, $\mathfrak{R}^c$ in the CSA closeout cash flow is defined the way $\mathfrak{R}$ was given in Chapter 3,

$$\mathfrak{R}^c = \Gamma_\tau + \mathbb{1}_{\tau=\tau_c}(R_c \chi^+ - \chi^-) - \mathbb{1}_{\tau=\tau_b}(R_b \chi^- - \chi^+) - \mathbb{1}_{\tau_b=\tau_c} \chi \tag{4.35}$$

(see (3.2), to which we refer the reader for the detailed notation and comments).

4.5.3 TVA Representation

With $\mathfrak{R}^c$ thus specified in $\mathfrak{R} = \mathfrak{R}^c + \mathbb{1}_{\tau=\tau_b} \mathfrak{R}^f$, we are now ready to introduce the TVA process Θ of the bank. Recall from the definition 4.4.5 that, unless $\bar{R}_b = 1$, the terminal condition $\mathfrak{R} = \mathfrak{R}^c + \mathbb{1}_{\tau=\tau_b} \mathfrak{R}^f$ in a solution (Π, ζ) to the price BSDE (4.20) implicitly depends on $(\Pi_{\tau-}, \zeta_{\tau-})$, via $\mathfrak{R}^f = (1-\bar{R}_b)\mathfrak{X}^{\star,+}_{\tau_b-}$, where $\mathfrak{X}^\star_t$ is used as a shorthand for $\mathfrak{X}_t(\Pi_t, \zeta_t)$. Note that (compare this with (3.13))

$$\begin{aligned}\xi &= P_\tau + \Delta_\tau - \mathfrak{R}\\ &= P_\tau - Q_\tau + \chi - \mathbb{1}_{\tau=\tau_c}\big(R_c\chi^+ - \chi^-\big)\\ &\qquad + \mathbb{1}_{\tau=\tau_b}\big(R_b\chi^- - \chi^+\big) + \mathbb{1}_{\tau_b=\tau_c}\chi - \mathbb{1}_{\tau=\tau_b}(1-\bar{R}_b)\mathfrak{X}^{\star,+}_{\tau_b-}\\ &= P_\tau - Q_\tau + \mathbb{1}_{\tau=\tau_c}(1-R_c)\chi^+ - \mathbb{1}_{\tau=\tau_b}\Big((1-R_b)\chi^- + (1-\bar{R}_b)\mathfrak{X}^{\star,+}_{\tau-}\Big).\end{aligned} \tag{4.36}$$

Definition 4.5.1 *Given a solution (Π, ζ) to the price BSDE (4.20), the corresponding (cumulative) TVA process Θ is defined by:*

$$\Theta_t = P_t - \Pi_t + \mathbb{1}_{t\geq\tau}\mathbb{1}_{\tau<T}\Delta_\tau,\ t \in [0, \bar{\tau}]. \tag{4.37}$$

In particular, $\Theta_{\bar{\tau}} = \mathbb{1}_{\tau<T}\xi$.

As in the previous chapter (see the remark 3.2.6) this TVA process is cumulative in the sense that, at time τ, it also includes any contractual promised dividend Δ_τ.

The following result extends to the multi-curve setup the single-curve bilateral CVA representation result of Proposition 3.2.8. Note that in a multi-curve setup this representation, in the form of the equation (4.38) below, is implicit, i.e. the right-hand side of (4.38) involves Θ and ζ, via $(1-\bar{R}_b)\mathfrak{X}^{\star,+}_{\tau_b-}$ in ξ and via the integrand g_t. This is at least the case unless $\bar{R}_b = 1$ and a funding coefficient $g_t(\pi, \varsigma) = g_t(\pi)$ is linear in π, so that one can get rid of these dependencies by a suitable adjustment of the discount factor (more on this in the remark 6.2.6).

Proposition 4.5.2 *Let there be given a hedge ζ and $\mathbb{G}$-semimartingales Π and Θ such that $\Theta = P - \Pi$ on $[0, \bar{\tau}]$. The pair-process (Π, ζ) is a solution to the price BSDE (4.20) if and only if Θ satisfies, for $t \in [0, \bar{\tau}]$,*

$$\beta_t\Theta_t = \mathbb{E}_t\Big[\beta_{\bar{\tau}}\mathbb{1}_{\tau<T}\xi + \int_t^{\bar{\tau}} \beta_s g_s(P_s - \Theta_s, \zeta_s)ds\Big], \tag{4.38}$$

with $\Pi_{\tau-} = P_{\tau-} - \Theta_{\tau-}$ implicitly in ξ.

Proof. This is due to the martingale property of $d(\beta_t\Theta_t) + \beta_t g_t(P_t - \Theta_t, \zeta_t)dt$ with $\beta_{\bar{\tau}}\Theta_{\bar{\tau}} = \beta_{\bar{\tau}}\mathbb{1}_{\tau<T}\xi$, as it results from (4.22), (4.33) and (4.37). □

In a multi-curve setup we still get an interpretation of the TVA as the price of a contingent credit default swap (CCDS), which is an option on the debt χ (that is embedded in ξ) of the counterparty to the bank at time τ. However, this CCDS is a , paying not only the amount ξ at time $\tau < T$, but also dividends $g_t(P_t - \Theta_t, \zeta_t)dt$ until $\bar{\tau}$.

Example 4.5.3 In the fully swapped hedge case (4.18) and under the funding specification of Sect. 4.2.2, we have that

$$g_t(\pi, \varsigma) = g_t(\pi) = \bar{b}_t\Gamma_t^+ - b_t\Gamma_t^- + \bar{\lambda}_t(\pi - \Gamma_t)^+ - \lambda_t(\pi - \Gamma_t)^-. \tag{4.39}$$

The equation (4.38) can then be rewritten as

$$\begin{aligned}
\beta_t\Theta_t = \mathbb{E}_t\Big[&\mathbb{1}_{\tau=\tau_c<T}\beta_\tau(1-R_c)(Q_\tau+\Delta_\tau-\Gamma_\tau)^+\Big]\\
-\ \mathbb{E}_t\Big[&\mathbb{1}_{\tau=\tau_b<T}\beta_\tau(1-R_b)(Q_\tau+\Delta_\tau-\Gamma_\tau)^-\Big]\\
+\ \mathbb{E}_t\Big[&\int_t^{\bar\tau}\beta_s\Big(\bar b_s\Gamma_s^+ - b_s\Gamma_s^- + \bar\lambda_s\big(P_s-\Theta_s-\Gamma_s\big)^+\\
&-\lambda_s\big(P_s-\Theta_s-\Gamma_s\big)^-\Big)ds\ -\ \mathbb{1}_{\tau=\tau_b<T}\beta_\tau(1-\bar R_b)(P_{\tau-}-\Theta_{\tau-}-\Gamma_{\tau-})^+\Big]\\
+\ \mathbb{E}_t\big[&\mathbb{1}_{\tau<T}\beta_\tau\,(P_\tau-Q_\tau)\big].
\end{aligned}\tag{4.40}$$

From the perspective of the bank, the four terms (conditional expectations) in this representation can respectively be interpreted as a costly credit value adjustment (CVA), a beneficial debt valuation adjustment (DVA), a liquidity valuation adjustment (LVA) and a replacement cost/benefit (depending on the sign of $P_\tau - Q_\tau$). We will dwell more on such decompositions in the example 5.2.6 and in Chapter 6. See also Sect. 4.6 for a decomposition that arises in the context of an externally funded hedge (as opposed to the most common case of a fully swapped hedge with $g(\pi,\varsigma) = g(\pi)$ in (4.39)).

Let us call the TVA BSDE of the bank the BSDE on the random time interval $[0,\bar\tau]$ with terminal condition $\mathbb{1}_{\tau<T}\xi$ at $\bar\tau$ and with driver coefficient $g_t(P_t-\vartheta,\varsigma)-r_t\vartheta$, $\vartheta\in\mathbb{R}, \varsigma\in\mathbb{R}^d$. The following statement rephrases Proposition 4.5.2 in BSDE terms.

Lemma 4.5.4 *Given a hedge* ζ *and semimartingales* Π *and* Θ *summing-up to* P, (Π,ζ) *solves the price BSDE if and only if* (Θ,ζ) *solves the corresponding TVA BSDE.*

Passing from the price BSDE in (Π,ζ) to the TVA BSDE in (Θ,ζ), we get rid of the contract's dividend dC_t in (4.20).

4.5.3.1 CCDS Static Hedging Interpretation

Let us temporarily assume, for the sake of the argument, that the clean contract with price process P and the corresponding CCDS are traded assets. Given the price process Π solving the price BSDE (4.20) for $\zeta=0$, a static replication scheme for the bank buying the contract to the counterparty consists of:

- at time 0, funding the costs Π_0 of the contract and $\Theta_0 = P_0 - \Pi_0$ of a CCDS by the sales of the clean contract at price P_0;
- on the time interval $(0,\bar\tau)$, holding $(-P)$ and Θ, transferring the dividends dD_t perceived on the Π-position (contract) to the $(-P+\Theta)$-position, the dt-benefits at rate $r_tP_t + g(\Pi_t,0) - r_t\Theta_t$ on the $(-P+\Theta)$-position offsetting the dt-costs at rate $g_t(\Pi_t,0)+r_t\Pi_t$ on the Π-position.

Thus, at time $\bar\tau$:

- if $\bar{\tau} = \tau < T$, the bank is left with an amount $-P_\tau - \Delta_\tau + \Theta_\tau = -P_\tau - \Delta_\tau + \xi = -\mathfrak{R}$, but it also gets $\mathfrak{R}$ on its Π-position;

- if $\bar{\tau} = T$, there are no cash flows at $\bar{\tau}$.

In both cases the bank is left break-even at $\bar{\tau}$.

In conclusion, based on this static hedge argument, the definition 4.4.5 of a price-and-hedge (Π, ζ) also makes sense in this $\zeta = 0$ case. But of course this static buy-and-hold replication strategy is not practical, since neither the clean contract nor the CCDS are traded assets. We are thus led to active management of the hedging error of the trading strategy through dynamic hedging. Here a question arises whether one should try to hedge the contract globally or, if any freedom in this is left by the internal organization of the bank, to hedge the clean contract P separately from the TVA component Θ of Π. In order to address these issues we need to extend the analysis of the cost process $d\varepsilon = d\nu - \zeta d\mathcal{M}$ of a price-and-hedge (Π, ζ). This will be done in the next chapter.

4.6 Example

This section illustrates our approach on a Black-Scholes case considered in Burgard-Kjaer Burgard and Kjaer (2011a, 2011b). However, these papers do not consider defaultability of the bank regarding its funding debt. So their setup corresponds to the special case where $\bar{R}_b = 1$. With bilateral counterparty risk, defaultability of the bank regarding its funding debt is an important issue. We thus treat the general case where $\bar{R}_b \leq 1$.

4.6.1 Setup

We consider a European option with payoff $\phi(S_T)$ on a Black-Scholes stock S. Burgard and Kjaer (2011b) consider an option sold by the bank to the counterparty at time 0. Under our sign convention we can see this as the bank buying an option with payoff $(-\phi(S_T))$ from the counterparty. Both parties are defaultable but they cannot default simultaneously. The option position is hedged by the bank with the stock S and zero-recovery risky bonds B^b and B^c issued by the bank and by the counterparty, respectively. Repo markets with zero repo bases for notational simplicity are assumed to exist for S, B^b and B^c. Assuming a constant risk-free rate r, the gain process $\mathcal{M}$ of a buy-and-hold position into the hedging assets traded in swapped form is:

$$d\mathcal{M}_t = \begin{pmatrix} dS_t - rS_t dt, \\ dB_t^c - rB_t^c dt, \\ dB_t^b - rB_t^b dt \end{pmatrix}.$$

Consistent with the martingale requirement of the assumption 4.4.1 on $\mathcal{M}$, we assume the following model for (S, B^c, B^b):

$$\begin{cases} dS_t - rS_t dt = \sigma S_t dW_t \\ dB_t^c - rB_t^c dt = B_{t-}^c \left(dJ_t^c + \gamma_c dt \right) \\ dB_t^b - rB_t^b dt = B_{t-}^b \left(dJ_t^b + \gamma_b dt \right), \end{cases} \tag{4.41}$$

where W_t is a $\mathbb{Q}$-Brownian motion and where $J_t^b = \mathbb{1}_{t<\tau_b}, J_t^c = \mathbb{1}_{t<\tau_c}$ are the survival indicator processes of the bank and the counterparty, with constant $\mathbb{Q}$-default intensities γ_b and γ_c. We assume that the bank trades S and B^c on repo markets and that it externally funds B^b. The external borrowing basis of the bank is a constant $\bar{\lambda}$ and its external lending rate is simply r. Moreover, there is no collateralization ($\Gamma = 0$). In this case, (4.17) and (4.35) with $\Gamma = 0$ and $\Delta_\tau = 0$ on $\tau < T$ (since the only promised cash flow of the contract is the option's payoff at T) yield:

$$\begin{aligned} &\mathfrak{X}_t(\pi, \varsigma) = \pi - \varsigma^b B_t^b,, \;\; g_t(\pi, \varsigma) = \bar{\lambda}\big(\pi - \varsigma^b B_t^b\big)^+ \\ &\mathfrak{R}^c = \mathbb{1}_{\tau=\tau_c}\big(R_c \chi^+ - \chi^-\big) - \mathbb{1}_{\tau=\tau_b}\big(R_b \chi^- - \chi^+\big), \end{aligned} \tag{4.42}$$

where $\chi = Q(\tau, S_\tau)$ for a CSA closeout pricing function Q and where R_b and R_c denote constant recovery rates of the bank and the counterparty relative to each other. Note that $\mathfrak{X}$ and g depend on $\varsigma = (\varsigma^S, \varsigma^c, \varsigma^b)^\mathsf{T}$ through the hedging position ς^b of the bank in its own bond. This is thus a case of an externally funded, non fully swapped hedge.

Letting $\zeta = (\zeta^S, \zeta^c, \zeta^b)$, the price BSDE (4.20) is written as follows:

$$\begin{aligned} &\Pi_{\bar{\tau}} = \mathbb{1}_{\tau<T}\mathfrak{R} \text{ and, for } t \in [0, \bar{\tau}] : \\ &d\Pi_t + \mathbb{1}_{T<\tau}\boldsymbol{\delta}_T(dt)\phi(S_T) = \big(r\Pi_t + \bar{\lambda}\big(\Pi_t - \zeta_t^b B_t^b\big)^+\big)dt + \zeta_t d\mathcal{M}_t + d\varepsilon_t, \end{aligned} \tag{4.43}$$

where

$$\begin{aligned} \mathfrak{R} &= \mathfrak{R}^c + \mathbb{1}_{\tau=\tau_b}\mathfrak{R}^f = \mathfrak{R}^c + \mathbb{1}_{\tau=\tau_b}(1 - \bar{R}_b)\mathfrak{X}_{\tau-}^+(\Pi_{\tau-}, \zeta_{\tau-}) \\ &= \mathbb{1}_{\tau=\tau_c}\big(R_c\chi^+ - \chi^-\big) - \mathbb{1}_{\tau=\tau_b}\Big(R_b\chi^- - \chi^+ - (1 - \bar{R}_b)\left(\Pi_{\tau-} - \zeta_{\tau-}^b B_{\tau-}^b\right)^+\Big), \end{aligned} \tag{4.44}$$

for some recovery rate $\bar{R}_b$ (a positive constant) of the bank toward its external funder.

4.6.2 Analysis of a Solution

In this simple, complete market case (note there are three independent sources of randomness W, τ_c, τ_b in the model and three hedging assets S, B^c and B^b, plus an external funding source), a solution (Π, ζ) to the price BSDE can be guessed intuitively. Moreover, this will be a solution with vanishing cost process $\varepsilon = 0$, hence $\varrho = 0$ (see Sect. 4.4.3), in other words a replication strategy for the bank selling the option to the counterparty.

The Markov structure of the problem leads us to seek a solution (Π, ζ) to (4.43) such that, for $t \in [0, \bar{\tau}]$:

$$\begin{aligned} &\Pi_t + \mathbb{1}_{t=T<\tau}\phi(S_T) = u(t, S_t, J_t^c, J_t^b), \\ &\zeta_t = \delta(t, S_t), \end{aligned} \tag{4.45}$$

for suitable pricing and "delta" functions u and $\delta = (\delta^S, \delta^c, \delta^b)$. We denote $\widetilde{u}(t, S) = u(t, S, 1, 1)$, $\widetilde{B}^c(t) = B_0^c e^{(r+\gamma_c)t}$ and $\widetilde{B}^b(t) = B_0^b e^{(r+\gamma_b)t}$. The equation (4.44) can be rewritten as:

$$\begin{aligned} \mathfrak{R} =& \mathbb{1}_{\tau=\tau_c}\Big(R_c Q(\tau, S_\tau)^+ - Q(\tau, S_\tau)^-\Big) \\ &- \mathbb{1}_{\tau=\tau_b}\Big(R_b Q(\tau, S_\tau)^- - Q(\tau, S_\tau)^+ - (1-\bar{R}_b)\left(\widetilde{u}(\tau, S_\tau) - \delta^b(\tau, S_\tau)\widetilde{B}^b(\tau)\right)^+\Big). \end{aligned}$$

In view of the first lines in (4.43) and (4.45), it follows, for $(t, S) \in [0, T] \times (0, \infty)$, that

$$\begin{aligned} u(t, S, 0, 1) &= R_c Q(t, S)^+ - Q(t, S)^-, \\ u(t, S, 1, 0) &= -R_b Q(\tau, S_\tau)^- + Q(\tau, S_\tau)^+ \\ &\qquad + (1-\bar{R}_b)\left(\widetilde{u}(\tau, S_\tau) - \delta^b(\tau, S_\tau)\widetilde{B}^b(\tau)\right)^+. \end{aligned} \tag{4.46}$$

Besides, on one hand, the first line in (4.45) and the second line in (4.43) (assuming $\varepsilon = 0$) yield, for $t \in [0, \bar{\tau}]$, that

$$du(t, S_t, J_t^c, J_t^b) = \Big(r\widetilde{u}(t, S_t) + \bar{\lambda}\big(\widetilde{u}(t, S_t) - \delta^b(t, S_t)\widetilde{B}^b(t)\big)^+\Big)dt + \delta(t, S_t)d\mathcal{M}_t. \tag{4.47}$$

On the other hand, assuming $\widetilde{u}$ of class $\mathcal{C}^{1,2}$, we have, in view of the model dynamics for (S_t, J_t^c, J_t^b), the following Itô formula on $[0, \bar{\tau}]$:

$$\begin{aligned} du(t, S_t, J_t^c, J_t^b) =& \Big(\partial_t \widetilde{u}(t, S_t) + \mathcal{A}^{bs}\widetilde{u}(t, S_t)\Big)\, dt + \partial_S \widetilde{u}(t, S_t)\sigma S_t\, dW_t \\ &- (u(t, S_t, 0, 1) - \widetilde{u}(t, S_t))\, dJ_t^c - (u(t, S_t, 1, 0) - \widetilde{u}(t, S_t))\, dJ_t^b, \end{aligned}$$

where $\mathcal{A}^{bs} = rS\partial_S + \frac{\sigma^2 S^2}{2}\partial^2_{S^2}$ is the Black-Scholes generator and where the right-hand side can be rewritten as

$$\begin{aligned} &\Big(\partial_t \widetilde{u}(t, S_t) + \mathcal{A}^{bs}\widetilde{u}(t, S_t) + \gamma_c\big(u(t, S_t, 0, 1) - \widetilde{u}(t, S_t)\big) \\ &\qquad + \gamma_b\big(u(t, S_t, 1, 0) - \widetilde{u}(t, S_t)\big)\Big)dt \\ &+ \partial_S \widetilde{u}(t, S_t)\sigma S_t\, dW_t \\ &- (u(t, S_t, 0, 1) - \widetilde{u}(t, S_t))\,(dJ_t^c + \gamma_c dt) \\ &- (u(t, S_t, 1, 0) - \widetilde{u}(t, S_t))\left(dJ_t^b + \gamma_b dt\right). \end{aligned} \tag{4.48}$$

Equating the martingale terms on the right-hand sides in (4.47) and (4.48) and using (4.46), we obtain that, for $(t, S) \in [0, T] \times (0, \infty)$:

$$\begin{aligned}
\delta^S(t, S) &= \partial_S \widetilde{u}(t, S) \\
-\delta^c(t, S)\widetilde{B}^c(t) &= R_c Q(t, S)^+ - Q(t, S)^- - \widetilde{u}(t, S) \\
-\delta^b(t, S)\widetilde{B}^b(t) &= -R_b Q(\tau, S_\tau)^- + Q(\tau, S_\tau)^+ \\
&\quad + (1 - \bar{R}_b)\left(\widetilde{u}(\tau, S_\tau) - \delta^b(\tau, S_\tau)\widetilde{B}^b(\tau)\right)^+ - \widetilde{u}(t, S).
\end{aligned} \tag{4.49}$$

The third line in (4.49) is equivalent to

$$\begin{aligned}
\bar{R}_b\big(\widetilde{u}(t, S) - \delta^b(t, S)\widetilde{B}^b(t)\big)^+ - \big(\widetilde{u}(t, S) - \delta^b(t, S)\widetilde{B}^b(t)\big)^- \\
= -R_b Q(\tau, S_\tau)^- + Q(\tau, S_\tau)^+,
\end{aligned} \tag{4.50}$$

i.e.

$$-\delta^b(t, S)\widetilde{B}^b(t) = \begin{cases} R_b Q(t, S) - \widetilde{u}(t, S), & Q(t, S) \le 0 \\ \frac{1}{\bar{R}_b} Q(t, S) - \widetilde{u}(t, S), & Q(t, S) \ge 0. \end{cases} \tag{4.51}$$

This, together with the first two lines in (4.49) explicitly yields $\delta(t, S)$ in terms of $\widetilde{u}(t, S)$. In particular, (4.50) yields that

$$\big(\widetilde{u}(t, S) - \delta^b(t, S)\widetilde{B}^b(t)\big)^+ = \frac{1}{\bar{R}_b} Q(t, S)^+. \tag{4.52}$$

Equating now the dt-terms on the right-hand sides in (4.47) and (4.48) and accounting also for (4.46), (4.52) and for the terminal payoff $\phi(S_T)$ in case $T < \tau$, one obtains that the pre-default pricing function $\widetilde{u}(t, S)$ should satisfy the following pricing equation over $[0, T] \times (0, \infty)$:

$$\begin{cases} \widetilde{u}(T, S) = \phi(S), \ S \in (0, \infty), \\ \left(\partial_t + \mathcal{A}^{bs}\right) \widetilde{u}(t, S) + k(t, S, \widetilde{u}(t, S)) = 0, \ t < T, S \in (0, \infty), \end{cases} \tag{4.53}$$

with, for any real number π,

$$\begin{aligned}
k(t,S,\pi) &= -r\pi - \bar{\lambda}\frac{1}{\bar{R}_b} Q(t, S)^+ + \gamma_c\Big(R_c Q(t, S)^+ - Q(t, S)^- - \pi\Big) \\
&\quad + \gamma_b\Big(- R_b Q(\tau, S_\tau)^- + Q(\tau, S_\tau)^+ + (1 - \bar{R}_b)\frac{1}{\bar{R}_b} Q(t, S)^+ - \pi\Big).
\end{aligned}$$

Or, equivalently, in terms of $\widetilde{\lambda} = \bar{\lambda} - (1 - \bar{R}_b)\gamma_b$ (a related "liquidity basis"; see the example 5.2.6) and $\widetilde{r} = r + \gamma_b + \gamma_c$:

$$\begin{aligned} k\,(t,S,\pi) &= -\frac{\widetilde{\lambda}}{\bar{R}_b} Q(t,S)^+ + \gamma_c \Big(R_c Q(t,S)^+ - Q(t,S)^- \Big) \\ &\quad + \gamma_b \Big(- R_b Q(\tau, S_\tau)^- + Q(\tau, S_\tau)^+ \Big) - \widetilde{r}\pi \\ &= -\frac{\widetilde{\lambda}}{\bar{R}_b} Q(t,S)^+ - \gamma_c \Big((1-R_c) Q(t,S)^+ - Q(t,S) \Big) \\ &\quad + \gamma_b \Big((1-R_b) Q(\tau, S_\tau)^- + Q(\tau, S_\tau) \Big) - \widetilde{r}\pi. \end{aligned} \tag{4.54}$$

This provides an analysis of a solution to the price BSDE (4.43), assumed to exist in the form (4.45) and with null cost process ε. Conversely, the linear PDE (4.53) is known to have a unique classical solution $\widetilde{u}(t,S)$ under mild conditions on the coefficients. Starting from the solution $\widetilde{u}(t,S)$ to the PDE (4.53) and the function $\delta(t,S)$ associated with it via (4.49), proceeding in opposite direction to the calculations done above shows that the processes Π and ζ obtained in terms of $\widetilde{u}(t,S)$ and $\delta(t,S)$ through (4.45) yield a solution to the price BSDE (4.43) with null cost process ε. We finally obtain a replicating price-and-hedge for the bank, accounting also for the defaultability of the latter regarding its funding debt, for any $0 < \bar{R}_b \leq 1$.

Remark 4.6.1 **(i)** Practicality of the solution only holds if $\delta^b \geq 0$, corresponding to the bank repurchasing its own bond. Otherwise $\delta^b \leq 0$ would mean that the bank should issue more bond for hedging its TVA, which is not practical (see Burgard and Kjaer (2011a, 2011b)).
(ii) The case $\bar{R}_b = 0$ can be dealt with similarly provided $Q \leq 0$, otherwise (4.50) reduces to

$$-\big(\widetilde{u}(t,S) - \delta^b(t,S)\widetilde{B}^b(t)\big)^- = Q(t,S_t),$$

which can have no solution given that the signs of both sides differ, so that replicability does not hold.

4.6.2.1 TVA

Let $v(t,S)$ denote the Black-Scholes (clean) pricing function of the option (price clean of counterparty risk and excess funding costs), i.e.

$$\begin{cases} v(T,S) = \phi(S), \; S \in (0,\infty), \\ \left(\partial_t + \mathcal{A}^{bs}\right) v(t,S) - r v(t,S) = 0, \; t < T, S \in (0,\infty). \end{cases} \tag{4.55}$$

Defining the pre-default TVA function $\widetilde{w} = v - \widetilde{u}$, the following pre-default TVA pricing equation follows by substraction of (4.53)-(4.54) from (4.55):

$$\begin{cases} \widetilde{w}(T,S) = 0, \; S \in (0,\infty), \\ \left(\partial_t + \mathcal{A}^{bs}\right) \widetilde{w}(t,S) + f(t,S,\widetilde{w}(t,S)) = 0, \; t < T, S \in (0,\infty), \end{cases} \tag{4.56}$$

where, for any real number ϑ,

$$\begin{aligned} f(t,S,\vartheta) + rv(t,S) &= -k\,(t,S,v(t,S)-\vartheta) \\ &= \frac{\widetilde{\lambda}}{\bar{R}_b} Q(t,S)^+ + \gamma_c\Big((1-R_c)Q(t,S)^+ - Q(t,S)\Big) \\ &\quad - \gamma_b\Big((1-R_b)Q(t,S_t)^- + Q(t,S_t)\Big) + \widetilde{r}(v(t,S)-\vartheta), \end{aligned}$$

i.e.

$$\begin{aligned} f(t,S,\vartheta) + r\vartheta &= \gamma_c(1-R_c)Q(t,S)^+ - \gamma_b(1-R_b)Q(t,S_t)^- \\ &\quad + \frac{\widetilde{\lambda}}{\bar{R}_b} Q(t,S)^+ + (\gamma_b+\gamma_c)(v(t,S)-\vartheta-Q(t,S_t)). \end{aligned} \tag{4.57}$$

4.6.2.2 CSA Close-Out Pricing Schemes

It is implicitly understood above that a CSA closeout valuation scheme Q is an exogenous process, as in the standard clean CSA closeout pricing scheme, where $Q(t,S) = v(t,S)$.

However, we can also see, by reverse-engineering in the above computations, that it is possible to likewise deal with the so-called pre-default CSA closeout pricing scheme $Q(t,S) = \widetilde{u}(t,S)$ (see Sect. 3.3.1). In this "implicit" case, where the data (in principle) Q depends on the solution $\widetilde{u}$ (see the remark 3.2.4), the pre-default price and TVA PDEs become semilinear via the resulting nonlinear dependence of k or f in their argument π or ϑ. This implies some viscosity (instead of classical) solution technicalities but it does essentially not change the flow of arguments.

Note finally that the solution by replication that we were able to obtain in the complete market Black-Scholes example of this section was only possible due to many specific features of the setup, including the use of zero-recovery bonds as hedging instruments. With nonzero recovery bonds, even though we have three hedging assets S, B^c and B^b in respect to three independent sources of randomness W, τ_c, τ_b, the nonlinearity (unless $\bar{R}_b = 1$) of the funding closeout cash flow (we take a positive part in $\mathfrak{X}^+$ in (4.44)) makes the replication equations nonlinear and therefore nontrivial to solve (even with three equations in three unknowns; see for instance the case made at the end of the remark 4.6.1(ii)).

4.6.3 Comparison with the Results of Burgard and Kjaer

In the special case where $\bar{R}_b = 1$, the results of Burgard and Kjaer (2011a, 2011b) coincide with the ones we just derived (with, in this case, $\widetilde{\lambda} = \bar{\lambda}$ representing a liquidity basis since, for $\bar{R}_b = 1$, $\bar{\lambda}$ should not incorporate any credit spread). There is a caveat however. Burgard and Kjaer implicitly disregard defaultability of the bank regarding its funding debt (i.e. in our notation $\bar{R}_b = 1$). But in the interpretation of their results, they dwell upon a case where, despite of the bank being practically risk-free with regard to its funding debt, the external borrowing basis $\bar{\lambda}$ (that they denote s_F) would be of the form

$(1-R_b)\gamma_b$. However, $\bar{\lambda}=(1-R_b)\gamma_b$ implicitly refers to a case (actually a sensible one) where $\bar{R}_b(=R_b)<1$, whereas in their case $\bar{R}_b$ is always (implicitly) equal to one. As a consequence, in this case, the simplifications that they find in the coefficient f in (4.57) (for $Q=v$; see Sect. 4.6.2.2) and the conclusions that they draw regarding the appropriate internal organization of the bank for managing counterparty risk and funding costs may not be relevant (at least, in this situation $s_F=(1-R_b)\gamma_b$ and $Q=v$ that they consider among other cases). More precisely, in the equations (4.53)-(4.54) above, the coefficient of the first $Q(t,S)^+$-term that appears in the expression of k in (4.54) is $\frac{\widetilde{\lambda}}{\bar{R}_b}$[15], as opposed to s_F in Burgard and Kjaer. This $\frac{\widetilde{\lambda}}{\bar{R}_b}$ has no special relation to $\gamma_b(1-R_b)$, so no simplification occurs between the $\frac{\widetilde{\lambda}}{\bar{R}_b}Q(t,S)^+$-term and the $-\gamma_b(1-R_b)Q(t,S)^-$-term[16] in f in (4.57).

Note that in this section, in the presence of nonlinear funding costs, a martingale pricing approach is already useful in the context of a complete market model, allowing us to streamline the analysis of Burgard and Kjaer (2011a, 2011b). However, the corresponding computations were quite setup-dependent and a general study at the level of the price BSDE (4.43) needs to be conducted, as soon as, in particular, one leaves the realm of complete markets.

Observe in this example that, consistent with the general comment made after Lemma 4.5.4, passing from a price BSDE to a TVA BSDE allowed us to get rid of the coupon payments (the option payoff in the example) that blur the picture in the price BSDE. We also learned from this example that the valuation problem is essentially a pre-default one, so that a reduced-form approach should be a fruitful way to go. These avenues of research are explored systematically in Part III (see also Chapter 12 and Crépey and Song (2014)).

Already some take-away messages of the present chapter are that for properly valuing and hedging (bilateral in particular) counterparty risk in a multi-curve setup reflecting the presence of various funding costs, it is necessary to focus on a party of interest, called the bank in this book, and to consider the "system" consisting of the bank, the counterparty and the third-party (in practice, a group of various entities) which funds the bank. One must also have a clear view of the three equally important pillars of the bank's position, consisting of the contract itself, its hedging portfolio and its funding portfolio (as opposed to getting rid of the funding component of the position by risk-free discounting in a classical single-curve setup).

[15] which indeed coincides with $\bar{\lambda}$ alias s_F in Burgard and Kjaer in case $\bar{R}_b=1$, but in this case only.
[16] corresponding to the FCA- and the DVA-terms in Burgard and Kjaer (2011a).

Part III

Reduced-Form BSDE Modeling

Chapter 5

A Reduced-Form TVA BSDE Approach to Counterparty Risk under Funding Constraints

5.1 Introduction

In this chapter we develop a reduced-form backward stochastic differential equations (BSDE) approach to the problem of pricing and hedging the TVA, in a general framework of bilateral counterparty risk under funding constraints. The end results of Sect. 5.3.3 and 5.3.4 yield concrete recipes for risk-managing the contract or its TVA according to the following objective of the bank: minimizing the (risk-neutral) variance of the cost process (which is essentially the hedging error) of the contract or of its TVA, subject to various hedging policies for the jump-to-default exposure. The results of this chapter also shed more light on the structure of the TVA and on the debate about unilateral versus bilateral counterparty risk.

A take-away message is that the two-stage valuation and hedging methodology (counterparty risky price obtained as clean price minus TVA), which has emerged for practical reasons in banks, is also useful in the mathematical analysis of the problem. This makes the TVA not only a very important and legitimate financial object, but also a valuable mathematical tool.

5.1.1 Outline

In Sect. 5.2, we develop a practical reduced-form TVA BSDE approach to the problem of pricing and hedging bilateral counterparty risk under funding constraints. Counterparty risk and funding corrections to the clean price-and-hedge of the contract are represented as the solution to a pre-default TVA BSDE stated with respect to a reference filtration in which defaultability of the two parties only shows up through their default intensities. This representation allows us to deepen, in the example 5.2.6, the analysis of the TVA initiated in the example 4.5.3 in terms of credit cost/benefits and liquidity funding cost/benefits (as well as replacement cost/benefits). In the Markov setup of Sect. 5.3, explicit TVA pricing and hedging schemes are formulated in terms of semilinear pre-default TVA PDEs.

5.2 Pre-Default BSDE Modeling

We develop in this section a reduced-form TVA BSDE approach to the problem of pricing and hedging counterparty risk under funding constraints.

5.2.1 Bilateral Reduced Form Setup

We assume that the model filtration $\mathbb{G}$ can be decomposed as $\mathbb{G} = \mathbb{F} \vee \mathbb{H}^b \vee \mathbb{H}^c$, where $\mathbb{F}$ is some reference filtration and $\mathbb{H}^b$ and $\mathbb{H}^c$ stand for the natural filtrations of τ_b and τ_c.[1] Also let $\mathbb{G}^\tau = \mathbb{F} \vee \mathbb{H}$, where $\mathbb{H}$ is the natural filtration of $\bar{\tau}$ (or equivalently, of τ). We refer the reader to Sect. 13.7 for the background material regarding the reduced-form approach in credit risk modeling. The Azéma supermartingale associated with τ is the process A defined, for $t \in [0, T]$, by:

$$A_t = \mathbb{Q}(\tau > t \,|\, \mathcal{F}_t).$$

We assume that A is a positive, continuous and nonincreasing process. As explained in Sect. 13.7, this is a classical, slight relaxation of the so-called immersion or $(\mathcal{H})$-hypothesis of $\mathbb{F}$ into $\mathbb{G}^\tau$, corresponding to τ being an $\mathbb{F}$-pseudo stopping time (see Nikeghbali and Yor (2005)). In particular,

Lemma 5.2.1 **(i)** *An $\mathbb{F}$-martingale stopped at τ is a $\mathbb{G}^\tau$-martingale, and a $\mathbb{G}^\tau$-martingale stopped at τ is a $\mathbb{G}$-martingale.*
(ii) *An $\mathbb{F}$-adapted càdlàg process cannot jump at τ, i.e. $\Delta X_\tau = 0$ almost surely, for any $\mathbb{F}$-adapted càdlàg process X.*

Proof. **(i)** The first part follows immediately from Lemma 13.7.3(i) (where the filtration $\mathbb{G}$ corresponds to $\mathbb{G}^\tau$ here), so let us prove the second part. Repeated applications of the Lemma 13.7.1[2] yield that, for any $\mathbb{G}^\tau$-adapted integrable process M we have, for $0 \le s \le t \le T$:

$$\begin{aligned} \mathbb{E}\left(M_{t\wedge\tau} \,|\, \mathcal{G}_s\right) &= \mathbb{1}_{s\ge\tau} M_\tau + \mathbb{1}_{s<\tau} \frac{\mathbb{E}\left(M_{t\wedge\tau}\mathbb{1}_{s<\tau} \,|\, \mathcal{F}_s\right)}{A_s} \\ &= \mathbb{1}_{s\ge\tau} M_{s\wedge\tau} + \mathbb{1}_{s<\tau}\mathbb{E}\left(M_{t\wedge\tau} \,|\, \mathcal{G}^\tau_s\right), \end{aligned}$$

which, in case M is a $\mathbb{G}^\tau$-true martingale, reduces to $M_{s\wedge\tau}$. A $\mathbb{G}^\tau$-true martingale stopped at τ is thus a $\mathbb{G}$-true martingale. Then a standard localization argument yields that a $\mathbb{G}^\tau$-(local) martingale stopped at τ is a $\mathbb{G}$-(local) martingale.
(ii) This follows from Lemma 13.7.3(ii). □

[1] I.e. filtrations generated by the corresponding indicator process.

[2] See Sect. 13.7 and note that (13.63) is also valid for $\mathbb{G} = \mathbb{F} \vee \mathbb{H}^b \vee \mathbb{H}^c$ as here, not only for $\mathbb{G}^\tau$ as in the standard setup of Sect. 13.7.

Remark 5.2.2 (Immersion) A reduced-form approach draws its computational efficiency from, essentially, an immersion hypothesis between the reference filtration, that "ignores" the default times of the two parties, and the reference filtration progressively enlarged by the latter. Under immersion the Azéma supermartingale A of τ has no martingale component, as a predictable and finite variation process. Consequently, in most modeling, computations of various probabilities can be done efficiently as there is a nice connection between martingale hazard process and the conditional survival probability (in fact, for even more simplicity, we will soon assume that A is absolutely continuous, so that these computations can be carried out in terms of the so called intensity of τ).

The basic immersion hypothesis of this part of the book, where it will be additionally assumed below that the data D, Q and Γ are $\mathbb{F}$-adapted and hence don't jump at τ, implies a kind of weak or indirect dependence between the reference contract and the default times of the two parties (see Jeanblanc and Le Cam (2007) or Jamshidian (2002)). In the language of counterparty risk, this basic immersion setup precludes major right/wrong-way risk effects such as the ones that are observed for instance with counterparty risk on credit derivatives. On the contrary, in the case without strong dependence between the contract and the default of the parties, this "advantage" should be promoted in the model and this is precisely the object of a reduced-form approach. Moreover, with credit derivatives, a reduced-form approach to counterparty risk loses some of its computational appeal. With credit derivatives, the discontinuous and high-dimensional nature of the problem is such that the gain in tractability resulting from the above reduction of filtration is not so tangible. As a consequence, a reduced-form approach, at least in the basic form of this chapter, is not sufficient to deal with counterparty risk on credit derivatives. We refer the reader to Part IV or to Brigo and Chourdakis (2008), Brigo and Capponi (2008a), Lipton and Sepp (2009a), Blanchet-Scalliet and Patras (2008) regarding possible approaches to appropriately deal with counterparty risk on credit derivatives (or strong wrong-way risk more generally). Immersion may also be a concern with FX derivatives since it has been shown empirically that there can be some rather strong dependence between the default risk of an obligor and an exchange rate (see also the remark 5.2.4 about related aspects regarding the collateral). This being said, we will see in Chapter 12 that one can extend the basic immersion setup of this part of the book to cover in a unified framework the models of Parts III and IV.

Note that ideally counterparty risk should not be considered at the level of a specific class of assets, but at the level of all the contracts between two counterparties under a given CSA. The construction of a global model and methodology for valuing and hedging a CSA hybrid book of derivatives, including credit derivatives, remains a challenging modeling issue (see e.g. Albanese, Bellaj, Gimonet, and Pietronero (2011)).

5.2.2 Reduction of Filtration

We assume that the risk-free short rate process r (or equivalently the risk-free discount factor process $\beta = e^{-\int_0^{\cdot} r_t dt}$) and the clean dividend process D are $\mathbb{F}$-adapted. Then, consistent with (4.32) in the basic immersion setup of this part of the book, the clean price process P of the contract can be equivalently be defined, for $t \in [0, T]$, as:

$$\beta_t P_t = \mathbb{E}\left(\int_t^T \beta_s dD_s \,\Big|\, \mathcal{F}_t\right) \tag{5.1}$$

instead of the expectation given $\mathcal{G}_t$ in (4.32). Indeed, assuming (5.1), the discounted cumulative clean price process defined by:

$$\beta P + \int_{[0,\cdot]} \beta_t dD_t,\ t \in [0, T] \tag{5.2}$$

is an $\mathbb{F}$-martingale. By Lemma 5.2.1(i), this process

stopped at τ is then a $\mathbb{G}$-martingale, so (4.33) follows as before. With P understood as in (5.1) henceforth in this part, the corresponding clean $\mathbb{F}$-martingale M on $[0, T]$, to be compared with the $\mathbb{G}$-martingale component ν of Π in the price BSDE (4.20), is then defined, for $t \in [0, T]$, by:

$$dM_t = dP_t + dD_t - r_t P_t dt, \tag{5.3}$$

along with the terminal condition $P_T = 0$ (compare with the terminal condition at $\bar{\tau}$ for Π in the first line of (4.20)).

Lemma 5.2.3 *There is no promised dividend or jump of the clean price process at the default time τ, i.e. (recall Δ is the jump process of D) $\Delta_\tau = D_\tau - D_{\tau-} = 0$ and $P_\tau - P_{\tau-} = 0$.*

***Proof*.** Since all our semimartingales are taken in a càdlàg version, then by Lemma 5.2.1(ii) the $\mathbb{F}$-semimartingales D and P cannot jump at τ. □

In particular, since $\Delta_\tau = 0$, the expression (3.1) for χ reduces to

$$\chi = Q_\tau - \Gamma_\tau. \tag{5.4}$$

In the sequel we assume that Γ and Q are $\mathbb{F}$-adapted, so that, by Lemma 5.2.1(ii), they cannot jump at τ. In addition, we assume that the recovery rates R_c, R_b and $\bar{R}_b$ can be represented as $R^c_{\tau_c}$, $R^b_{\tau_b}$ and $\bar{R}^b_{\tau_b}$, for some $\mathbb{G}$-predictable processes R^c_t, R^b_t and $\bar{R}^b_t$.

Remark 5.2.4 Assuming $\mathbb{F}$-adaptedness of Q_t and Γ_t is sometimes too restrictive. For instance, in case of collateral posted in another currency, strongly dependent on the default of a party, one may wish to model a jump of Γ_t at τ. See Ehlers and Schönbucher (2006) for a model with a jump of a currency at a default time and see Chapter 12 and Crépey and Song (2014) for more developments in this regard.

We now introduce an equivalent pre-default TVA BSDE over $[0, T]$, relative to the pre-default filtration $\mathbb{F}$. By Theorem 67.b in Dellacherie and Meyer (1975) and Lemma 13.7.2, the $\mathcal{G}_{\tau-}$-measurable random variables $\mathbb{Q}(\tau = \tau_b \,|\, \mathcal{G}_{\tau-})$, $\mathbb{Q}(\tau = \tau_c \,|\, \mathcal{G}_{\tau-})$ and $\mathbb{Q}(\tau_b = \tau_c \,|\, \mathcal{G}_{\tau-})$ can be represented as p^b_τ, p^c_τ and q_τ, for some $\mathbb{F}$-predictable processes,

p^b, p^c and q. Likewise, since Q and Γ cannot jump at τ, there exist $\mathbb{F}$-predictable processes with the same values as Q and Γ at τ (take pre-default values, which exist by Lemma 13.7.2, of the left-limit processes Q_- and Γ_-), so that we can assume that these processes are in fact $\mathbb{F}$- predictable. As a consequence, the debt χ of the counterparty to the bank is the value at time τ of the $\mathbb{F}$-predictable process χ_t defined, for $t \in [0, T]$, by:

$$\chi_t = Q_t - \Gamma_t. \tag{5.5}$$

Again from Lemma 13.7.2, we may assume without loss of generality that the process $g_t(P_t - \vartheta, \varsigma)$ is $\mathbb{F}$-progressively measurable for $\vartheta \in \mathbb{R}$, $\varsigma \in \mathbb{R}^d$, and that the processes R^b and R^c are $\mathbb{F}$-predictable. Henceforth, we assume that the Azéma supermartingale A of τ is absolutely continuous (i.e. pathwise time-differentiable) and we define the hazard intensity $\gamma_t = -\frac{d \ln A_t}{dt}$ of τ, where $A_t = e^{-\int_0^t \gamma_s ds}$; also the credit-risk-adjusted interest-rate $\widetilde{r}$ and the credit-risk-adjusted discount factor α are defined by

$$\widetilde{r}_t = r_t + \gamma_t,\ \alpha_t = \beta_t A_t = \beta_t e^{-\int_0^t \gamma_s ds} = e^{-\int_0^t \widetilde{r}_s ds}.$$

For every $\pi \in \mathbb{R}$ and $\varsigma \in \mathbb{R}^d$, let $\widetilde{\xi}_t(\pi, \varsigma)$ represent the $\mathbb{F}$-progressively measurable process defined, for $t \in [0, T]$, by:

$$\widetilde{\xi}_t(\pi, \varsigma) = (P_t - Q_t) + p_t^c(1 - R_t^c)\chi_t^+ - p_t^b\Big((1 - R_t^b)\chi_t^- + (1 - \bar{R}_t^b)\mathfrak{X}_t^+(\pi, \varsigma)\Big), \tag{5.6}$$

where we recall from Sect. 4.4.1 that $(1 - \bar{R}_b)\mathfrak{X}_t^+(\pi, \varsigma)$ is used to model the funding closeout cash flow of the bank, for any $\pi \in \mathbb{R}$ and $\varsigma \in \mathbb{R}^d$, which respectively represent the contract's price and hedge in the financial interpretation.

Definition 5.2.5 The pre-default TVA BSDE of the bank is the $\mathbb{F}$-BSDE in $(\widetilde{\Theta}, \zeta)$ on $[0, T]$ with a null terminal condition at T and with driver coefficient

$$f_t(P_t - \vartheta, \varsigma) = g_t(P_t - \vartheta, \varsigma) + \gamma_t \widetilde{\xi}_t(P_t - \vartheta, \varsigma) - \widetilde{r}_t \vartheta, \tag{5.7}$$

i.e.

$$\begin{cases} \widetilde{\Theta}_T = 0 \text{ and, for } t \in [0, T], \\ -d\widetilde{\Theta}_t = f_t(P_t - \widetilde{\Theta}_t, \zeta_t)dt - d\widetilde{\mu}_t, \end{cases} \tag{5.8}$$

where $\widetilde{\Theta}$ is an $\mathbb{F}$-special semimartingale, ζ is a hedge and $\widetilde{\mu}$ is an $\mathbb{F}$-martingale (the $\mathbb{F}$-martingale component of $\widetilde{\Theta}$).

Or, equivalent to the second line in (5.8):

$$-d(\alpha_t \widetilde{\Theta}_t) = \alpha_t\Big(g_t(P_t - \widetilde{\Theta}_t, \zeta_t) + \gamma_t \widetilde{\xi}_t(P_t - \widetilde{\Theta}_t, \zeta_t)\Big)dt - \alpha_t d\widetilde{\mu}_t. \tag{5.9}$$

Or, equivalent to (5.8) but in integral form:

$$\alpha_t \widetilde{\Theta}_t = \mathbb{E}\Big[\int_t^T \alpha_s\big(g(s, P_s - \widetilde{\Theta}_s) + \gamma_s \widetilde{\xi}(s, P_s - \widetilde{\Theta}_s)\big)ds \,\Big|\, \mathcal{F}_t\Big],\ t \in [0, T]. \tag{5.10}$$

Example 5.2.6 Under the fully swapped hedge funding specification of the example 4.5.3, with $g(\pi,\varsigma) = g(\pi)$ given by (4.39), one obtains, by plugging (4.39) into (5.7) and reordering terms, that

$$\begin{aligned} f_t(P_t - \vartheta) + r_t\vartheta = \gamma_t p_t^c(1 - R_t^c)(Q_t - \Gamma_t)^+ \\ - \gamma_t p_t^b(1 - R_t^b)(Q_t - \Gamma_t)^- \\ + \bar{b}_t\Gamma_t^+ - b_t\Gamma_t^- + \widetilde{\lambda}_t\,(P_t - \vartheta - \Gamma_t)^+ - \lambda_t\,(P_t - \vartheta - \Gamma_t)^- \\ + \gamma_t\,(P_t - \vartheta - Q_t)\,, \end{aligned} \quad (5.11)$$

where the coefficient $\widetilde{\lambda}_t = \bar{\lambda}_t - \gamma_t p_t^b(1 - \bar{R}_t^b)$ of $(P_t - \vartheta - \Gamma_t)^+$ can be interpreted as an external borrowing basis net of credit risk. This basis represents the liquidity component of $\bar{\lambda}$. The four terms on the lines of this decomposition can be interpreted similarly as in (4.40).

Remark 5.2.7 (CSA Close-Out Pricing and Collateralization Schemes) From a mathematical point of view, the "implicit"(so called pre-default) CSA closeout pricing scheme, i.e. $Q_\tau = \Pi_{\tau-}$ (as opposed to the "explicit" clean CSA closeout pricing scheme $Q = P$; see Sect. 3.3.1), can be accounted or in a reduced-form TVA BSDE setup simply by letting $Q = P - \widetilde{\Theta}$ everywhere in the coefficient f_t of the pre-default TVA BSDE (5.8) (as will result from our modeling assumption $\Pi = P - \Theta$ in (5.16) below).

However, in order to meet ISDA requirements, a real-life collateralization scheme Γ is typically path-dependent in Q (see Sect. 3.3.2). Under the pre-default CSA closeout pricing scheme and in case of a path-dependent collateralization, one ends-up with a so-called time delayed BSDE with a coefficient depending on the past of $\widetilde{\Theta}$ (see Delong and Imkeller (2010)). This raises a mathematical difficulty of the pre-default CSA closeout pricing scheme since, as shown in the just-mentioned paper, even for a Lipschitz coefffi-cient, a time-delayed BSDE may only have a solution for T small enough, depending on the Lipschitz constant of the coefficient.

5.2.3 Modeling Assumption

In this section we work under the following hypothesis:

Assumption 5.2.8 The pre-default TVA BSDE (5.8) admits a solution $(\widetilde{\Theta}, \zeta)$.

Note that after specification of a jump-diffusion setup endowed with a martingale representation property, existence and uniqueness for a solution to (5.8) (or an equivalent BSDE (5.24)) holds under mild regularity and square-integrability conditions (see e.g. Crépey (2013)). At this stage we merely postulate existence, examining its consequences in terms of the existence of a solution (Π, ζ) to the price BSDE (4.20) in this subsection and of the analysis of the cost process ε of (Π, ζ) in Sect. 5.2.4. Letting

$$\begin{aligned} \xi_t(\pi,\varsigma) = (P_t - Q_t) + \mathbb{1}_{t\geq\tau_c}(1 - R_t^c)\chi_t^+ \\ - \mathbb{1}_{t\geq\tau_b}\Big((1 - R_t^b)\chi_t^- + (1 - \bar{R}_t^b)\mathfrak{X}_{t-}^+(\pi,\varsigma)\Big), \end{aligned} \quad (5.12)$$

we use $\xi^\star_t$, $\widetilde{\xi}^\star_t$ and $\mathfrak{X}^\star_t$ as shorthand notation for $\xi_t(P_t-\widetilde{\Theta}_t,\zeta_t)$, $\widetilde{\xi}_t(P_t-\widetilde{\Theta}_t,\zeta_t)$ and $\mathfrak{X}_t(P_t-\widetilde{\Theta}_t,\zeta_t)$. The following results (decomposition (5.13) of μ in particular, to be compared with Lemma 3.2.13 in the single-curve setup) are key in the sequel.

Lemma 5.2.9 *On $[0,\bar{\tau}]$, the $(\mathbb{G},\mathbb{Q})$-compensated martingale of $\xi^\star_t dJ_t$ is written as*

$$\xi^\star_t dJ_t + \gamma_t\widetilde{\xi}^\star_t dt.$$

Proof. Conditional on a first jump of a party at time $\tau = t$, an event with intensity γ_t, this jump represents a default of the counterparty alone with probability $p^{\{c\}}_t = p^c_t - q_t$, of the bank alone with probability $p^{\{b\}}_t = p^b_t - q_t$ and of both parties jointly with probability q_t. Moreover, the corresponding exposures can be represented respectively as:

$$\begin{aligned}
\xi^{\{c\}}_t &= P_t - Q_t + (1-R^c_t)\chi^+_t\\
\xi^{\{b\}}_t &= P_t - Q_t - \Big((1-R^b_t)\chi^-_t + (1-\bar{R}^b_t)\mathfrak{X}^{\star,+}_{t-}\Big)\\
\xi^{\{b,c\}}_t &= P_t - Q_t + (1-R^c_t)\chi^+_t - \Big((1-R^b_t)\chi^-_t + (1-\bar{R}^b_t)\mathfrak{X}^{\star,+}_{t-}\Big).
\end{aligned}$$

Therefore the compensator of $\xi^\star_t dJ_t$ is given by

$$\gamma_t\left(p^{\{c\}}_t\xi^{\{c\}}_t + p^{\{b\}}_t\xi^{\{b\}}_t + q_t\xi^{\{b,c\}}_t\right)dt = \gamma_t\widetilde{\xi}^\star_t dt,$$

in view of (5.6). □

Proposition 5.2.10 (Reduced-Form TVA Modeling) *Under the assumption 5.2.8, define $\Theta = \widetilde{\Theta}$ on $[0,\bar{\tau})$ and $\Theta_{\bar{\tau}} = \mathbb{1}_{\tau<T}\xi^\star_\tau$. Then the process Θ satisfies the TVA BSDE (4.38) on $[0,\bar{\tau}]$. Moreover, the $(\mathbb{G},\mathbb{Q})$-martingale component*

$$d\mu_t = d\Theta_t + (g_t(P_t-\Theta_t,\zeta_t) - r_t\Theta_t)dt$$

of Θ satisfies, for $t\in[0,\bar{\tau}]$:

$$d\mu_t = d\widetilde{\mu}_{t\wedge\tau} - \left((\xi^\star_t - \widetilde{\Theta}_t)dJ_t + \gamma_t(\widetilde{\xi}^\star_t - \Theta_t)dt\right), \tag{5.13}$$

where

$$d\widetilde{\mu}_t = d\widetilde{\Theta}_t + f_t(P_t-\widetilde{\Theta}_t,\zeta_t)dt$$

is the $(\mathbb{F},\mathbb{Q})$-martingale component of $\widetilde{\Theta}$.

Proof. Recall that $\mathbb{F}$-adapted càdlàg processes (including $\mathbb{F}$-adapted semimartingales, $\widetilde{\Theta}$ in particular) don't jump at τ, so each unnecesssary "$_-$" is omitted in the notation. By definition of Θ (a $(\mathbb{G},\mathbb{Q})$-semimartingale), for $t\in[0,\bar{\tau}]$, we have:

$$d(\beta_t\Theta_t) = d(J_t\beta_t\widetilde{\Theta}_t) + \beta_t\xi^\star_t\boldsymbol{\delta}_\tau(dt) = d\big(\beta_{t\wedge\tau}\widetilde{\Theta}_{t\wedge\tau}\big) + \beta_t\widetilde{\Theta}_t\,dJ_t - \beta_t\xi^\star_t dJ_t, \tag{5.14}$$

where by (5.8):

$$-d\widetilde{\Theta}_t = f_t(P_t - \widetilde{\Theta}_t, \zeta_t)dt - d\widetilde{\mu}_t. \tag{5.15}$$

Therefore, for $t \in [0, \bar{\tau}]$:

$$\begin{aligned}-\beta_t^{-1}d(\beta_t\Theta_t) &= \big(f_t(P_t - \widetilde{\Theta}_t, \zeta_t) + r_t\widetilde{\Theta}_t\big)dt - d\widetilde{\mu}_{t\wedge\tau} + (\xi_t^\star - \widetilde{\Theta}_t)dJ_t \\ &= g_t(P_t - \widetilde{\Theta}_t, \zeta_t)dt - d\widetilde{\mu}_{t\wedge\tau} + \Big((\xi_t^\star - \widetilde{\Theta}_t)dJ_t + \gamma_t(\widetilde{\xi}_t^\star - \Theta_t)dt\Big),\end{aligned}$$

by definition (5.7) of f. Moreover, by Lemma 5.2.1, $\widetilde{\mu}_{t\wedge\tau}$ is a $(\mathbb{G}, \mathbb{Q})$-martingale, as is

$$(\xi_t^\star - \widetilde{\Theta}_t)dJ_t + \gamma_t(\widetilde{\xi}_t^\star - \widetilde{\Theta}_t)dt,$$

by Lemma 5.2.9. This yields the decomposition (5.13) of the $(\mathbb{G}, \mathbb{Q})$-martingale component μ of Θ. In particular, the process Θ satisfies the TVA BSDE (4.38) on $[0, \bar{\tau}]$. □

Corollary 5.2.11 *Under the assumption 5.2.8, the pair* (Π, ζ), *where*

$$\Pi := P - \Theta = J(P - \widetilde{\Theta}) + (1 - J)\mathbb{1}_{\tau<T}\mathfrak{R} \tag{5.16}$$

on $[0, \bar{\tau}]$ *and* $\mathfrak{R} = P_\tau - \xi_\tau^\star$, *solves the price BSDE* (4.20). *Moreover, the* $\mathbb{G}$*-martingale component* ν *(cf.* (4.20)*) of* Π *satisfies, for* $t \in [0, \bar{\tau}]$,

$$d\nu_t = d\widetilde{\nu}_t - \Big((\mathfrak{R}_t - \widetilde{\Pi}_t)dJ_t + \gamma_t(\widetilde{\mathfrak{R}}_t - \widetilde{\Pi}_t)dt\Big), \tag{5.17}$$

where $\widetilde{\Pi} = P - \widetilde{\Theta}$ *is the pre-default value of* Π, $\widetilde{\nu} = M - \widetilde{\mu}$ *is the* $\mathbb{F}$*-martingale component of* $\widetilde{\Pi}$ *and the* $\mathbb{G}$*-progressively measurable process* $\mathfrak{R}_t$ *and the* $\mathbb{F}$*-progressively measurable process* $\widetilde{\mathfrak{R}}_t$ *are defined, for* $t \in [0, T]$, *by:*

$$\begin{aligned}\mathfrak{R}_t &= \Gamma_t + \mathbb{1}_{t\ge\tau_c}\big(R_t^c\chi_t^+ - \chi_t^-\big) \\ &\quad - \mathbb{1}_{t\ge\tau_b}\Big(\big(R_t^b\chi_t^- - \chi_t^+\big) - (1 - \bar{R}_t^b)\mathfrak{X}_{t-}^{\star,+}\Big) - \mathbb{1}_{t\ge\tau_b=\tau_c}\chi_t \\ \widetilde{\mathfrak{R}}_t &= \Gamma_t + p_t^c\big(R_t^c\chi_t^+ - \chi_t^-\big) - p_t^b\Big(\big(R_t^b\chi_t^- - \chi_t^+\big) - (1 - \bar{R}_t^b)\mathfrak{X}_{t-}^{\star,+}\Big) - q_t\chi_t.\end{aligned} \tag{5.18}$$

Proof. **(i)** Since (Θ, ζ) solves the TVA BSDE (4.38), it follows by Lemma 4.5.4 that the pair (Π, ζ), where $\Pi = P - \Theta$, solves the price BSDE (4.20). Also recall that $P_T = 0$, which justifies the right-hand side identity in (5.16).

(ii) By (5.13), for $t \in [0, \bar{\tau}]$, we have (recalling that M defined by (5.3) is the $\mathbb{F}$-martingale component of P):

$$\begin{aligned}d\nu_t = dM_t - d\mu_t &= (dM_t - d\widetilde{\mu}_t) + \Big((\xi_t^\star - \widetilde{\Theta}_t)dJ_t + \gamma_t(\widetilde{\xi}_t^\star - \widetilde{\Theta}_t)dt\Big) \\ &= d\widetilde{\nu}_t - \Big((\mathfrak{R}_t - \widetilde{\Pi}_t)dJ_t + \gamma_t(\widetilde{\mathfrak{R}}_t - \widetilde{\Pi}_t)dt\Big),\end{aligned}$$

where the last equality follows by algebraic manipulations similar to those in (4.36). This proves (5.17). □

Remark 5.2.12 The jump-to-default exposure corresponding to the dJ-term in (5.13) or (5.17) can be seen as a marked process, where the mark corresponds to the default of the counterparty alone, of the bank alone, or to a joint default. Consistent with this interpretation, the compensator of either dJ-term in (5.13) or (5.17) is given by the corresponding "intensity $\times$ average jump size", in turn given by the dt-term in the same line, where the average is taken with respect to the probabilities of the marks, conditionally on the occurrence of a jump at τ. Further enrichment the mark space of τ gives a generalization of the basic immersion setup of this chapter; see Chapter 12 and Crépey and Song (2014).

From now on our approach for dealing with the price BSDE (4.20) will consist in modeling the counterparty risky price process Π as in (5.16), via the corresponding pre-default TVA process $\widetilde{\Theta}$.

5.2.4 Cost Processes Analysis

Let us now postulate, for the $\mathbb{G}$-martingale component $\mathcal{M}$ of the primary risky assets price process $\mathcal{P}$, with pre-default value process denoted by $\widetilde{\mathcal{P}}$, a structure analogous to the one that is apparent in (5.17) for the $\mathbb{G}$-martingale component ν of Π. Thus we assume that on $[0, \bar{\tau}]$

$$d\mathcal{M}_t = d\widetilde{\mathcal{M}}_t - \Big((\mathcal{R}_t - \widetilde{\mathcal{P}}_t) dJ_t + \gamma_t (\widetilde{\mathcal{R}}_t - \widetilde{\mathcal{P}}_t) dt \Big) \tag{5.19}$$

for an $\mathbb{F}$-martingale $\widetilde{\mathcal{M}}$, a $\mathbb{G}$-progressively measurable primary recovery process $\mathcal{R}_t$ and an $\mathbb{F}$-progressively measurable process $\widetilde{\mathcal{R}}_t$ such that $\gamma_t(\widetilde{\mathcal{R}}_t - \widetilde{\mathcal{P}}_t)dt$ compensates $(\mathcal{R}_t - \widetilde{\mathcal{P}}_t) dJ_t$ over $[0, \bar{\tau}]$.

For hedges ϕ and ζ, understood respectively as hedges of the contract clean price P and price Π, let $\eta = \phi - \zeta$ denote the corresponding hedge of the TVA component Θ of Π. Then let the cost processes $\varepsilon^{P,\phi}$, $\varepsilon^{\Theta,\eta}$ and $\varepsilon^{\Pi,\zeta}$ be defined by $\varepsilon_0^{P,\phi} = \varepsilon_0^{\Theta,\eta} = \varepsilon_0^{\Pi,\zeta} = 0$ and, for $t \in [0, \bar{\tau}]$,

$$\begin{aligned} d\varepsilon_t^{P,\phi} &= dM_t - \phi_t d\mathcal{M}_t \\ d\varepsilon_t^{\Theta,\eta} &= d\mu_t - \eta_t d\mathcal{M}_t \\ d\varepsilon_t^{\Pi,\zeta} &= d\varepsilon_t^{P,\phi} - d\varepsilon_t^{\Theta,\eta} = d\nu_t - \zeta_t d\mathcal{M}_t. \end{aligned} \tag{5.20}$$

In particular, we retrieve $\varepsilon^{\Pi,\zeta} = \varepsilon$, the cost process of a price-and-hedge (Π, ζ) in (4.21). Application of (5.13), (5.17) and (5.19) yields the following:

Proposition 5.2.13 *For* $t \in [0, \bar{\tau}]$,

$$d\varepsilon_t^{P,\phi} = \Big(dM_t - \phi_t d\widetilde{\mathcal{M}}_t\Big) + \phi_t\big(\mathcal{R}_t - \widetilde{\mathcal{P}}_t\big)dJ_t + \gamma_t\phi_t\big(\widetilde{\mathcal{R}}_t - \widetilde{\mathcal{P}}_t\big)dt \tag{5.21}$$

$$\begin{aligned} d\varepsilon_t^{\Theta,\eta} = & \Big(d\widetilde{\mu}_t - \eta_t d\widetilde{\mathcal{M}}_t\Big) - \Big(\big(\xi_t^\star - \widetilde{\Theta}_t\big) - \eta_t\big(\mathcal{R}_t - \widetilde{\mathcal{P}}_t\big)\Big)dJ_t \\ & -\gamma_t\Big(\big(\widetilde{\xi}_t^\star - \widetilde{\Theta}_t\big) - \eta_t\big(\widetilde{\mathcal{R}}_t - \widetilde{\mathcal{P}}_t\big)\Big)dt \end{aligned} \tag{5.22}$$

$$\begin{aligned} d\varepsilon_t^{\Pi,\zeta} = & \Big(d\widetilde{\nu}_t - \zeta_t d\widetilde{\mathcal{M}}_t\Big) - \Big(\big(\mathfrak{R}_t - \widetilde{\Pi}_t\big) - \zeta_t\big(\mathcal{R}_t - \widetilde{\mathcal{P}}_t\big)\Big)dJ_t \\ & -\gamma_t\Big(\big(\widetilde{\mathfrak{R}}_t - \widetilde{\Pi}_t\big) - \zeta_t\big(\widetilde{\mathcal{R}}_t - \widetilde{\mathcal{P}}_t\big)\Big)dt. \end{aligned} \tag{5.23}$$

Thus we get decompositions of the different cost processes as $\mathbb{F}$-martingales stopped at τ, hence $\mathbb{G}$-martingales, plus $\mathbb{G}$-compensated jump-to-default exposures. These decompositions can be used for devising specific pricing and hedging schemes, such as pricing at the cost of hedging by replication (whenever possible), hedging only pre-default risk, hedging only the jump-to-default risk (dJ-terms), min-variance hedging, among others. This will now be made practical in a Markov setup.

5.3 Markov Case

In a Markov setup, explicit TVA pricing and hedging schemes can be formulated in terms of semilinear pre-default TVA PDEs. More precisely, in this section we will relate suitable notions of solutions of the pre-default TVA BSDE to:

- from a financial point of view, corresponding min-variance hedging strategies of the bank, based on the cost processes analysis of Sect. 5.2.4;
- from a mathematical point of view, classical Markov BSDEs driven by an explicit set of fundamental martingales (Brownian motions and/or a compensated jump counting measure).

These Markov BSDEs will be well-posed under mild conditions, yielding related notions of orthogonal solutions to the pre-default TVA BSDE and of min-variance hedges. This approach will be developed for three different min-variance hedging objectives, respectively considered in Sect. 5.3.2, 5.3.3 and 5.3.4. In the end, the preferred criterion (we mainly see the analysis of Sect. 5.3.2 as preparatory to those of Sect. 5.3.3 and 5.3.4) can be optimized by solving (numerically if need be) the related Markov BSDE or an equivalent semilinear parabolic PDE. This methodology is applicable to the risk-management of either the contract as a whole or of its TVA component in isolation. But in all cases the pre-default TVA BSDE will be key to the mathematical analysis. The end results of Sect. 5.3.3 (respectively 5.3.4) yield concrete recipes for risk-managing the contract as a whole or its TVA component, according to the following objective of the bank: minimizing the variance of the cost process of the contract or of its TVA component, subject to the constraint of a perfect hedge of the jump-to-default exposure (respectively of the jump-to-counterparty-default exposure).

As explained in the preface, a clean price-and-hedge (P, ϕ) is typically determined by a business trading desk of the bank. The central TVA desk is left with the task of devising a TVA price-and-hedge (Θ, η). Henceforth, following this logic, for a clean price-and-hedge (P, ϕ), a solution $(\widetilde{\Theta}, \zeta)$ to the pre-default TVA BSDE is sought in the form $(\widetilde{\Theta}, \phi - \eta)$, where an $\mathbb{F}$-adapted triplet $(\widetilde{\Theta}, \eta, \epsilon)$ solves

$$\begin{cases} \widetilde{\Theta}_T = 0, \text{ and for } t \in [0, T]: \\ -d\widetilde{\Theta}_t = f_t(P_t - \widetilde{\Theta}_t, \phi_t - \eta_t)dt - \big(\eta_t d\widetilde{\mathcal{M}}_t + d\epsilon_t\big), \end{cases} \tag{5.24}$$

for an $\mathbb{F}$-predictable[3] integrand η and an $(\mathbb{F}, \mathbb{Q})$-martingale ϵ. The pre-default TVA BSDE in the form (5.24) is indeed equivalent to the original pre-default TVA BSDE (5.8), with $\eta = \phi - \zeta$ and ϵ defined by the second line of (5.24) (and $\epsilon_0 = 0$). Accordingly:

Definition 5.3.1 Given a solution $(\widetilde{\Theta}, \eta, \epsilon)$ to (5.24), meaning in particular that ϵ, defined through $(\widetilde{\Theta}, \eta)$ by the second line in (5.24) (and $\epsilon_0 = 0$), is an $(\mathbb{F}, \mathbb{Q})$-martingale, we call TVA price-and hedge the pair-process (Θ, η), where $\Theta = \widetilde{\Theta}$ on $[0, \bar{\tau})$ and $\Theta_{\bar{\tau}} = \mathbb{1}_{\tau < T}\xi^{\star}_{\bar{\tau}}$, in which $\xi^{\star}_t$ is used as shorthand notation for $\xi_t(P_t - \widetilde{\Theta}_t, \phi_t - \eta_t)$.

5.3.1 Factor Process

We assume that the pre-default TVA BSDE, thus redefined as (5.24), is Markov in the sense that any of its input data of the form $\mathcal{D}_t$ is given as a measurable function $\mathcal{D}(t, X_t)$ of an $\mathbb{F}$-Markov factor process X. In particular, $(P_t, \phi_t) = (P(t, X_t), \phi(t, X_t))$. Consequently, we have:

$$f_t(P_t - \widetilde{\Theta}_t, \phi_t - \eta_t)dt = f\big(t, X_t, P(t, X_t) - \widetilde{\Theta}_t, \phi(t, X_t) - \eta_t\big)dt, \tag{5.25}$$

where the meaning of the function $f(t, x, \pi, \varsigma)$ is clear from the above context.

We will use as factor process the jump-diffusion of Sect. 13.3,

$$dX_t = b(t, X_t)\,dt + \sigma(t, X_t)\,dW_t + j(t, X_{t-}) \cdot dm_t, \;\; X_0 = x, \tag{5.26}$$

driven by an $\mathbb{R}^q$-valued $\mathbb{F}$-Brownian motion W and an $\mathbb{F}$-compensated jump measure m on $[0, T] \times \mathbb{R}^q$, for some integer q, with a (random) disintegrated jump intensity measure given by $c(t, X_t, dx)$ (e.g. Lévy measure $c(dx)$).

Further analysis of the cost processes (5.21)–(5.23) depends on a hedging criterion of the bank. In the following sections we will propose three tractable approaches, all of them involving, to some extent, min-variance hedging. In case of a complete primary

[3] Typically left-continuous in a Markov setup.

market, min-variance hedging of course reduces to hedging by replication. Moreover, we will consider the two issues of hedging the contract globally, or only hedging its TVA. In all cases the mathematical analysis will ultimately rely on the pre-default TVA BSDE (5.24).

Note that in this chapter min-variance hedging is performed with respect to the reference filtration $\mathbb{F}$, on the top of a given choice of a hedging strategy regarding the jump-to-default exposure of the bank: no hedge in Sect. 5.3.2, perfect hedge in Sect. 5.3.3 and hedge of a default of the counterparty alone in Sect. 5.3.4.

We assume further that $\widetilde{\mathcal{P}}_t = \widetilde{\mathcal{P}}(t, X_t)$ for some pre-default primary risky assets pricing function $\widetilde{\mathcal{P}}$, so that the dynamics of the $\mathbb{F}$-martingale component $\widetilde{\mathcal{M}}$ of $\mathcal{M}$ in (5.19) is written as

$$d\widetilde{\mathcal{M}}_t = (\partial\widetilde{\mathcal{P}}\sigma)(t, X_t)dW_t + \delta\widetilde{\mathcal{P}}(t, X_{t-}) \cdot dm_t.$$

Given n- and m-dimensional vector functions $u = u(t,x)$ and $v = v(t,x)$ on $[0,T] \times \mathbb{R}^q$, we denote by $\mathcal{C}(u,v)$ the $\mathbb{R}^{n\times m}$-valued *carré du champ* matrix function with entry (function of (t,x)) $\mathcal{C}(u_i, v_j)$ defined through the formula (13.41). In the probabilistic interpretation, $\mathcal{C}(u,v)(t,X_t)dt$ represents "the $\mathcal{F}_t$-conditional covariance matrix of $du(t,X_t)$ and $dv(t,X_t)$".

5.3.2 Min-Variance Hedging of Market Risk

Our first objective is to min-variance hedge the market risk corresponding to the term $(d\widetilde{\mu}_t - \eta_t d\widetilde{\mathcal{M}}_t)$ in the TVA cost process $\varepsilon^{\Theta,\eta}$ in (5.22), or $(d\widetilde{\nu}_t - \zeta_t d\widetilde{\mathcal{M}}_t)$ in the overall contract cost process $\varepsilon^{\Pi,\zeta} = \varepsilon$ of (5.23).

Regarding (5.22), this is tantamount to seeking for a solution $(\widetilde{\Theta}, \eta, \epsilon)$ to the pre-default TVA BSDE (5.24) in which ϵ is $\mathbb{F}$-orthogonal to $\widetilde{\mathcal{M}}$ (cf. Proposition 5.2 in El Karoui, Peng, and Quenez (1997)). Given such an orthogonal solution $(\widetilde{\Theta}, \eta, \epsilon)$ to (5.24) and moreover if $\widetilde{\Theta}_t = \widetilde{\Theta}(t, X_t)$, we have by a standard min-variance sharp bracket regression formula[4], using the above-introduced matrix-form $\mathcal{C}$ of (13.41): carré du champ matrix-form $\mathcal{C}$ of (13.41):

$$\eta_t = \frac{d\langle\widetilde{\mu},\widetilde{\mathcal{M}}\rangle_t}{dt}\left(\frac{d\langle\widetilde{\mathcal{M}}\rangle_t}{dt}\right)^{-1} = \left(\mathcal{C}\left(\widetilde{\Theta},\widetilde{\mathcal{P}}\right)\Lambda\right)(t, X_{t-}) =: \eta(t, X_{t-}), \tag{5.27}$$

where we let $\Lambda = \left(\mathcal{C}\left(\widetilde{\mathcal{P}},\widetilde{\mathcal{P}}\right)\right)^{-1}$. Here invertibility of the *carré du champ* matrix $\mathcal{C}(\widetilde{\mathcal{P}},\widetilde{\mathcal{P}})$ of $\widetilde{\mathcal{P}}$ is assumed. We are led to the following Markov BSDE over $[0,T]$ in

[4]See e.g. Sect. 4.2.3.1 in Crépey (2013); note that by regression we mean multilinear regression everywhere in this chapter.

$(\widetilde{\Theta}(t,X_t),(\partial\widetilde{\Theta}\sigma)(t,X_t),\delta\widetilde{\Theta}(t,X_{t-},\cdot))$:

$$\begin{cases} \widetilde{\Theta}(T,X_T)=0 \text{ and, for } t\in[0,T]: \\ -\,d\widetilde{\Theta}(t,X_t)=\widehat{f}\left(t,X_t,\widetilde{\Theta}(t,X_t),(\partial\widetilde{\Theta}\sigma)(t,X_t),\left((\delta\widetilde{\Theta}\delta\widetilde{\mathcal{P}}^{\mathsf{T}})\cdot c\right)(t,X_t)\right)dt \\ \qquad -(\partial\widetilde{\Theta}\sigma)(t,X_t)dW_t-\delta\widetilde{\Theta}(t,X_{t-})\cdot dm_t, \end{cases} \tag{5.28}$$

with, for $(t,x,\vartheta,z,w)\in[0,T]\times\mathbb{R}^q\times\mathbb{R}\times\mathbb{R}^q\times\mathbb{R}^d$ (for row-vectors z,w),

$$\widehat{f}\,(t,x,\vartheta,z,w)=f\,(t,x,P(t,x)-\vartheta,\phi(t,x)-\widehat{\eta}(t,x,\vartheta,z,w))$$

where

$$\widehat{\eta}(t,x,\vartheta,z,w)=\left(z(\partial\widetilde{\mathcal{P}}\sigma)^{\mathsf{T}}(t,x)+w\right)\Lambda(t,x)$$

so that, in view of (13.41) and (5.27):

$$\begin{aligned} &\widehat{\eta}(t,X_t,\widetilde{\Theta}(t,X_t),(\partial\widetilde{\Theta}\sigma)(t,X_t),\left((\delta\widetilde{\Theta}\delta\widetilde{\mathcal{P}}^{\mathsf{T}})\cdot c\right)(t,X_t))=\eta(t,X_t),\\ &\widehat{f}\left(t,X_t,\widetilde{\Theta}(t,X_t),(\partial\widetilde{\Theta}\sigma)(t,X_t),\left((\delta\widetilde{\Theta}\delta\widetilde{\mathcal{P}}^{\mathsf{T}})\cdot c\right)(t,X_t)\right)dt=\\ &\qquad f\left(t,X_t,P(t,x)-\widetilde{\Theta}(t,X_t),\phi(t,X_t)-\eta(t,X_t)\right)dt. \end{aligned}$$

As is classical (see e.g. Chapters 12 and 13 of Crépey (2013)), under mild regularity and square-integrability conditions on the coefficient $\widehat{f}$, the Markov BSDE (5.28) has a unique square-integrable solution $\widetilde{\Theta}_t=\widetilde{\Theta}(t,X_t)$. Moreover, the pre-default TVA function $\widetilde{\Theta}=\widetilde{\Theta}(t,x)$ is the unique solution in suitable spaces to the following semilinear partial integro-differential equation (PDE for short):

$$\begin{cases} \widetilde{\Theta}(T,x)=0,\ x\in\mathbb{R}^q, \\ (\partial_t+\mathcal{A})\,\widetilde{\Theta}(t,x)+\widehat{f}(t,x,\widetilde{\Theta}(t,x),(\partial\widetilde{\Theta}\sigma)(t,x),((\delta\widetilde{\Theta}\delta\widetilde{\mathcal{P}}^{\mathsf{T}})\cdot c)(t,x)) \\ \qquad =0 \text{ on } [0,T)\times\mathbb{R}^q, \end{cases} \tag{5.29}$$

where $\mathcal{A}$ stands for the infinitesimal generator of X.

Remark 5.3.2 A BSDE comparison theorem is key in the connection between a BSDE and a PDE approach to a semilinear parabolic equation (see for instance Chapter 14 of Crépey (2013)). Note that for BSDE with jumps, such a comparison theorem is subject to a monotonicity condition of the coefficient $\widehat{f}$ with respect to the jump variable w, i.e. in our case

$$\widehat{f}\,(t,x,\vartheta,z,w)\le\widehat{f}\,(t,x,\vartheta,z,w') \text{ if } w_i\le w'_i,\ i=1,\ldots,d$$

(see the remark 12.1.11 in (Crépey 2013) or Royer (2006)). Of course these technicalities disappear in the most common case of a fully swapped hedge satisfying (4.18) so that $f(t,x,\pi,\varsigma)=f(t,x,\pi)$ (see also the remark 5.3.8 regarding the corresponding Markov BSDEs and semilinear PDEs).

Proposition 5.3.3 *Assuming invertibility of the carré du champ matrix of the hedging assets, $\mathcal{C}(\widetilde{\mathcal{P}},\widetilde{\mathcal{P}})$, the solution $\widetilde{\Theta} = \widetilde{\Theta}(t,x)$ to* (5.29) *yields, via* (5.27) *for η and then* (5.24) *for ϵ, an orthogonal solution $(\widetilde{\Theta},\eta,\epsilon)$ to the pre-default TVA BSDE* (5.24).

The TVA-market-risk-min-variance hedge is given by the formula (5.27)*, i.e.*

$$\eta_t = \eta(t, X_{t-}) = \left(\mathcal{C}\left(\widetilde{\Theta},\widetilde{\mathcal{P}}\right)\left(\mathcal{C}\left(\widetilde{\mathcal{P}},\widetilde{\mathcal{P}}\right)\right)^{-1}\right)(t, X_{t-}).$$

The process ϵ in the solution to (5.24) *represents the residual TVA market risk under the TVA hedge η.*

Hedging of the Contract as a Whole We now consider hedging of the market risk $(d\widetilde{\nu}_t - \zeta_t d\widetilde{\mathcal{M}}_t)$ of the overall contract cost process $\varepsilon^{\Pi,\zeta} = \varepsilon$ of (5.23). Let the clean hedge ϕ be specifically given here as the coefficient of regression in an $\mathbb{F}$-orthogonal decomposition $dM = \phi d\widetilde{\mathcal{M}} + de$. By the min-variance sharp bracket formula, it follows, for $t \in [0,\bar{\tau}]$, that

$$\phi_t = \frac{d<M,\widetilde{\mathcal{M}}>_t}{dt}\left(\frac{d<\widetilde{\mathcal{M}}>_t}{dt}\right)^{-1} = \left(\mathcal{C}\left(P,\widetilde{\mathcal{P}}\right)\Lambda\right)(t, X_{t-}) =: \phi(t, X_{t-}). \tag{5.30}$$

Moreover, in view of (5.21)–(5.23), we have:

$$d\widetilde{\nu}_t - \zeta_t d\widetilde{\mathcal{M}}_t = \left(dM_t - \phi_t d\widetilde{\mathcal{M}}_t\right) - \left(d\widetilde{\mu}_t - \eta_t d\widetilde{\mathcal{M}}_t\right).$$

Since $dM - \phi d\widetilde{\mathcal{M}}$ and $d\widetilde{\mu} - \eta d\widetilde{\mathcal{M}}$ are $\mathbb{F}$-orthogonal to $d\widetilde{\mathcal{M}}$, so is $d\widetilde{\nu} - \zeta d\widetilde{\mathcal{M}}$. In conclusion, Proposition 5.3.3 implies the following:

Corollary 5.3.4 *For a clean hedge ϕ given as the regression coefficient of M against $\widetilde{\mathcal{M}}$, the strategy $\zeta_t = (\phi - \eta)(t, X_{t-})$ is a min-variance hedge of the market risk component $(d\widetilde{\nu} - \zeta d\widetilde{\mathcal{M}})$ of the contract cost process $\varepsilon^{\Pi,\zeta} = \varepsilon$. The residual market risk of the contract hedged in this way is given by $e - \epsilon$.*

5.3.3 Min-Variance Hedging Constrained to Perfect Hedging of Jump-to-Default Risk

The previous approach disregards the jump-to-default risk corresponding to the dJ-terms in (5.22) or (5.23). Now we want to min-variance hedge the market risk corresponding to the term $(d\widetilde{\mu}_t - \eta_t d\widetilde{\mathcal{M}}_t)$ in the TVA cost process $\varepsilon^{\Theta,\eta}$ of (5.22) (respectively $(d\widetilde{\nu}_t - \zeta_t d\widetilde{\mathcal{M}}_t)$ in the overall contract cost process $\varepsilon^{\Pi,\zeta} = \varepsilon$ of (5.23)), under the constraint of perfectly hedging the jump-to-default risk corresponding to the dJ-term in (5.22) (respectively (5.23)). Note that in view of the marked point process interpretation provided in the remark 5.2.12, cancelation of the dJ-term in any of the equation (5.21)–(5.23) implies cancelation of the dt-driven process which compensates it in the same equation. We are thus equivalently minimizing the variance of the cost processes $\varepsilon^{\Theta,\eta}$ or $\varepsilon^{\Pi,\zeta} = \varepsilon$, under the constraint of perfectly hedging the jump-to-default exposure.

We let a superscript "0" refer to the subset of the hedging instruments with price processes which cannot jump at time τ, so $\mathcal{R}^0 = \widetilde{\mathcal{R}}^0 = \widetilde{\mathcal{P}}^0$, and we let "1" refer to the subset, complement of "0", of the hedging instruments with price processes which can jump at time τ. The TVA cost equation (5.22) can thus be rewritten, for $t \in [0, \bar{\tau}]$ as:

$$\begin{aligned} d\varepsilon_t^{\Theta,\eta} =& \Big(d\widetilde{\mu}_t - \eta_t^0 d\widetilde{\mathcal{M}}_t^0 - \eta_t^1 d\widetilde{\mathcal{M}}_t^1\Big) - \Big(\big(\xi_t^\star - \widetilde{\Theta}_t\big) - \eta_t^1\big(\mathcal{R}_t^1 - \widetilde{\mathcal{P}}_t^1\big)\Big) dJ_t \\ & - \gamma_t\Big(\big(\widetilde{\xi}_t^\star - \widetilde{\Theta}_t\big) - \eta_t^1\big(\widetilde{\mathcal{R}}_t^1 - \widetilde{\mathcal{P}}_t^1\big)\Big) dt. \end{aligned} \tag{5.31}$$

The condition that a TVA price-and-hedge (Θ, η) perfectly hedges the dJ-term in (5.31) is written as

$$\xi_t^\star - \widetilde{\Theta}_{t-} = \eta_t^1\big(\mathcal{R}_t^1 - \widetilde{\mathcal{P}}_t^1\big),\ t \in [0, \bar{\tau}], \tag{5.32}$$

where it should be noted in view of (5.12) that $\xi_t^\star$ is, via $(1 - \bar{R}_t^b)\mathfrak{X}_{t-}^{\star,+}$, a random function of $\widetilde{\Theta}_{t-}$ and $\zeta_{t-} = \phi_{t-} - \eta_{t-}$. Condition (5.32) is thus implicitly a nonlinear equation in η_t^1, except in the special case where (in the present Markov setup)

$$(1 - \bar{R}_b(t,x))\mathfrak{X}^+(t,x,\pi,\varsigma) = (1 - \bar{R}_b(t,x))\mathfrak{X}^+(t,x,\pi) \tag{5.33}$$

(i.e. the left-hand side does not depend on ς), so that $\xi_t^\star$ does not depend on η_{t-}. In this case, in view of the expression derived from (5.12) for $\xi_t^\star$, depending on whether one considers a model of unilateral counterparty risk ($\tau_b = \infty$), of bilateral counterparty risk without joint default of the bank and of the counterparty ($\tau_b, \tau_c < \infty$ with $\tau_b \neq \tau_c$ almost surely), or of bilateral counterparty risk with a possible joint default of the bank and of the counterparty, we have that the equation (5.32) respectively reduces to a system of one, two or three linear equations in η_t^1.

Remark 5.3.5 (Discussion of Condition (5.33)**)** Condition (5.33) holds in the most common case of a fully swapped hedge, as well as in the partial default case where $\bar{R}_t^b = 1$ (including the case of unilateral counterparty risk). See also Sect. 4.6 for the example, based on Burgard and Kjaer (2011a, 2011b), where a solution to the equation (5.32) may be found without condition (5.33). If condition (5.33) does not hold, a possible idea to recover it (if need be) could be to forget about the closeout funding cash flow $\mathfrak{R}^f = (1 - \bar{R}_{\tau_b}^b)\mathfrak{X}_{\tau_b-}^{\star,+}$ in $\mathfrak{R}$, thus working everywhere as if $\bar{R}_b$ was equal to one and using a dt-funding coefficient $g_t(\pi,\varsigma)$ adjusted to

$$g_t^\sharp(\pi,\varsigma) = g_t(\pi,\varsigma) - \gamma_t p_t^b(1 - \bar{R}_t^b)\mathfrak{X}_t^+(\pi,\varsigma). \tag{5.34}$$

The modified problem satisfies (5.33). The adjusted funding cost coefficient $g_t^\sharp(\pi,\varsigma)$ represents a pure liquidity (as opposed to credit risk) funding cost coefficient. Using this approach also allows one to decouple the credit risk ingredients in the model, represented by τ_b and τ_c, from the liquidity funding ingredients, represented by the adjusted funding coefficient $g^\sharp$. Note that simply ignoring the closeout funding cash flow $\mathfrak{R}^f$ without adjusting g in (5.34) would introduce a valuation and hedging bias. In contrast, accordingly adjusting g in (5.34) makes it at least correct from the valuation point of view, for any fixed ζ. But this correctness in value is only for a given hedge process ζ. Since a central point in all this (particularly without (5.33)) is precisely how to choose ζ, we believe this adjustment approach is, in the end, fallacious.

Henceforth in this subsection we work under the assumption that the equation (5.32) has a solution of the form

$$\eta_t^1 = \eta_t^1(\widetilde{\Theta}_{t-}) = \eta^1(t, X_{t-}, \widetilde{\Theta}_{t-}). \tag{5.35}$$

Again, under condition (5.33), this is satisfied under a mild nonredundancy condition on the hedging instruments in group "1", with η^1 typically univariate in case $\tau_b = \infty$, bivariate in case $\tau_b, \tau_c < \infty$ with $\tau_b \neq \tau_c$, and trivariate otherwise; and we also refer the reader to Sect. 4.6 for a case where this holds without condition (5.33).

For any TVA hedge η with the components η^1 of η in group "1" given as $\eta_t^1(\widetilde{\Theta}_{t-})$ in (5.35), the TVA cost process (5.31) reduces to

$$d\varepsilon_t^{\Theta,\eta} = d\widetilde{\mu}_t - \eta_t^0 d\widetilde{\mathcal{M}}_t^0 - \eta_t^1 d\widetilde{\mathcal{M}}_t^1. \tag{5.36}$$

This leads us to seek a solution (Θ, η) to our problem (min-variance hedging of the TVA constrained to perfect hedging of TVA jump-to-default risk) with η_t of the form

$$\eta_t = \left(\eta_t^0, \eta_t^1(\widetilde{\Theta}_{t-})\right) \tag{5.37}$$

and with $(\widetilde{\Theta}, \eta, \epsilon)$ solving the pre-default TVA BSDE (5.24). Note in view of the pre-default TVA BSDE (5.24) that $d\epsilon_t$ then reduces to $d\varepsilon_t^{\Theta,\eta}$ in (5.36), the variance of which we want to minimize. Now, in order to minimize the variance of $d\varepsilon_t^{\Theta,\eta} = d\epsilon_t$ among all solutions $(\widetilde{\Theta}, \eta, \epsilon)$ of (5.24) such that $\eta_t^1 = \eta_t^1(\widetilde{\Theta}_{t-})$, we must choose η^0 as the coefficient of regression of $d\bar{\mu}_t = d\widetilde{\mu}_t - \eta_t^1(\widetilde{\Theta}_{t-})d\widetilde{\mathcal{M}}_t^1$ against $d\widetilde{\mathcal{M}}_t^0$. In other words, we are now looking for a solution $(\widetilde{\Theta}, \eta, \epsilon)$ to the pre-default TVA BSDE (5.24), with $\eta_t^1 = \eta_t^1(\widetilde{\Theta}_{t-})$ and with $d\widetilde{\mu}_t - \eta_t^1(\widetilde{\Theta}_{t-})d\widetilde{\mathcal{M}}_t^1 - \eta_t^0 d\widetilde{\mathcal{M}}_t^0$ orthogonal to $d\widetilde{\mathcal{M}}_t^0$. In such a solution, additionally assuming $\widetilde{\Theta}_t = \widetilde{\Theta}(t, X_t)$ for some Borel function $\widetilde{\Theta}(t,x)$, the min-variance sharp bracket regression formula yields:

$$\begin{aligned}
\eta_t^0 &= \frac{d\langle\bar{\mu}, \widetilde{\mathcal{M}}^0\rangle_t}{dt}\left(\frac{d < \widetilde{\mathcal{M}}^0 >_t}{dt}\right)^{-1} \\
&= \left(\mathcal{C}\left(\widetilde{\Theta}, \widetilde{\mathcal{P}}^0\right)\Lambda^0\right)(t, X_{t-}) - \eta^1(t, X_{t-}, \widetilde{\Theta}(t, X_{t-}))\left(\mathcal{C}\left(\widetilde{\mathcal{P}}^1, \widetilde{\mathcal{P}}^0\right)\Lambda^0\right)(t, X_{t-}) \\
&=: \eta^0(t, X_{t-}),
\end{aligned} \tag{5.38}$$

where we write" $\Lambda^0 = \left(\mathcal{C}\left(\widetilde{\mathcal{P}}^0, \widetilde{\mathcal{P}}^0\right)\right)^{-1}$, assumed to exist. This leads us to the following Markov BSDE in $(\widetilde{\Theta}(t, X_t), (\partial\widetilde{\Theta}\sigma)(t, X_t), \delta\widetilde{\Theta}(t, X_{t-}, \cdot))$ over $[0, T]$:

$$\begin{cases}
\widetilde{\Theta}(T, X_T) = 0 \text{ and, for } t \in [0, T], \\
-d\widetilde{\Theta}(t, X_t) = \bar{f}\left(t, X_t, \widetilde{\Theta}(t, X_t), (\partial\widetilde{\Theta}\sigma)(t, X_t), ((\delta\widetilde{\Theta}\delta(\widetilde{\mathcal{P}}^0)^{\mathsf{T}}) \cdot c)(t, X_t)\right) dt \\
\qquad - (\partial\widetilde{\Theta}\sigma)(t, X_t)dW_t - \delta\widetilde{\Theta}(t, X_{t-}) \cdot dm_t,
\end{cases} \tag{5.39}$$

with for $(t, x, \vartheta, z, w) \in [0, T] \times \mathbb{R}^q \times \mathbb{R} \times \mathbb{R}^q \times \mathbb{R}^{d_0}$, in which d_0 is the number of assets in group "0":

$$\bar{f}(t, x, \vartheta, z, w) = f\left(t, x, P(t, x) - \vartheta, \phi(t, x) - \left(\bar{\eta}^0(t, x, \vartheta, z, w), \eta^1(t, x, \vartheta)\right)\right),$$

where we let

$$\bar{\eta}^0(t, x, \vartheta, z, w) = \left(z(\partial\widetilde{\mathcal{P}}^0\sigma)^{\mathsf{T}}(t, x) + w\right)\Lambda^0(t, x) - \eta^1(t, x, \vartheta)\left(\mathcal{C}\left(\widetilde{\mathcal{P}}^1, \widetilde{\mathcal{P}}^0\right)\Lambda^0\right)(t, x).$$

Indeed, we have in view of (13.41) and (5.38):

$$\bar{\eta}^0\left(t, X_t, \widetilde{\Theta}(t, X_t), (\partial\widetilde{\Theta}\sigma)(t, X_t), ((\delta\widetilde{\Theta}\delta(\widetilde{\mathcal{P}}^0)^{\mathsf{T}}) \cdot c)(t, X_t)\right) = \eta^0(t, X_t),$$

$$\begin{aligned}
&\bar{f}\left(t, X_t, \widetilde{\Theta}(t, X_t), (\partial\widetilde{\Theta}\sigma)(t, X_t), ((\delta\widetilde{\Theta}\delta(\widetilde{\mathcal{P}}^0)^{\mathsf{T}}) \cdot c)(t, X_t)\right) dt \\
&\quad = f\left(t, X_t, P(t, X_t)\widetilde{\Theta}(t, X_t), \phi(t, X_t) - \left(\eta^0(t, X_t), \eta^1(t, X_t, \widetilde{\Theta}(t, X_t))\right)\right) dt.
\end{aligned}$$

Now, under mild technical conditions, the Markov BSDE (5.39) has a unique solution[5] and the pre-default TVA function $\widetilde{\Theta} = \widetilde{\Theta}(t, x)$ in this solution can be characterized as the unique solution to the following semilinear PDE:

$$\begin{cases}
\widetilde{\Theta}(T, x) = 0, \; x \in \mathbb{R}^q, \\
(\partial_t + \mathcal{A})\,\widetilde{\Theta}(t, x) + \bar{f}(t, x, \widetilde{\Theta}(t, x), (\partial\widetilde{\Theta}\sigma)(t, x), ((\delta\widetilde{\Theta}\delta(\widetilde{\mathcal{P}}^0)^{\mathsf{T}}) \cdot c)(t, x)) \\
\qquad = 0 \text{ on } [0, T) \times \mathbb{R}^q.
\end{cases} \tag{5.40}$$

We can summarize the above analysis as follows.

Proposition 5.3.6 *Let us assume existence of a solution* $\eta_t^1 = \eta^1(\widetilde{\Theta}_{t-})$ *to the equation* (5.32) *and invertibility of the carré du champ matrix* $\mathcal{C}\left(\widetilde{\mathcal{P}}^0, \widetilde{\mathcal{P}}^0\right)$ *of the hedging assets in group* "0". *Then the solution* $\widetilde{\Theta} = \widetilde{\Theta}(t, x)$ *to* (5.40) *yields, via* (5.37)-(5.38) *for* η *and* (5.24) *for* ϵ, *a solution* $(\widetilde{\Theta}, \eta, \epsilon)$ *to the pre-default TVA BSDE* (5.24) *such that* $\eta_t^1 = \eta_t^1(\widetilde{\Theta}_{t-})$ *and* $(d\widetilde{\mu} - \eta_t^1(\widetilde{\Theta}_{t-})d\widetilde{\mathcal{M}}_t^1 - \eta_t^0 d\widetilde{\mathcal{M}}_t^0)$ *is orthogonal to* $d\widetilde{\mathcal{M}}_t^0$.

The min-variance hedge of the TVA, constrained to a perfect hedge of the TVA jump-to-default risk, is given as $\eta_t = (\eta_t^0, \eta_t^1(\widetilde{\Theta}_{t-}))$, *where* $\eta_t^0 = \eta^0(t, X_{t-})$ *is given by the formula* (5.38), *i.e.*

$$\begin{aligned}
\eta_t^0 = &\left(\mathcal{C}\left(\widetilde{\Theta}, \widetilde{\mathcal{P}}^0\right)\left(\mathcal{C}\left(\widetilde{\mathcal{P}}^0, \widetilde{\mathcal{P}}^0\right)\right)^{-1}\right)(t, X_{t-}) \\
&- \eta_t^1(\widetilde{\Theta}_{t-})\left(\mathcal{C}\left(\widetilde{\mathcal{P}}^1, \widetilde{\mathcal{P}}^0\right)\left(\mathcal{C}\left(\widetilde{\mathcal{P}}^0, \widetilde{\mathcal{P}}^0\right)\right)^{-1}\right)(t, X_{t-}).
\end{aligned}$$

The process $\epsilon = \varepsilon^{\Theta,\eta}$ *represents the residual TVA risk (in this case a pure market risk) under this TVA hedge* η.

[5] Up to the monotonicity condition of the remark 5.3.2, applying here to $\bar{f}$.

Hedging of the Contract as a Whole We now consider the constrained min-variance hedging problem of the contract as a whole, rather than simply of its TVA component. We assume further that the hedge ϕ of the contract clean price P only involves the primary assets in group "0" and that ϕ^0 is given as the coefficient of regression in an $\mathbb{F}$-orthogonal decomposition $dM = \phi^0 d\widetilde{\mathcal{M}}^0 + d\bar{e}$, so that

$$\phi_t^0 = \frac{d<M,\widetilde{\mathcal{M}}^0>_t}{dt}\left(\frac{d<\widetilde{\mathcal{M}}^0>_t}{dt}\right)^{-1} = \left(\mathcal{C}\left(P,\widetilde{\mathcal{P}}^0\right)\Lambda^0\right)(t,X_{t-}) =: \phi^0(t,X_{t-}).$$

For $(\widetilde{\Theta},\eta,\epsilon)$ as in Proposition 5.3.6 and for $\zeta = \phi - \eta$, the cost equations (5.21)-(5.23) reduce to

$$\begin{aligned}
d\varepsilon_t^{P,\phi} &= dM_t - \phi_t^0 d\widetilde{\mathcal{M}}_t^0 = d\bar{e}_t,\\
d\varepsilon_t^{\Theta,\eta} &= d\widetilde{\mu}_t - \eta_t^0 d\widetilde{\mathcal{M}}_t^0 - \eta_t^1(\widetilde{\Theta}_{t-})d\widetilde{\mathcal{M}}_t^1 = d\epsilon_t,\\
d\varepsilon_t^{\Pi,\zeta} &= d\varepsilon_t = d\varepsilon_t^{P,\phi} - d\varepsilon_t^{\Theta,\eta}\\
&= d\widetilde{\nu}_t - \zeta_t^0 d\widetilde{\mathcal{M}}_t^0 + \eta_t^1(\widetilde{\Theta}_{t-})d\widetilde{\mathcal{M}}_t^1.
\end{aligned}$$

Since $dM_t - \phi_t^0 d\widetilde{\mathcal{M}}_t^0$ and $d\widetilde{\mu}_t - \eta_t^0 d\widetilde{\mathcal{M}}_t^0 - \eta_t^1(\widetilde{\Theta}_{t-})d\widetilde{\mathcal{M}}_t^1$ are $\mathbb{F}$-orthogonal to $d\widetilde{\mathcal{M}}_t^0$, so is $d\widetilde{\nu}_t - \zeta_t^0 d\widetilde{\mathcal{M}}_t^0 + \eta_t^1(\widetilde{\Theta}_{t-})d\widetilde{\mathcal{M}}_t^1$. Proposition 5.3.6 thus admits the following:

Corollary 5.3.7 *For ϕ^0 given as the regression coefficient of M against $\widetilde{\mathcal{M}}^0$, the strategy $\zeta_t = \left(\phi^0(t,X_{t-}) - \eta^0(t,X_{t-}), -\eta_t^1(\widetilde{\Theta}_{t-})\right)$ is a min-variance hedge of the contract (market risk), under the contract jump-to-default perfect hedge constraint (i.e. $\zeta_t^1 = -\eta_t^1(\widetilde{\Theta}_{t-})$). The residual (market) risk of the contract hedged in this way is given by $\varepsilon^{\Pi,\zeta} = \varepsilon = \bar{e} - \epsilon$.*

Remark 5.3.8 Under the fully swapped hedge condition (4.18), which in the current Markov setup implies (5.33) through a more specific $f(t,x,\pi,\varsigma) = f(t,x,\pi)$ and with $f(t,x,P(t,x)-\vartheta) = \widetilde{f}(t,x,\vartheta)$, the Markov BSDEs (5.28) and (5.39) both reduce to:

$$\begin{cases}\widetilde{\Theta}(T,X_T) = 0 \text{ and, for } t\in[0,T],\\ -\,d\widetilde{\Theta}(t,X_t) = \widetilde{f}\left(t,X_t,\widetilde{\Theta}(t,X_t)\right)dt - (\partial\widetilde{\Theta}\sigma)(t,X_t)dW_t - \delta\widetilde{\Theta}(t,X_{t-})\cdot dm_t,\end{cases} \tag{5.41}$$

with a related semilinear PDE given as

$$\begin{cases}\widetilde{\Theta}(T,x) = 0,\ x\in\mathbb{R}^q,\\ (\partial_t + \mathcal{A})\,\widetilde{\Theta}(t,x) + \widetilde{f}(t,x,\widetilde{\Theta}(t,x)) = 0 \text{ on } [0,T)\times\mathbb{R}^q.\end{cases} \tag{5.42}$$

Note that even though the value $\widetilde{\Theta}$ of the TVA is uniquely defined through (5.41)–(5.42) (assumed well-posed), the hedges η that follow via Propositions 5.3.3 or 5.3.6 (respectively ζ via Corollaries 5.3.4 or 5.3.7) typically differ, since they solve different hedging problems.

5.3.4 Unilateral or Bilateral in the End?

The importance of hedging counterparty risk in terms not only of market risk, but also of jump-to-default exposure, was revealed in the 2007–09 subprime crisis. But, since selling protection on oneself via a CDS contract is not really doable (who would buy it?), whether it is practically possible to hedge one's own jump-to-default exposure is rather dubious due to absence of suitable hedging instruments – apart from the possibility of repurchasing one's own bond considered in Burgard and Kjaer (2011a, 2011b), which we discussed in Sect. 4.6. (A contrario, hedging the spread risk of DVA is possible, by selling CDS protection on peers, but this strategy is very risky and also poses a systemic risk issue; one may also consider that DVA, like CVA, is a natural hedge to other things, being countercyclical).

Alternatively (in particular, in case of unilateral counterparty risk), the bank can resort to a variant of the approach of Sect. 5.3.3 consisting in min-variance hedging of market risk constrained to perfect hedging of the counterparty's jump-to-default risk, leaving ones own default unhedged. Only hedging the counterparty's jump-to-default risk means hedging $\mathbb{1}_{\{\tau_c<\tau_b\wedge T\}}\times$ (the dJ-term) in (5.31). In view of the TVA cost equation (5.31) and given the specification (5.5) of ξ_t, this reduces to the following explicit (as opposed to (5.32)) univariate linear equation to be satisfied by a scalar process η^1:

$$P_t - Q_t + (1-R_t^c)\chi_t^+ - \widetilde{\Theta}_{t-} = \eta_t^1\left(\mathcal{R}_t^1 - \widetilde{\mathcal{P}}_t^1\right),\ t\in[0,T]. \tag{5.43}$$

Therefore min-variance hedging of the market risk of the TVA (or of the contract as a whole), subject to perfect hedge of the jump-to-default of the counterparty alone reduces to min-variance hedging of the market risk of the TVA (or of the contract as a whole) subject to $\eta_t^1 = \eta_t^1(\widetilde{\Theta}_{t-})$ solving (5.43). This min-variance hedging can be implemented as in Sect. 5.3.3, yielding easily derived analogs of Proposition 5.3.6 and Corollary 5.3.7. Note that this involves no technical condition like (5.33).

In this approach the bank does not hedge its own jump-to-default and therefore cannot monetize it (as opposed again to the case where, for instance, the bank could make instant money by selling upfront CDS protection on itself). If the bank wants to be consistent in this regard and disregard such "fake benefit", it can set $R_b = \bar{R}_b = 1$. Then the equations (5.7)–(5.6) for f reduce to

$$\begin{aligned} f_t(P_t-\vartheta,\varsigma) + r_t\vartheta = {} & \gamma_t p_t^c(1-R_t^c)(Q_t-\Gamma_t)^+ \\ & + g_t\left(P_t-\vartheta,\varsigma\right) \\ & + \gamma_t\left(P_t-\vartheta-Q_t\right), \end{aligned} \tag{5.44}$$

where there is no longer a beneficial debt valuation adjustment and where the funding coefficient g is interpreted as a pure liquidity cost (see Sect. 6.3.5 for a more concrete specification, in the framework of the example 5.2.6).

Such an "asymmetrical TVA approach", in contrast with the "symmetrical TVA approach" of Sect. 5.3.3, even though still bilateral, allows one to get rid of many concerns, including the arbitrage issue that arises for $\bar{R}_b < 1$, the hypothetical and paradoxical benefit of the bank at its own default time and the puzzle of the bank having to hedge its own jump-to-default risk in order to monetize it. This approach is also justified with regard to the fact that the windfall benefits at own default are in effect cashflows to senior bond holders, whereas only the interest of the shareholders should be considered in the optimization (or hedging) process of the bank (see Albanese, Brigo, and Oertel (2013) and Albanese and Iabichino (2013)).

Chapter 6

The Four Wings of the TVA

6.1 Introduction

The intricacy issues regarding the CVA, DVA, LVA and RC, jointly referred to as TVA (for total valuation adjustment), were investigated in the previous chapters, to which this one is a numerical companion. Sect. 6.2 provides an executive summary of Chapters 4 and 5. Sect. 6.3 describes various CSA specifications. In Sects. 6.4 and 6.5 we present clean valuation (clean of counterparty risk and funding costs) and TVA computations in two simple models for interest rate derivatives. We show in a series of practical examples how CVA, DVA, LVA and RC, the four "wings" of the TVA, can be computed in various situations, and we assess the related model risk.

6.2 TVA Representations

6.2.1 Setup

The setup is the same as in Chapter 5. The full model filtration is given as $\mathbb{G} = \mathbb{F} \vee \mathbb{H}^b \vee \mathbb{H}^c$, where $\mathbb{F}$ is a reference (or background) filtration. A risk-neutral pricing measure $\mathbb{Q}$ is fixed throughout on $\mathcal{G}_T$. As in Chapter 5, the meaning of a risk-neutral pricing measure in a nonlinear funding setup is specified by martingale conditions, stated below in the form of suitable pricing backward stochastic differential equations (BSDEs).

Remark 6.2.1 Even though it will not appear explicitly in this chapter, a pricing measure must also satisfy the assumption 4.4.1, so that the gain processes on the hedging assets follow martingales.

Moreover, we assume that an $\mathbb{F}$-martingale stopped at τ is a $\mathbb{G}$-martingale, i.e. τ is an $(\mathbb{F}, \mathbb{G})$-pseudo stopping time. As discussed in the remark 5.2.2, this basic assumption precludes major wrong-way risk effects such as the ones which occur with counterparty risk on credit derivatives. In particular, under these assumptions an $\mathbb{F}$-adapted càdlàg process cannot jump at τ, e.g. $\Delta_\tau = 0$. Readers are referred to Part IV as well as to Chapter 12 and Crépey and Song (2014) for various developments in this regard.

We denote by r_t an OIS rate (OIS stands for an overnight indexed swap), which is the best market proxy of a risk-free rate. By $\widetilde{r}_t = r_t + \gamma_t$ we denote the credit-risk adjusted rate, where γ_t is the $\mathbb{F}$-hazard intensity of τ, assumed to exist. Let $\beta_t = \exp(-\int_0^t r_s ds)$ and $\alpha_t = \exp(-\int_0^t \widetilde{r}_s ds)$ denote the corresponding discount factors. Furthermore, $\mathbb{E}_t$ and $\widetilde{\mathbb{E}}_t$ denote the conditional expectations given $\mathcal{G}_t$ and $\mathcal{F}_t$, respectively. The clean value process P_t of the contract with promised dividends dD_t is defined, for $t \in [0, \bar{\tau}]$, by (cf. (5.1))

$$\beta_t P_t = \widetilde{\mathbb{E}}_t \left(\int_t^T \beta_s dD_s \right). \tag{6.1}$$

Let an $\mathbb{F}$-predictable process Q represent the CSA value process of the contract and an $\mathbb{F}$-adapted process Γ stand for the value process of a CSA (cash) collateralization scheme. We denote by π a real number, meant to represent the wealth of the hedging portfolio of the bank in the financial interpretation. For simplicity we choose to conform to the most common situation where the hedge is securely funded as swapped and/or traded via repo markets. As follows from Chapter 5, the $\mathcal{G}_\tau$-measurable exposure at default ξ will then be defined in terms of (cf. (5.12))

$$\begin{aligned}\xi_t(\pi) = P_t - Q_t + \mathbb{1}_{t \geq \tau_c}(1 - R_c)\chi_t^+ \\ - \mathbb{1}_{t \geq \tau_b}\Big((1 - R_b)\chi_t^- + (1 - \bar{R}_b)\mathfrak{X}_{t-}^+(\pi)\Big),\end{aligned} \tag{6.2}$$

where (cf. (5.5), (4.17))

$$\chi_t = Q_t - \Gamma_t \text{ and } \mathfrak{X}_t(\pi) = \pi - \Gamma_t \tag{6.3}$$

represent the algebraic debt of the counterparty to the bank and of the bank to its funder, respectively, and where R_b and R_c stand for recovery rates between the two parties, whereas $\bar{R}_b$ stands for a recovery rate of the bank to its funder.

Remark 6.2.2 Under some of the specifications considered below, Q_t and Γ_t also depend on π. Such additional dependence does not essentially alter the flow of arguments and is therefore omitted for notational simplicity.

We now consider the cash flows required for funding the bank's position (the contract and its hedging portfolio together). The OIS rate r_t is used as a reference for all other funding rates, which are defined in terms of corresponding bases to r_t. Given such bases b_t and $\bar{b}_t$ related to the collateral posted and received by the bank and λ_t and $\bar{\lambda}_t$ related to external lending and borrowing by the bank, the funding cost coefficient $g_t(\pi)$ is defined as in the example 5.2.6 by

$$g_t(\pi, \varsigma) = g_t(\pi) = \bar{b}_t \Gamma_t^+ - b_t \Gamma_t^- + \bar{\lambda}_t (\pi - \Gamma_t)^+ - \lambda_t (\pi - \Gamma_t)^-. \tag{6.4}$$

Then $(r_t \pi + g_t(\pi))dt$ represents the bank's funding cost over $(t, t + dt)$, depending on the contract's value[1] π.

[1] Or, more precisely, minus the value of the collateralization, hedging and funding portfolio.

Remark 6.2.3 A funding basis is interpreted as a combination of liquidity and credit risk (see Chapter 2). Collateral posted in different currencies and related optionalities can be accounted for by suitable amendments to b and $\bar{b}$ (see Fujii and Takahashi (2011a) and Piterbarg (2012)).

6.2.2 BSDEs

With the data ξ and g_t thus specified, the TVA process Θ is implicitly defined on $[0, \bar{\tau}]$ as a solution, assumed to exist, to the following BSDE (in integral form), posed over the random time interval $[0, \bar{\tau}]$ (cf. 4.38):

$$\beta_t \Theta_t = \mathbb{E}_t\Big[\beta_{\bar{\tau}} \mathbb{1}_{\tau<T} \xi_\tau (P_\tau - \Theta_{\tau-}) + \int_t^{\bar{\tau}} \beta_s g_s(P_s - \Theta_s) ds\Big], \ t \in [0, \bar{\tau}]. \tag{6.5}$$

See Proposition 4.5.2 for the derivation of the TVA BSDE (6.5).

Remark 6.2.4 For $\bar{R}_b = 1$, the exposure at default ξ_t does not depend on π, and in the linear case of the funding coefficient given as $g_t(P_t - \vartheta) = g_t^0(P_t) - \lambda_t^0 \vartheta$ (for some g^0 and λ^0), the TVA equation (6.5) reduces to the explicit representation

$$\beta_t^0 \Theta_t = \mathbb{E}_t\Big[\beta_{\bar{\tau}}^0 \mathbb{1}_{\tau<T} \xi + \int_t^{\bar{\tau}} \beta_s^0 g_s^0(P_s) ds\Big], \tag{6.6}$$

for the adjusted discount factor

$$\beta_t^0 = \exp(-\int_0^t (r_s + \lambda_s^0) ds).$$

For $t \in [0, T]$ and $\pi \in \mathbb{R}$, let

$$\begin{aligned} \widetilde{\xi}_t(\pi) &= (P_t - Q_t) + p_t^c (1 - R_t^c) \chi_t^+ \\ &\quad - p_t^b \Big((1 - R_t^b) \chi_t^- + (1 - \bar{R}_t^b)(\pi - \Gamma_t)^+ \Big) \end{aligned} \tag{6.7}$$

for predictable processes p^b and p^c such that

$$p_\tau^b = \mathbb{Q}(\tau = \tau_b \,|\, \mathcal{G}_{\tau-}), \ p_\tau^c = \mathbb{Q}(\tau = \tau_c \,|\, \mathcal{G}_{\tau-})$$

(note that, in case of unilateral counterparty risk, we have $\tau_b = \infty$ and, consequently, $p_\tau^b = 0$, $p_\tau^c = 1$). We saw in Proposition 5.2.10 that we obtain a solution Θ to the full TVA equation (6.5) by setting $\Theta = \widetilde{\Theta}$ on $[0, \bar{\tau})$ and $\Theta_{\bar{\tau}} = \mathbb{1}_{\tau<T} \xi$, provided an $\mathbb{F}$-semimartingale $\widetilde{\Theta}$ satisfies the following $\mathbb{F}$-BSDE:

$$\alpha_t \widetilde{\Theta}_t = \widetilde{\mathbb{E}}_t\Big[\int_t^T \alpha_s \big(g_s(P_s - \widetilde{\Theta}_s) + \gamma_s \widetilde{\xi}_s (P_s - \widetilde{\Theta}_s)\big) ds\Big], \tag{6.8}$$

for $t \in [0, T]$.

Remark 6.2.5 This assumes that the data of (6.8) are $\mathbb{F}$-adapted, a mild condition which can always be met by passing to $\mathbb{F}$-representatives (or pre-default values) of the original data (see Sect. 5.2.2).

The pre-default TVA BSDE (6.8) can be written in differential form as (we recover (5.8))

$$\begin{cases} \widetilde{\Theta}_T = 0 \text{ and, for } t \in [0,T], \\ -d\widetilde{\Theta}_t = f_t(P_t - \widetilde{\Theta}_t)dt - d\widetilde{\mu}_t, \end{cases} \tag{6.9}$$

where $\widetilde{\mu}$ is the $\mathbb{F}$-martingale component of $\widetilde{\Theta}$ and where (cf. (5.7))

$$f_t(P_t - \vartheta) = g_t(P_t - \vartheta) + \gamma_t \widetilde{\xi}_t(P_t - \vartheta) - \widetilde{r}_t \vartheta. \tag{6.10}$$

Remark 6.2.6 In the linear pre-default case, where

$$g_t(P_t - \vartheta) + \gamma_t\big(\widetilde{\xi}_t(P_t - \vartheta) - \vartheta\big) = \widetilde{g}^0_t(P_t) - \widetilde{\lambda}^0_t \vartheta$$

for some $\widetilde{g}^0$ and $\widetilde{\lambda}^0$, the pre-default TVA equation (6.8) reduces to the explicit representation

$$\widetilde{\beta}^0_t \widetilde{\Theta}_t = \widetilde{\mathbb{E}}_t\Big[\int_t^T \widetilde{\beta}^0_s \widetilde{g}^0_s(P^0_s)ds\Big], \tag{6.11}$$

for the adjusted discount factor

$$\widetilde{\beta}^0_t = \exp(-\int_0^t (r_s + \widetilde{\lambda}^0_s)ds).$$

On the numerical side, explicit representations such as (6.6) or (6.12) allow one to estimate the corresponding "linear TVA" by a standard Monte Carlo loop (provided P_t and Q_t can be computed explicitly). Otherwise nonlinear TVA computations can only be done by more advanced schemes: nonlinear regressions as in Cesari, Aquilina, Charpillon, Filipovic, Lee, and Manda (2010), expansions (see Fujii and Takahashi (2011b)) or branching particles (see Henry-Labordère (2012)). Deterministic schemes for the corresponding semilinear TVA PDEs can only be used in low dimensional Markov models.

Remark 6.2.7 From (6.1), P satisfies the following $\mathbb{F}$-BSDE

$$\begin{cases} P_T = 0 \text{and, for } t \in [0,T], \\ -dP_t = dD_t - r_t P_t dt - dM_t, \end{cases} \tag{6.12}$$

for some $\mathbb{F}$-martingale M. The following pre-default $\mathbb{F}$-BSDE for $\widetilde{\Pi} := P - \widetilde{\Theta}$ follows by (6.9) and (6.12):

$$\begin{cases} \widetilde{\Pi}_T = 0 \text{ and, for } t \in [0,T], \\ -d\widetilde{\Pi}_t = dD_t - \Big(f_t(\widetilde{\Pi}_t) + r_t P_t\Big)\, dt - d\widetilde{\nu}_t, \end{cases} \tag{6.13}$$

where $d\widetilde{\nu}_t = dM_t - d\widetilde{\mu}_t$. As the pre-default price BSDE (6.13) involves the contractual promised cash flows dD_t, it is less convenient than the pre-default TVA BSDE (6.9). This mathematical incentive comes on top of the financial justification presented in the preface for adopting a "clean price P minus TVA Θ" valuation and hedging approach.

6.2.2.1 Pre-default Markov Setup

Assume as in the remark 5.3.8 (cf. also (5.25)) that

$$f_t(P_t - \vartheta) = \widetilde{f}(t, X_t, \vartheta) \tag{6.14}$$

for some measurable function $\widetilde{f}(t, x, \theta)$ and an $\mathbb{R}^d$-valued $\mathbb{F}$-Markov pre-default factor process X. Then $\widetilde{\Theta}_t = \widetilde{\Theta}(t, X_t)$, where the pre-default TVA pricing function $\widetilde{\Theta}(t, x)$ is the solution to a related pre-default pricing PDE. However, as mentioned in the remark 6.2.6, from the point of view of numerical solution deterministic PDE schemes can only be used provided the dimension of X is less than 3 or 4, otherwise simulation schemes for (6.9) are the only viable choices.

6.2.3 CVA, DVA, LVA and RC

Substituting (6.7) into (6.10) and reordering terms (cf. (5.11)),

$$\begin{aligned} f_t(P_t - \vartheta) + r_t\vartheta = {} & \gamma_t p_t^c (1 - R_c)(Q_t - \Gamma_t)^+ \\ & - \gamma_t p_t^b ((1 - R_b)(Q_t - \Gamma_t)^- \\ & + \bar{b}_t \Gamma_t^+ - b_t \Gamma_t^- + \widetilde{\lambda}_t (P_t - \vartheta - \Gamma_t)^+ - \lambda_t (P_t - \vartheta - \Gamma_t)^- \\ & + \gamma_t (P_t - \vartheta - Q_t), \end{aligned} \tag{6.15}$$

where the coefficient $\widetilde{\lambda}_t = \bar{\lambda}_t - \gamma_t p_t^b (1 - \bar{R}_b)$ multiplying $(P_t - \vartheta - \Gamma_t)^+$ in the third line can be interpreted as an external borrowing basis net of credit spread. This coefficient represents the liquidity component of $\bar{\lambda}$. From the perspective of the bank, the four terms in this decomposition of the TVA coefficient can respectively be interpreted as a costly (nonalgebraic, strict) credit value adjustment (CVA) component, a beneficial debt valuation adjustment (DVA) component, a liquidity funding cost/benefit (LVA) component and a replacement benefit/cost (RC) component. In particular, the time-0 TVA can be represented as

$$\begin{aligned} \Theta_0 = {} & \mathbb{E}\Big[\int_0^T \beta_t \gamma_t p_t^c (1 - R_c)(Q_t - \Gamma_t)^+ \big) dt\Big] \\ & - \mathbb{E}\Big[\int_0^T \beta_t \gamma_t p_t^b \big((1 - R_b)(Q_t - \Gamma_t)^- \big) dt\Big] \\ & + \mathbb{E}\Big[\int_0^T \beta_t \big(\bar{b}_t \Gamma_t^+ - b_t \Gamma_t^- + \widetilde{\lambda}_t \Big(P_t - \widetilde{\Theta}_t - \Gamma_t\Big)^+ - \lambda_t \Big(P_t - \widetilde{\Theta}_t - \Gamma_t\Big)^- \big) dt\Big] \\ & + \mathbb{E}\Big[\int_0^T \beta_t \gamma_t (P_t - \widetilde{\Theta}_t - Q_t) dt\Big]. \end{aligned} \tag{6.16}$$

The CVA and the $\bar{\lambda}_t(P_t - \widetilde{\Theta}_t - \Gamma_t)^+$ components of the LVA (for $\bar{\lambda}_t$ positive) are "deal adverse" as they increase the TVA and therefore decrease the "buying price" , at which the bank can consider buying the contract from the counterparty (cost of the related hedge).

Conversely, the DVA and the $-\gamma_t p_t^b(1-\bar{R}_b)(P_t - \widetilde{\Theta}_t - \Gamma_t)^+$ component of the LVA are "deal friendly", as they decrease the TVA and therefore increase the price the bank can afford. Likewise the other terms of (6.16) can be interpreted as "deal friendly or adverse" depending on their sign, positive or negative.

Remark 6.2.8 A possible application of the analysis of Chapter 2 would be to calibrate (or, at least, gauge) the volatility of the liquidity funding borrowing basis $\widetilde{\lambda}_t$ in a stochastic model for the latter. In view of the LOIS formula (2.12), this volatility could be identified, at a first level of approximation, with the LOIS market spread between Libor and OIS rates (see also the remark 6.4.2). This would correspond to a so called market approach to funding costs, as opposed to a funding approach based on an analysis of the funding transfer process of the bank.

6.3 CSA Specifications

In this section we detail various specifications of the general form (6.15) for f, depending on the CSA data: the closeout valuation scheme Q, the collateralization scheme Γ and the collateral remuneration bases b and $\bar{b}$.

6.3.1 Clean CSA Recovery Scheme

In case of a clean CSA recovery scheme $Q = P$, (6.15) takes the form

$$\begin{aligned} f_t(P_t - \vartheta) + \widetilde{r}_t\vartheta = \gamma_t p_t^c(1-R_c)(P_t - \Gamma_t)^+ \\ - \gamma_t p_t^b(1-R_b)(P_t - \Gamma_t)^- \\ + b_t\Gamma_t^+ - \bar{b}_t\Gamma_t^- + \lambda_t(P_t - \vartheta - \Gamma_t)^+ - \widetilde{\lambda}_t(P_t - \vartheta - \Gamma_t)^- \end{aligned} \quad (6.17)$$

(note $\widetilde{r}_t$ on the left-hand side, as opposed to r_t in (6.15)). In case of no collateralization ($\Gamma = 0$), the right-hand-side of (6.17) reduces to

$$\gamma_t p_t^c(1-R_c)P_t^+ - \gamma_t p_t^b(1-R_b)P_t^- + \lambda_t(P_t - \vartheta)^+ - \widetilde{\lambda}_t(P_t - \vartheta)^-; \quad (6.18)$$

whereas in case of continuous collateralization with $\Gamma = Q = P$, it reduces to

$$b_t P_t^+ - \bar{b}_t P_t^- + \lambda_t\vartheta^- - \widetilde{\lambda}_t\vartheta^+. \quad (6.19)$$

Remark 6.3.1 (Symmetrical funding liquidity case) If $\lambda = \widetilde{\lambda}$ (equal external borrowing and lending liquidity bases), then the TVA is linear (cf. the remark 6.2.6) for every collateralization scheme of the form $P_t - \Gamma_t = \varepsilon_t$ that yields an exogenous residual exposure ε_t (not depending on $\widetilde{\Theta}_t$, e.g. for null or continuous collateralization). Setting

$$f_t^\lambda = (\gamma_t p_t^c(1-R_c) + \lambda_t)\,\varepsilon^+ - \left(\gamma_t p_t^b(1-R_b) + \lambda_t\right)\varepsilon^- + b_t\Gamma_t^+ - \bar{b}_t\Gamma_t^-, \quad (6.20)$$

we end up, similar to (6.11), with

$$\widetilde{\beta}_t^{\lambda}\widetilde{\Theta}_t = \widetilde{\mathbb{E}}_t \int_t^T \widetilde{\beta}_s^{\lambda} f_s^{\lambda} ds, \tag{6.21}$$

for the adjusted discount factor

$$\widetilde{\beta}_t^{\lambda} = \exp\left(-\int_0^t (\widetilde{r}_s + \lambda_s) ds\right). \tag{6.22}$$

6.3.2 Pre-Default CSA Recovery Scheme

In case of a pre-default CSA recovery scheme $Q = \widetilde{\Pi} = P - \widetilde{\Theta}$, (6.15) takes the form

$$\begin{aligned} f_t(\vartheta) + r_t\vartheta = & \left(\gamma_t p_t^c(1 - R_c) + \widetilde{\lambda}_t\right)(P_t - \vartheta - \Gamma_t)^+ \\ & - \left(\gamma_t p_t^b(1 - R_b) + \lambda_t\right)(P_t - \vartheta - \Gamma_t)^- \\ & + \bar{b}_t\Gamma_t^+ - b_t\Gamma_t^-. \end{aligned} \tag{6.23}$$

In case of no collateralization ($\Gamma = 0$) the right-hand-side reduces to

$$\left(\gamma_t p_t^c(1 - R_c) + \widetilde{\lambda}_t\right)(P_t - \vartheta)^+ - \left(\gamma_t p_t^b(1 - R_b) + \lambda_t\right)(P_t - \vartheta)^-, \tag{6.24}$$

whereas the continuous collateralization with $\Gamma = Q = P - \widetilde{\Theta}$ yields

$$\bar{b}_t(P_t - \vartheta)^+ - b_t(P_t - \vartheta)^-. \tag{6.25}$$

Remark 6.3.2 If $b = \bar{b}$ (case of equal collateral borrowing and lending liquidity bases), the TVA is linear for every collateralization scheme of the form $P_t - \widetilde{\Theta}_t - \Gamma_t = \varepsilon_t$ that yields an exogenous residual exposure ε_t (not depending on $\widetilde{\Theta}_t$, e.g. for continuous collateralization). Setting

$$f_t^b = -\left(\gamma_t p_t^c(1 - R_c) + \widetilde{\lambda}_t\right)\varepsilon_t^- + b_t(P_t - \varepsilon_t) + \left(\gamma_t p_t^b(1 - R_b) + \lambda_t\right)\varepsilon_t^+, \tag{6.26}$$

we end up, again similar to (6.11), with

$$\widetilde{\beta}_t^b\widetilde{\Theta}_t = \widetilde{\mathbb{E}}_t\left[\int_t^T \widetilde{\beta}_s^b f_s^b ds\right], \tag{6.27}$$

for the adjusted discount factor

$$\widetilde{\beta}_t^b = \exp(-\int_0^t (r_s + b_s) ds). \tag{6.28}$$

Remark 6.3.3 If $b = \bar{b} = 0$, the two continuous collateralization schemes corresponding to equations (6.19)) and (6.25) both reduce to $\widetilde{\Theta} = 0$ and $\widetilde{\Pi} = P = \Gamma = Q$.

6.3.3 Full Collateralization CSA

Observe that (6.25) only involves the collateral remuneration bases, which are exchanged between the two parties. Therefore, in the corresponding case, $\Gamma = Q = \widetilde{\Pi}$, where $\widetilde{\Pi}$ is the bank's buying price of the contract is also the counterparty's selling price. Consequently, this specification deserves the name of full collateralization. Note, however, that for b or $\bar{b} \neq 0$, this specification, even though fully collateralized, entails some TVA.

If $b = \bar{b}$ then, in view of (6.15), $f_t(\widetilde{\Pi}_t)+r_t P_t$ reduces to $(r_t+b_t)\widetilde{\Pi}_t$ in the BSDE (6.13) for the fully collateralized price $\widetilde{\Pi} = \Gamma$. This BSDE is thus equivalent to the following explicit expression for $\widetilde{\Pi}$:

$$\widetilde{\beta}_t^b \widetilde{\Pi}_t = \widetilde{\mathbb{E}}_t \Big[\int_t^T \widetilde{\beta}_s^b dD_s \Big], \tag{6.29}$$

where $\widetilde{\beta}^b$ has been defined in (6.28). In the special case $b = \bar{b} = 0$, we have $\widetilde{\Theta} = 0$ and $\widetilde{\Pi} = P = Q = \Gamma$, which is the situation already considered in the remark 6.3.3. This case, where all the counterparty risk and excess-funding costs are mitigated out, justifies the status of the formula (6.1) as the master clean valuation formula for a fully collateralized price at an OIS collateral funding rate r_t. With such a fully collateralized CSA there is no need for pricing-and-hedging a (null) TVA. The problem reduces to the computation of a clean price P and a related hedge.

6.3.4 Pure Funding

In case $\gamma = 0$, i.e. in the case of default-free counterparties, which we refer to as the pure funding case, the CSA value process Q plays no actual role. The dt-coefficient of the BSDE (6.13) for $\widetilde{\Pi}$ is given by

$$f_t(\widetilde{\Pi}_t) + r_t P_t = (r_t + \bar{b}_t)\Gamma_t^+ - (r_t + b_t)\Gamma_t^- + (r_t + \widetilde{\lambda}_t)(\widetilde{\Pi} - \Gamma_t)^+ - (r_t + \lambda_t)(\widetilde{\Pi} - \Gamma_t)^-,$$

for an external borrowing basis $\widetilde{\lambda} = \bar{\lambda}$ that should be interpreted as a pure liquidity funding basis (as there is no counterparty risk involved). In case $\Gamma = 0$ and $\lambda = \widetilde{\lambda}$, this results in the following explicit expression for $\widetilde{\Pi}$:

$$\widetilde{\beta}_t^\lambda \widetilde{\Pi}_t = \widetilde{\mathbb{E}}_t \int_t^T \widetilde{\beta}_s^\lambda dD_s, \tag{6.30}$$

where $\widetilde{\beta}^\lambda$ has been defined in (6.22). In the special case $\widetilde{\lambda} = \bar{\lambda} = 0$, we recover the classical valuation formula (6.1) for $\widetilde{\Pi} = P$.

6.3.5 Asymmetrical TVA Approach

In an asymmetrical TVA approach with $R_b = \bar{R}_b = 1$ (see Sect. 5.3.4), the equation (6.15) reduces to (cf. (5.44))

$$\begin{aligned} f_t(P_t - \vartheta) + r_t\vartheta = {} & \gamma_t p_t^c (1 - R_c)(Q_t - \Gamma_t)^+ \\ & + \bar{b}_t \Gamma_t^+ - b_t \Gamma_t^- + \bar{\lambda}_t (P_t - \vartheta - \Gamma_t)^+ - \lambda_t (P_t - \vartheta - \Gamma_t)^- \\ & + \gamma_t (P_t - \vartheta - Q_t), \end{aligned} \tag{6.31}$$

where there is no longer any beneficial debit valuation adjustment (for $R_b = 1$ the second line of (6.15) vanishes) and where the borrowing funding basis $\bar{\lambda}_t$ is interpreted as a pure liquidity cost.

6.4 Clean Valuations

In the numerical section 6.5, we will resort to two univariate short rate models, presented in Sect. 6.4.2 and 6.4.3, that will be used for TVA computations in case of an interest rate swap. Our motivation for considering two different models is twofold. First, we want to emphasize that from an implementation point of view the BSDE schemes that we use for TVA computations are quite model-independent (at least the backward nonlinear regression stage, after a forward simulation of the model in a first stage). Second, comparing the results obtained in two different models calibrated to the same data allows us to assess the TVA model risk.

Remark 6.4.1 As mentioned earlier, the basic credit risk reduced-form methodology of Part III is suitable for situations of reasonable dependence between the reference contract and the two parties (reasonable or unknown, e.g. the dependence between interest rates and credit is not liquidly priced on the market). For cases of strong dependence such as counterparty risk on credit derivatives, additional tools are necessary (see Part IV).

Remark 6.4.2 (Multi-curve interest-rate modeling) Talking about interest rate derivatives, one should also mention "systemic" counterparty risk, referring to various significant spreads which emerged since August 2007 between quantities that were very similar before, such as the OIS rates and the Libor rates of different tenors. As developed in Chapter 2 (see also the remark 6.2.8), these spreads are a consequence of banks' counterparty risk too, but at a macro level (counterparty risk of the banking sector as a whole as opposed to the particular counterparty risk of "the bank" in this book). Through its discounting implications, such a systemic counterparty risk has impacted all derivative markets and should therefore be represented in an interest-rate derivative model, which otherwise would not calibrate jointly to current OIS and Libor markets. However, as shown in Crépey, Grbac, Ngor, and Skovmand (2013) (where the focus is precisely on the just-mentioned joint calibration exercise), in order to adapt to a multi-curve setup the general TVA methodology of this chapter, all we need do is use multi-curve clean value models of interest rate derivatives in the TVA computations, i.e. clean interest-rate derivative models where the cash flows are based on interest-rates distinct from those which are used in the discounting. In other words, all we need do is switch from a classical single-curve clean valuation interest-rates derivative model, such as (6.38) or (6.45), to a multi-curve model where, in the generic clean pricing formula (6.1), there may be a spread between the OIS rates underlying the risk-neutral discount factor β and the Libor rates underlying the fixings (promised cash flows) dD. For different multi-curve interest-rate models, see e.g. Kijima, Tanaka, and Wong (2009), Kenyon (2010), Henrard (2007, 2010), Bianchetti (2010), Mercurio (2010a, 2010b), Fujii, Shimada, and Takahashi (2011, 2011a), Moreni and Pallavicini (2013a)). Note however that all these are clean (even though multi-curve) models, which, for the purpose of this book, should also be considered from the point of view of their tractability regarding TVA computations. Beyond calibrability, in particular, a short rate process r_t and a parsimonious Markov structure are required. In this last regard the most suitable models are Crépey, Grbac, Ngor, and Skovmand (2013), the HJM multi-curve model of Fujii and Takahashi (2011a) and the "parsimonious" (in reference to the above Markov concern) models of Moreni and Pallavicini (2013a, 2013b).

6.4.1 Products

We will deal with the following interest rate derivatives: forward rate agreements (FRAs), IR swaps (portfolios of FRAs) and caps (that are used in Sect. 6.5 for calibration purposes), whose definitions we provide below. The underlying rate for all these derivatives is the Libor rate. We work under the usual convention that the Libor rate is set in advance and the payments are made in arrears. As pointed out in the remark 6.4.2, we do not pursue here the multi-curve issue. Thus we use the classical definition of the forward Libor rate $L_t(T, T+\delta)$, fixed at time $t \le T$, for the future time interval $[T, T+\delta]$:

$$L_t(T, T+\delta) = \frac{1}{\delta}\left(\frac{B_t(T)}{B_t(T+\delta)} - 1\right),$$

where $B_t(T)$ denotes the time-t price of a zero coupon bond with maturity T.

Definition 6.4.3 A forward rate agreement (FRA) is a financial contract which fixes the interest rate K which will be applied to a future time interval. Denote by $T > 0$ the future inception date, by $T+\delta$ the maturity of the contract (where $\delta \ge 0$) and by N the notional amount. The payoff of the payer FRA at maturity $T+\delta$ is equal to

$$P^{fra}(T+\delta; T, T+\delta, K, N) = N\delta(L_T(T, T+\delta) - K).$$

Since $P^{fra}(t; T, T+\delta, K, N)/B_t(T+\delta)$ and $L_t(T, T+\delta)$ are martingales under the forward measure $\mathbb{Q}^{T+\delta}$ (see e.g Brigo and Mercurio (2006)), the value at time $t \in [0, T]$ of the payer FRA is given by

$$\begin{aligned} P^{fra}(t; T, T+\delta, K, N) &= NB_t(T+\delta)\mathbb{E}^{T+\delta}\left[(L_T(T, T+\delta) - K)\,\Big|\,\mathcal{F}_t\right] \\ &= N(B_t(T) - \bar{K}B_t(T+\delta)), \end{aligned} \tag{6.32}$$

where $\mathbb{E}^{T+\delta}$ denotes the expectation with respect to the forward measure $\mathbb{Q}^{T+\delta}$ and $\bar{K} = 1 + \delta K$.

Definition 6.4.4 An interest rate (IR) swap is a financial contract between two parties to exchange one stream of future interest payments for another, based on a specified notional amount N. A fixed-for-floating swap is a swap in which fixed payments are exchanged for floating payments linked to the Libor rate. Denote by $T_0 \geq 0$ the inception date, by $T_1 < \cdots < T_n$, where $T_1 > T_0$, a collection of the payment dates and by K the fixed rate. The time-t clean price P_t of the payer swap is given, for $t \leq T_0$, by:

$$P_t = P^{sw}(t; T_1, T_n) = N\left(B_t(T_0) - B_t(T_n) - K\sum_{k=1}^{n}\delta_{k-1}B_t(T_k)\right), \tag{6.33}$$

where $\delta_{k-1} = T_k - T_{k-1}$. The swap rate K_t, i.e. the fixed rate K making the value of the swap at time t equal to zero, is given by:

$$K_t = \frac{B_t(T_0) - B_t(T_n)}{\sum_{k=1}^{n}\delta_{k-1}B_t(T_k)}. \tag{6.34}$$

The value of the swap from initiation onward, i.e. the time-t value, for $T_0 \leq t < T_n$, of the swap, is given by:

$$P^{sw}(t; T_1, T_n) = \\ N\left(\left(\frac{1}{B_{T_{k_t-1}}(T_{k_t})} - K\delta_{k_t-1}\right)B_t(T_{k_t}) - B_t(T_n) - K\sum_{k=k_t+1}^{n}\delta_{k-1}B_t(T_k)\right), \tag{6.35}$$

where T_{k_t} is the smallest T_k greater than t. The clean price $\bar{P}_t$ of the receiver swap is given by $\bar{P}_t = -P^{sw}(t; T_1, T_n)$.

Definition 6.4.5 An interest rate cap (respectively floor) is a financial contract in which the buyer receives payments at the end of each period in which the interest rate exceeds (respectively falls below) a mutually agreed strike. The payment that the seller has to make covers exactly the positive part of the difference between the interest rate and the strike K (respectively between the strike K and the interest rate) at the end of each period. Every cap (respectively floor) is a series of caplets (respectively floorlets). The payoff of a caplet with strike K and exercise date T, which is settled in arrears, is given by

$$P^{cpl}(T; T, K) = \delta\left(L_T(T, T+\delta) - K\right)^+.$$

The time-t price of a caplet with strike K and maturity T is given, with $\bar{K} = 1 + \delta K$, by:

$$\begin{aligned} P^{cpl}(t; T, K) &= \delta\, B_t(T+\delta)\mathbb{E}^{T+\delta}\left[(L_T(T, T+\delta) - K)^+ \,\middle|\, \mathcal{F}_t\right] \\ &= B_t(T+\delta)\mathbb{E}^{T+\delta}\left[\left(\frac{1}{B_T(T+\delta)} - \bar{K}\right)^+ \,\middle|\, \mathcal{F}_t\right] \\ &= \bar{K}B_t(T)\mathbb{E}^{T}\left[\left(\frac{1}{\bar{K}} - B_T(T+\delta)\right)^+ \,\middle|\, \mathcal{F}_t\right] \\ &= \bar{K}\mathbb{E}\left[e^{-\int_t^T r_s ds}\left(\frac{1}{\bar{K}} - B_T(T+\delta)\right)^+ \,\middle|\, \mathcal{F}_t\right]. \end{aligned} \tag{6.36}$$

The next-to-last equality is due to the fact that, for $t \leq T$, a payoff $\left(\frac{1}{B_T(T+\delta)} - \bar{K}\right)^+$ at time $T+\delta$ has the same time t-value as a payoff $B_T(T+\delta)\left(\frac{1}{B_T(T+\delta)} - \bar{K}\right)^+ = \bar{K}\left(\frac{1}{\bar{K}} - B_T(T+\delta)\right)^+$ at time T. The last equality is obtained by changing from the forward measure $\mathbb{Q}^T$ to the spot martingale measure $\mathbb{Q}$ (cf. Musiela and Rutkowski (2005, Definition 9.6.2)). The above equalities say that a caplet can be seen as a put option on a zero coupon bond.

In the two models considered in the next subsections, for all vanilla interest rate derivatives including IR swaps, caps/floors and swaptions, the counterparty clean price P of an interest rate derivative satisfies, as required for (6.14) to hold:

$$P_t = P(t, X_t), \tag{6.37}$$

where the factor process X_t is given as r_t, augmented if need be by a few additional auxiliary processes in order to account for some path dependence of the product at hand (see the remark 6.4.6 for an example below).

6.4.2 Gaussian Vasicek Short Rate Model

In the Vasicek model the evolution of the short rate r is described by the following SDE:

$$dr_t = a(k - r_t)dt + dW_t^\sigma, \tag{6.38}$$

where $a, k > 0$ and W^σ is an $\mathbb{F}$-Brownian motion with volatility $\sigma > 0$. The unique solution to this SDE is given by

$$r_t = r_0 e^{-at} + k(1 - e^{-at}) + \int_0^t e^{-a(t-u)} dW_u^\sigma.$$

The zero coupon bond price $B_t(T)$ in this model can be written as an exponential-affine function of the current level of the short rate r. We have

$$B_t(T) = e^{m_{va}(t,T) + n_{va}(t,T) r_t}, \tag{6.39}$$

where

$$m_{va}(t,T) = R_\infty \left(\frac{1}{a}\left(1 - e^{-a(T-t)}\right) - T + t\right) - \frac{\sigma^2}{4a^3}\left(1 - e^{-a(T-t)}\right)^2, \tag{6.40}$$

with $R_\infty = k - \frac{\sigma^2}{2a^2}$ and

$$n_{va}(t,T) = -e^{at} \int_t^T e^{-au} du = \frac{1}{a}\left(e^{-a(T-t)} - 1\right). \tag{6.41}$$

Inserting the expression (6.39) for the bond price $B_t(T)$ intothe equations (6.33) and (6.35) from the definition 6.4.4, the clean price P for interest rate swaps can be written as

$$P_t = P(t,r_t,r'_t), \qquad t \in [0,T], \tag{6.42}$$

where r'_t is a shorthand notation for $r_{T_{k_t-1}}$. In particular, the time-t price of the payer swap is given, for $T_0 \le t < T_n$, by

$$P_t = N\Big(\big(e^{-(m_{va}(T_{k_t-1},T_{k_t})+n_{va}(T_{k_t-1},T_{k_t})r_{T_{k_t-1}})} - K\delta_{k_t-1}\big)e^{m_{va}(t,T_{k_t})+n_{va}(t,T_{k_t})r_t}$$
$$- e^{m_{va}(t,T_n)+n_{va}(t,T_n)r_t} - K\sum_{k=k_t+1}^{n}\delta_{k-1}e^{m_{va}(t,T_k)+n_{va}(t,T_k)r_t}\Big),$$

which follows from (6.35) and (6.39). In the above equation $m_{va}(t,T_k)$ and $n_{va}(t,T_k)$ are given by (6.40) and (6.41).

Remark 6.4.6 In (6.42) as in (6.52) below, the dependence of the swap price on r'_t expresses a mild path-dependence which is intrinsic to payments in arrears (independent of the model).

6.4.2.1 Caplet

To price a caplet at time 0 in the Vasicek model, we use (6.36) with

$$B_T(T+\delta) = e^{m_{va}(T,T+\delta)+n_{va}(T,T+\delta)r_T}$$

for $m_{va}(T,T+\delta)$ and $n_{va}(T,T+\delta)$ given by (6.40) and (6.41). Combining this with Proposition 11.3.1 and the formula on the bottom of page 354 in Musiela and Rutkowski (2005) [2], one has (recall $\bar{K} = 1 + \delta K$):

$$P^{cpl}(0;T,K) = B_0(T)\Phi(-d_-) - \bar{K}B_0(T+\delta)\Phi(-d_+), \tag{6.43}$$

where Φ is the Gaussian distribution function and

$$d_\pm = \frac{\ln\left(\frac{B_0(T+\delta)}{B_0(T)}\bar{K}\right)}{\Xi\sqrt{T}} \pm \frac{1}{2}\Xi\sqrt{T},$$

with

$$\Xi^2 T = \frac{\sigma^2}{2a^3}\left(1-e^{-2aT}\right)\left(1-e^{-a\delta}\right)^2. \tag{6.44}$$

[2] Up to the typo in this formula: it should be $(U-T)$, consistent with $(-a\delta)$ in (6.44), instead of $(U-t)$ in the last term.

6.4.3 Lévy Hull-White Short Rate Model

In this section we recall a one-dimensional Lévy Hull-White model obtained within the HJM framework. Contrary to the Vasicek model, this model fits automatically the initial bond term structure $B_0(T)$. As in the example 3.5 of Crépey, Grbac, and Nguyen (2012), we consider the Lévy Hull–White extended Vasicek model for the short rate r given by

$$dr_t = \alpha(\kappa(t) - r_t)dt + dZ_t^{\varsigma}, \tag{6.45}$$

where $\alpha > 0$ and Z^{ς} denotes the Lévy process described below. Furthermore,

$$\begin{aligned}\kappa(t) = f_0(t) + \frac{1}{\alpha}\partial_t f_0(t) + \psi_{\varsigma}\left(\frac{1}{\alpha}\left(e^{-\alpha t} - 1\right)\right) \\ - \psi_{\varsigma}'\left(\frac{1}{\alpha}\left(e^{-\alpha t} - 1\right)\right)\frac{1}{\alpha}e^{-\alpha t},\end{aligned} \tag{6.46}$$

where $f_0(t) = -\partial_t \log B_0(t)$ and ψ_{ς} denotes the cumulant function of Z^{ς}, i.e. (see e.g. Crépey, Grbac, and Nguyen (2012))

$$\psi_{\varsigma}(z) = \ln \mathbb{E}[e^{zZ_1^{\varsigma}}] \tag{6.47}$$

(for any complex argument z for which it is well defined). We refer the reader to the example 3.5 in Crépey, Grbac, and Nguyen (2012) for the volatility specification $\sigma(s,T) = e^{-\alpha(T-s)}, 0 \le s \le T$.

We will use an inverse Gaussian (IG) process $Z^{\varsigma} = (Z_t^{\varsigma})_{t\ge 0}$, which is a pure-jump, infinite activity, subordinator (nonnegative Lévy process), providing an explicit control on the sign of the short rates (see Crépey, Grbac, and Nguyen (2012)). This IG process is obtained from a standard Brownian motion W by setting

$$Z_t^{\varsigma} = \inf\{s > 0 : W_s + \varsigma s > t\},$$

where $\varsigma > 0$. Its Lévy measure (see Sect. 13.1.2) is given by

$$F_{\varsigma}(dx) = \frac{1}{\sqrt{2\pi x^3}}e^{-\frac{\varsigma^2 x}{2}}\mathbb{1}_{\{x>0\}}\,dx.$$

The distribution of Z_t^{ς} is $IG(\frac{t}{\varsigma}, t^2)$. The cumulant function ψ_{ς} exists for all $z \in [-\frac{\varsigma^2}{2}, \frac{\varsigma^2}{2}]$ (actually for all $z \in (-\infty, \frac{\varsigma^2}{2}]$ since F_{ς} is concentrated on $(0,\infty)$) and is given by

$$\psi_{\varsigma}(z) = \left(1 - \sqrt{1 - 2\frac{z}{\varsigma^2}}\right)\varsigma. \tag{6.48}$$

Similar to the Gaussian Vasicek model, the bond price $B_t(T)$ in the Lévy Hull-White short rate model can be written as an exponential-affine function of the current level of the short rate r:

$$B_t(T) = e^{m_{le}(t,T)+n_{le}(t,T)r_t}, \tag{6.49}$$

where

$$m_{le}(t,T) = \log\left(\frac{B_0(T)}{B_0(t)}\right) - n_{le}(t,T)\left[f_0(t) + \psi_\varsigma\left(\frac{1}{\alpha}\left(e^{-\alpha t} - 1\right)\right)\right] \tag{6.50}$$

and

$$n_{le}(t,T) = -e^{\alpha t}\int_t^T e^{-\alpha u}du = \frac{1}{\alpha}\left(e^{-\alpha(T-t)} - 1\right). \tag{6.51}$$

By combining the exponential-affine representation (6.49) of the bond price $B_t(T)$ and the definition 6.4.4, the clean price P for interest rate swaps in the Lévy Hull-White model can be written as

$$P_t = P(t,r_t,r'_t), \qquad t \in [0,T]. \tag{6.52}$$

In particular, the time-t price, for $T_0 \le t < T_n$, of the payer swap is given by

$$\begin{aligned} P_t = N\Big(&\left(e^{-(m_{le}(T_{k_t-1},T_{k_t})+n_{le}(T_{k_t-1},T_{k_t})r_{T_{k_t-1}})} - K\delta_{k_t-1}\right)e^{m_{le}(t,T_{k_t})+n_{le}(t,T_{k_t})r_t} \\ &- e^{m_{le}(t,T_n)+n_{le}(t,T_n)r_t} - K\sum_{k=k_t+1}^{n}\delta_{k-1}e^{m_{le}(t,T_k)+n_{le}(t,T_k)r_t}\Big), \end{aligned} \tag{6.53}$$

which follows from (6.35) and (6.49). In the above equations $m_{le}(t,T_k)$ and $n_{le}(t,T_k)$ are given by (6.50) and (6.51).

6.4.3.1 Caplet

To calculate the price of a caplet at time 0 in the Lévy Hull-White model, we can replace $\bar{B}^\star$ with B, $\bar{\Sigma}^\star$ with Σ, $\bar{A}^\star$ with A and insert $\Sigma^\star = 0$ and $A^\star = 0$ in Subsection 4.4 of Crépey, Grbac, and Nguyen (2012), thus obtaining the time-0 price of the caplet

$$\begin{aligned} P^{cpl}(0;T,K) &= B_0(T+\delta)\mathbb{E}^{T+\delta}\left[\left(\frac{1}{B_T(T+\delta)} - \bar{K}\right)^+\right] \\ &= B_0(T+\delta)\mathbb{E}^{T+\delta}\left[\left(e^Y - \bar{K}\right)^+\right], \end{aligned}$$

with

$$\begin{aligned} Y = \log\frac{B_0(T)}{B_0(T+\delta)} &+ \int_0^T (A(s,T+\delta) - A(s,T))ds \\ &+ \int_0^T (\Sigma(s,T+\delta) - \Sigma(s,T))dZ_s^\varsigma, \end{aligned}$$

where $\Sigma(s,t) = \frac{1}{\alpha}\left(1 - e^{-\alpha(t-s)}\right)$ and $A(s,t) = \psi_\varsigma(-\Sigma(s,t))$, for $0 \le s \le t$. The time-0 price of the caplet is now given by (cf. Crépey, Grbac, and Nguyen (2012, Proposition 4.5))

$$P^{cpl}(0;T,K) = \frac{B_0(T+\delta)}{2\pi}\int_{\mathbb{R}} \frac{\bar{K}^{1+iv-R}M_Y^{T+\delta}(R-iv)}{(iv-R)(1+iv-R)}dv, \tag{6.54}$$

for $R > 1$ such that $M_Y^{T+\delta}(R) < \infty$. The moment generating function $M_Y^{T+\delta}$ of Y under the measure $\mathbb{Q}^{T+\delta}$ is provided by

$$M_Y^{T+\delta}(z) = \mathbb{E}^{T+\delta}\left[e^{zY}\right] = \exp\left(-\int_0^T \psi_\varsigma(-\Sigma(s,T+\delta))ds\right)$$
$$\times \exp\left(z\left(\log\frac{B_0(T)}{B_0(T+\delta)} + \int_0^T \left(\psi_\varsigma(-\Sigma(s,T+\delta)) - \psi_\varsigma(-\Sigma(s,T))\right)ds\right)\right)$$
$$\times \exp\left(\int_0^T \psi_\varsigma\left((z-1)\Sigma(s,T+\delta) - z\Sigma(s,T)\right)ds\right),$$

for $z \in \mathbb{C}$ such that the above expectation is finite.

As an alternative, the time-0 price of the caplet can be computed as the following expectation (cf. the formula (6.36)):

$$P^{cpl}(0;T,K) = \bar{K}\mathbb{E}\left[\exp^{-\int_t^T r_s ds}\left(\frac{1}{\bar{K}} - B_T(T+\delta)\right)^+\right], \qquad (6.55)$$

where r is given by (6.45) and $B_T(T+\delta)$ by (6.49).

6.4.4 Numerics

In Sect. 6.5 we will present TVA computations on an interest rate swap with ten years maturity, where the bank exchanges the swap rate K against a floating Libor at the end of each year 1 to 10. In order to get an idea of the TVA model risk, this will be done in the Vasicek model and in the Lévy Hull-White model calibrated to the same data, in the sense that they share a common initial zero-bond discount curve $B_0^\star(T)$ below and produce the same price for the cap with payments at years 1 to 10 struck at K (hence there is the same level of Black-implied volatility in both models at the strike level K). Specifically, we set $r_0 = 2\%$ and the following Vasicek parameters:

$$a = 0.25, \ \ k = 0.05, \ \ \sigma = 0.004,$$

with related zero-coupon rates and discount factors denoted by $R_0^\star(T)$ and $B_0^\star(T) = e^{-TR_0^\star(T)}$. It follows from (6.39) after some simple calculations:

$$\begin{aligned}
R_0^\star(T) &= R_\infty - (R_\infty - r_0)\frac{1}{aT}\left(1-e^{-aT}\right) + \frac{\sigma^2}{4a^3T}\left(1-e^{-aT}\right)^2, \\
f_0^\star(T) &= \partial_T\left(TR_0^\star(T)\right) = k + e^{-aT}(r_0 - k) - \frac{\sigma^2}{2a^2}\left(1-e^{-aT}\right)^2, \\
\partial_T f_0^\star(T) &= -ae^{-aT}(r_0-k) - \frac{\sigma^2}{a}\left(1-e^{-aT}\right)e^{-aT}.
\end{aligned}$$

An application of formula (6.34) at time 0 yields, for the corresponding swap rate, the value $K = 3.8859\%$. We choose a swap notional of $N = 310.13\$$ so that the fixed leg of the swap is worth 100\$ at inception.

In the Lévy Hull-White model we use $\alpha = a = 0.25$ (same speed of mean-reversion as in the Vasicek model), an initial bond term-structure $B_0(T)$ fitted to $B_0^\star(T)$ by using $f_0(T) = f_0^\star(T)$ above in (6.46), and a value of $\varsigma = 17.57$ obtained by calibration to the Vasicek price of the cap with payments at years 1 to 10 struck at K. The calibration is done by least square minimization based on the explicit formulas for caps in both of the models reviewed in Sect. 6.4.2.1 and 6.4.3.1. After calibration the price of the cap in both models is 20.161$ (for the above notional N that yields a value of 100$ for the fixed leg of the swap).

The top panels of Fig. 6.1 show 20 paths, along with expectations and 2.5/97.5-percentiles over 10^4 paths, simulated in the two models by an Euler scheme $\widehat{r}$ for the short-rate r on a uniform time grid with 200 time steps over $[0, 10]$yrs. Note that one doesn't see the jumps on the right panel because we used interpolation between the points so that we can identify better the twenty paths.

The top panels of Fig. 6.2 show the initial zero-coupon rate term structure $R_0^\star(T)$ and the corresponding forward curves $f_0^\star(T)$, while the corresponding discount factors $B_0^\star(T)$ can be seen on the lower left panel. This displays an increasing term structure of interest rates, meaning that the bank will on average be in-the-money with a positive $P_t = P_t^{sw}$ in (6.35) in the case of the payer swap, respectively out-of-the-money with a negative $\bar{P}_t = -P_t^{sw}$ in the case of the receiver swap (see the bottom panels of Fig. 6.1). Note that the swap price processes have quite distinct profiles in the two models even though these are co-calibrated. The bottom right panel of Fig. 6.2 shows the Lévy Hull-White mean-reversion function $\kappa(t)$ corresponding to $f_0^\star(T)$ through (6.46).

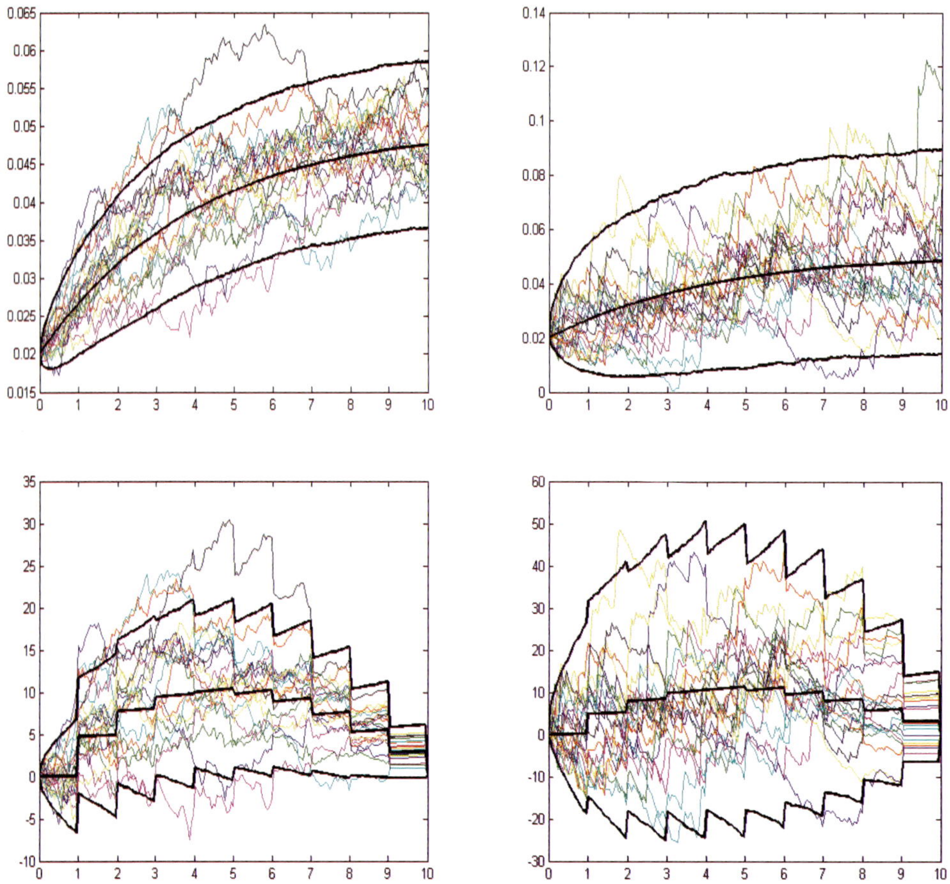

FIGURE 6.1: Clean Valuations. *Top panels:* short rate process r_t; *Bottom panels:* clean price process P_t of the payer swap; *Left panels:* results in the Vasicek model; *Right panels:* results in the Lévy model. Each panel shows twenty paths simulated with two hundred time points, along with the process mean and 2.5/97.5-percentiles (as function of time) computed over 10^4 simulated paths.

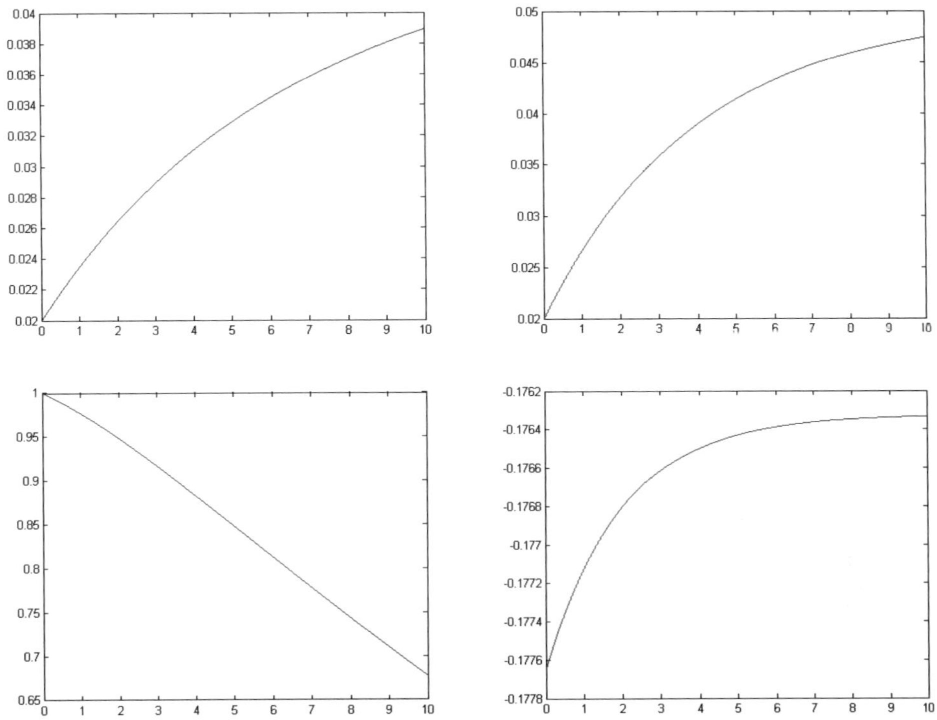

FIGURE 6.2: Initial term structures. *Upper left panel*: zero-coupon rates $R_0^\star(T)$; *Upper right panel*: forward rates $f_0^\star(T)$; *Bottom left panel*: discount factors $B_0^\star(T)$; *Bottom right panel*: Lévy Hull-White mean-reversion function $\kappa(t)$.

6.5 TVA Computations

In this section we show with practical examples how the CVA, DVA, LVA and RC components of the TVA can be computed for various CSA specifications of the coefficient f in Sect. 6.3. The computations are done for an IR swap in the two models of Sect. 6.4.

6.5.1 TVA Equations

Note that, in both Vasicek and Lévy Hull-White models, the pair $X_t = (r_t, r'_t)$, where $r'_t = r_{T_{k_t-1}}$, is a Markov process. Assuming (6.14) with $X_t = (r_t, r'_t)$, where r'_t is needed in view of the path-dependence induced by the payments in arrears (see the remark 6.4.6), the pre-default TVA Markov BSDE in both models is written:

$$\widetilde{\Theta}(t, X_t) = \widetilde{\mathbb{E}}_t\left(\int_t^T f(s, X_s, \widetilde{\Theta}(s, X_s))ds\right), \quad t \in [0, T]. \tag{6.56}$$

Even though finding numerical solutions to the corresponding PDEs would be possible in the above univariate setups, we nevertheless favor BSDE schemes, as they are more generic – in high-dimensional applications deterministic schemes can no longer be used. We solve (6.56) by backward regression over the simulated time-space grids of Subsection 6.4.4 (see the top panels of Fig. 6.1). We thus approximate $\widetilde{\Theta}(t, x)$ in (6.56) by $\widehat{\Theta}_i^j$ on the corresponding time-space grid, where the time-index i runs from 1 to $n = 200$ and the space-index j runs from 1 to $m = 10^4$. Denoting by $\widehat{\Theta}_i = (\widehat{\Theta}_i^j)_{1 \le j \le m}$ the vector of TVA values on the space grid at time i, we have $\widehat{\Theta}_n = 0$, and then for $i = n-1, \cdots, 0$ and $j = 1, \cdots, m$:

$$\widehat{\Theta}_i^j = \widehat{\mathbb{E}}_i^j\left(\widehat{\Theta}_{i+1} + f_{i+1}(\widehat{X}_{i+1}, \widehat{\Theta}_{i+1})\mathrm{h}\right)$$

for the time-step $\mathrm{h} = \frac{T}{n} = 0.05\mathrm{y}$ (see Crépey (2013) for the related discretization issues). The conditional expectations in space at every time-step are computed by a d-nearest neighbor average non-parametric regression estimate (see e.g. Hastie, Tibshirani, and Friedman (2009)). Moreover, as we found very little impact of the second factor r' in the regressions, we eventually only used one regressor, r, with $d = 5$, in our numerical experiments below.

6.5.2 Numerics

We set the following TVA parameters: $\gamma = 10\%$, $b = \bar{b} = \lambda = 1.5\%$, $\bar{\lambda} = 4.5\%$, $p^b = 50\%$, $p^c = 70\%$ and we consider five possible CSA specifications (in the following order):

$$\begin{array}{lll} (\bar{R}_b, R_b, R_c) = (40, 40, 40)\%, & Q = P, & \Gamma = 0 \\ (\bar{R}_b, R_b, R_c) = (100, 40, 40)\%, & Q = P, & \Gamma = 0, \\ (\bar{R}_b, R_b, R_c) = (100, 100, 40)\%, & Q = P, & \Gamma = 0, \\ (\bar{R}_b, R_b, R_c) = (100, 100, 40)\%, & Q = \Pi, & \Gamma = 0, \\ (\bar{R}_b, R_b, R_c) = (100, 40, 40)\%, & Q = P, & \Gamma = Q = P. \end{array} \tag{6.57}$$

Note that under the first specification, we have $\widetilde{\lambda} = 4.5\% - 0.6 \times 0.5 \times 10\% = 1.5\% = \lambda$, so this is a linear TVA special case of the remark 6.3.1, where the TVA at time 0 can be computed through a straight Monte Carlo.

Moreover, we will study the TVA in the two co-calibrated Vasicek and Lévy models and for the payer and receiver swaps. We thus consider twenty cases (5 CSA specifications $\times$ 2 models $\times$ payer versus receiver swap).

Table 6.1 shows the time-0 TVAs and the corresponding CVA/DVA/LVA/RC decompositions (four terms on the right-hand side of (6.16)) in each of the twenty cases. In the four cases corresponding to the first (linear) CSA specification (upper left "TVA" cells of the four parts of Table 6.1), the Monte Carlo 95%-confidence intervals for the time-0 TVA BSDEs are, respectively,

$$[1.8705, 1.9024], \ [2.0003, 2.1021], \ [-1.4767, -1.4505] \text{ and } [-1.3639, -1.2711].$$

In each of the four cases the BSDE time-0 value of the TVA is close to the middle of the confidence interval.

The numbers of Table 6.1 are fully consistent with the CVA/DVA/LVA/RC interpretation of the four terms in the TVA decomposition on the right-hand side of (6.16), given the increasing term structure of the data discussed in Sect. 6.4.4. For instance, a "large negative" DVA (-1.75 in the Vasicek model and -2.34 in the Lévy model) for the receiver swap in the first row of the bottom part of Table 6.1 is consistent with the fact that with an increasing term structure of rates, the bank is on average out-of-the-money on the receiver swap with a negative $\bar{P}_t = -P_t^{sw}$ (see Fig. 6.1). The CVA, on the contrary, is moderate and higher in the Lévy than in the Vasicek model (0.90 versus 0.06). Some of the numbers in Table 6.1 are not negligible at all in view of the initial value of 100$ of the fixed leg of the swap; see also Cont, Mondescu, and Yu (2011), where non-negligible differences in the valuation of swaps under various clearing conventions are found. In particular, the LVA terms are quite significant in case of the payer swap with R_b and/or $\bar{R}_b = 100\%$: see e.g. the corresponding terms in rows 3 and 4 in the two upper parts of Table 6.1. The choice of $\bar{R}_b$ and R_b thus has tangible practical consequences, given the "deal adverse" ("deal friendly") effect of the positive (negative) TVA terms as explained at the end of Sect. 6.2.3. It is worthwhile noting that all this happens in a simplistic TVA model, in which credit risk is independent from interest rates. These numbers could be significantly bigger in a model accounting for potential wrong-way risk dependence effects between interest rates and credit risk (see the remark 6.4.1).

TVA	CVA	DVA	LVA	RC	TVA	CVA	DVA	LVA	RC
1.90	2.45	-0.04	0.68	-1.17	2.08	3.28	-0.64	0.66	-1.25
2.64	2.45	-0.04	1.92	-1.67	3.17	3.28	-0.64	2.41	-1.92
2.67	2.45	0.00	1.92	-1.68	3.59	3.28	0.00	2.38	-2.11
3.59	1.77	0.00	1.83	0.00	4.80	2.49	0.00	2.26	0.00
0.50	0.00	0.00	0.81	-0.31	0.51	0.00	0.00	0.81	-0.31
TVA	CVA	DVA	LVA	RC	TVA	CVA	DVA	LVA	RC
-1.47	0.06	-1.75	-0.71	0.92	-1.34	0.90	-2.34	-0.72	0.85
-1.40	0.06	-1.75	-0.64	0.91	-0.93	0.90	-2.34	-0.15	0.68
-0.40	0.06	0.00	-0.76	0.29	0.45	0.90	0.00	-0.32	-0.12
-0.66	0.08	0.00	-0.74	0.00	0.43	0.76	0.00	-0.32	0.00
-0.43	0.00	0.00	-0.72	0.29	-0.44	0.00	0.00	-0.72	0.29

TABLE 6.1: Time-0 TVA and their decompositions into time-0 CVA, DVA, LVA and RC. *Top tables*: payer swap; *Bottom tables*: receiver swap. *Left*: Vasicek model; Right: Lévy model.

Figs. 6.3 and 6.4 (payer swap in the Vasicek and Lévy model, respectively) and Fig. 6.5 and 6.6 (receiver swap in the Vasicek and Lévy models, respectively) show the "expected exposures" corresponding to the four right-hand side terms of the "local" TVA decomposition (6.15) with ϑ replaced by $\widetilde{\Theta}_t$ there. These exposures are computed as the related space-averages over 10^4 paths, as a function of time t. Each time-0 "integrated" term of the TVA in Table 6.1 corresponds to the algebraic surface (time-integral) enclosed by the related curve in Figs. 6.5 to 6.4 (with correspondence between, respectively: Fig. 6.3 and the upper left panel of Table 6.1, Fig. 6.4 and the upper right panel of Table 6.1, Fig. 6.5 and the lower left panel of Table 6.1, Fig. 6.6 and the lower right panel of Table 6.1).

Finally, Fig. 6.7 (payer swap) and 6.8 (receiver swap) show the TVA processes in the same format as the swap clean prices at the bottom of Fig. 6.1.

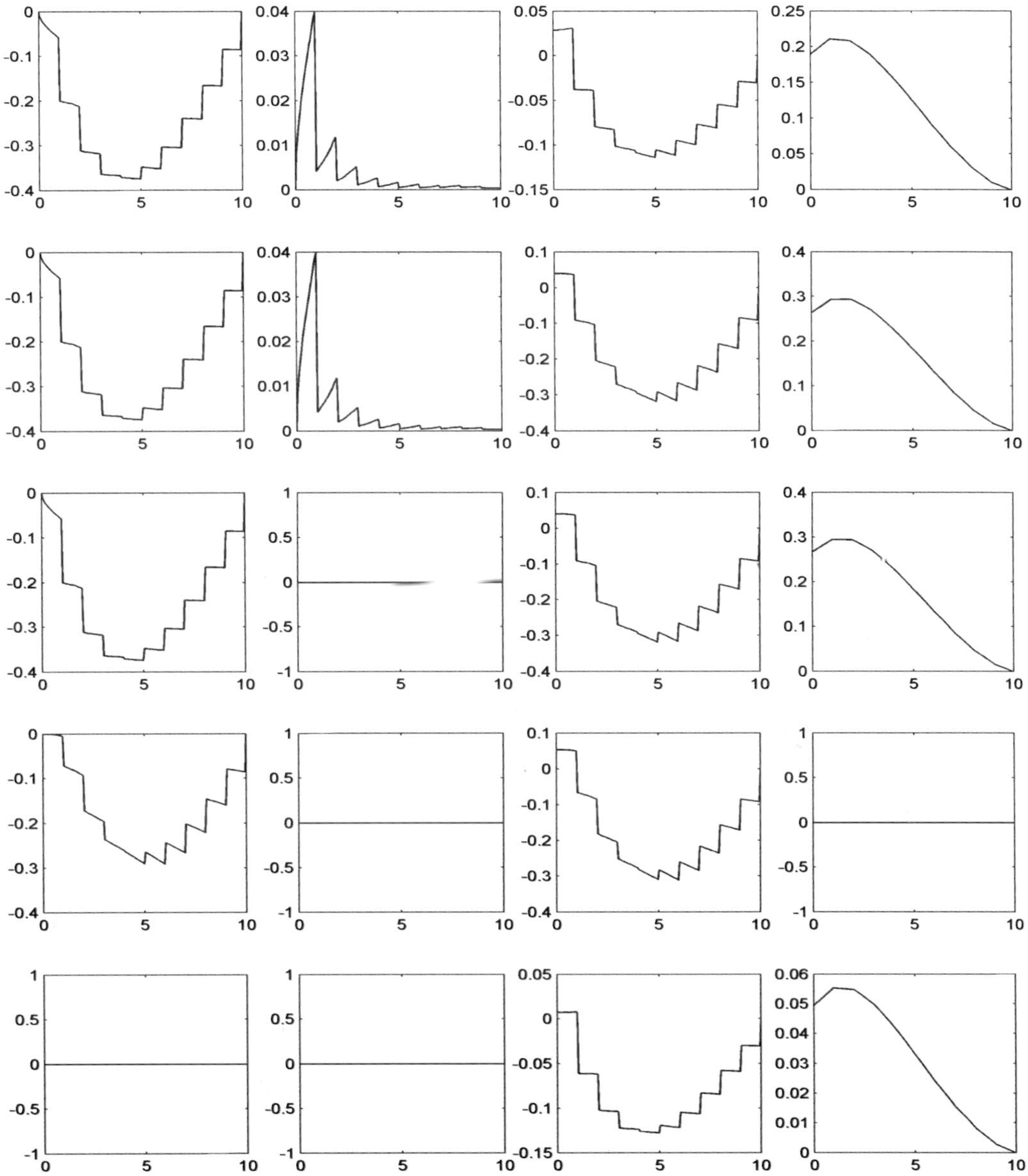

FIGURE 6.3: Expected exposures of the TVA components of a payer swap in the Vasicek model. *Columns (from left to right)*: exposures related to the CVA, the DVA, the LVA and the RC components of the TVA process. *Rows (from top to bottom)*: results corresponding to the five CSA specifications listed in (6.57).

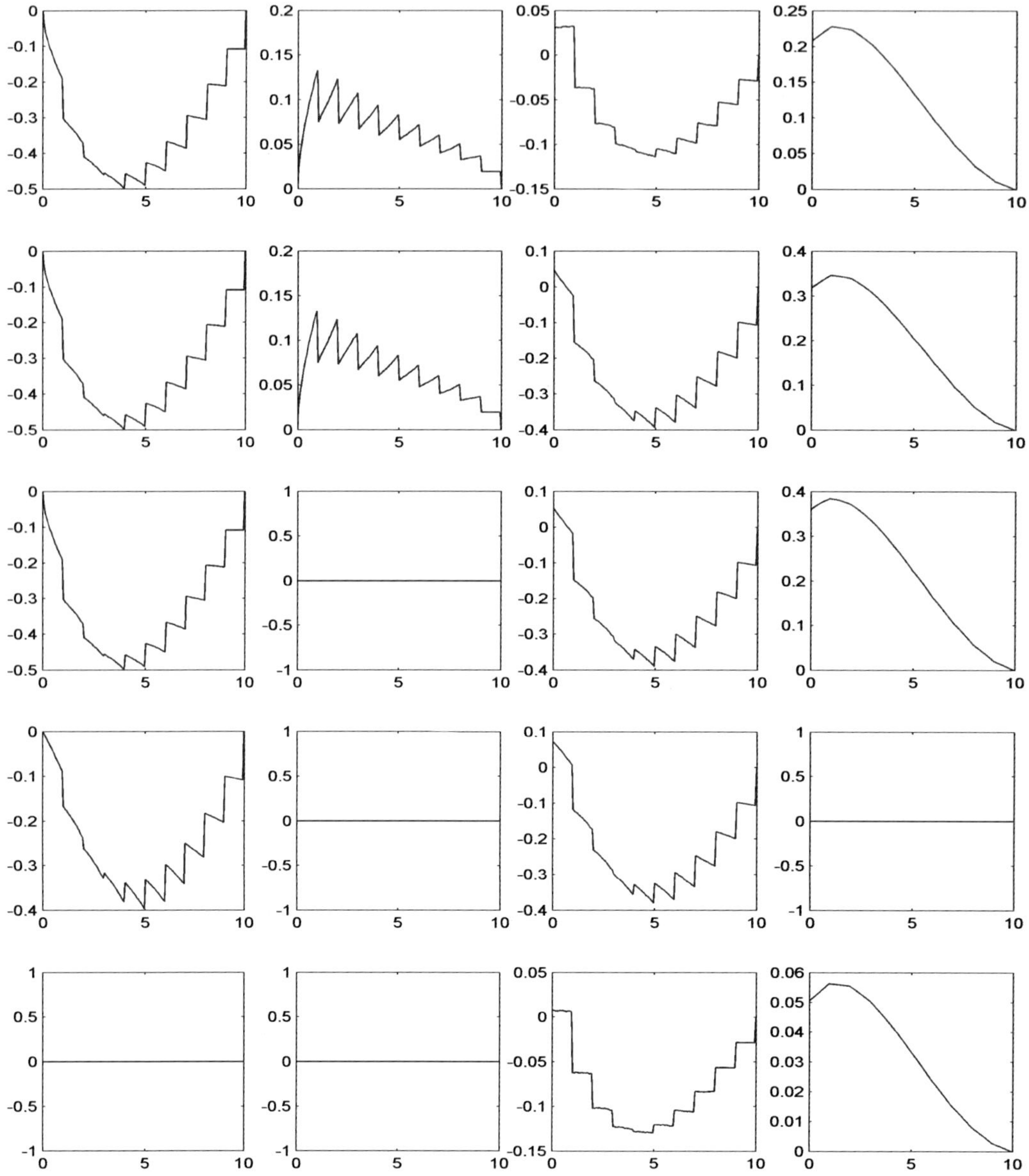

FIGURE 6.4: Expected exposures of the TVA components of a payer swap in the Lévy model. *Columns (from left to right)*: exposures related to the CVA, the DVA, the LVA and the RC components of the TVA process. *Rows (from top to bottom)*: results corresponding to the five CSA specifications listed in (6.57).

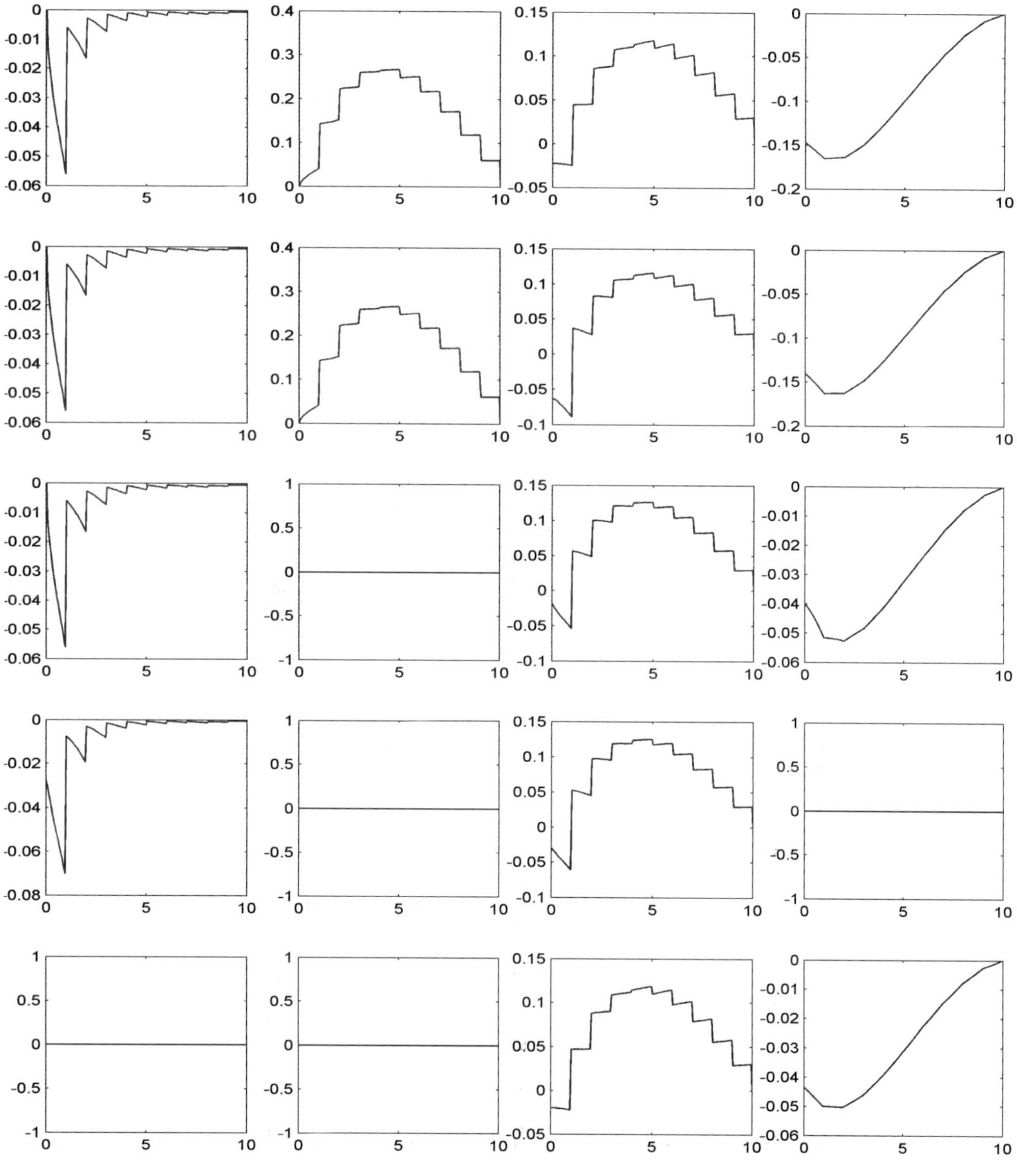

FIGURE 6.5: Expected exposures of the TVA components of a receiver swap in the Vasicek model. *Columns (from left to right)*: exposures related to the CVA, the DVA, the LVA and the RC components of the TVA process. *Rows (from top to bottom)*: results corresponding to the five CSA specifications listed in (6.57).

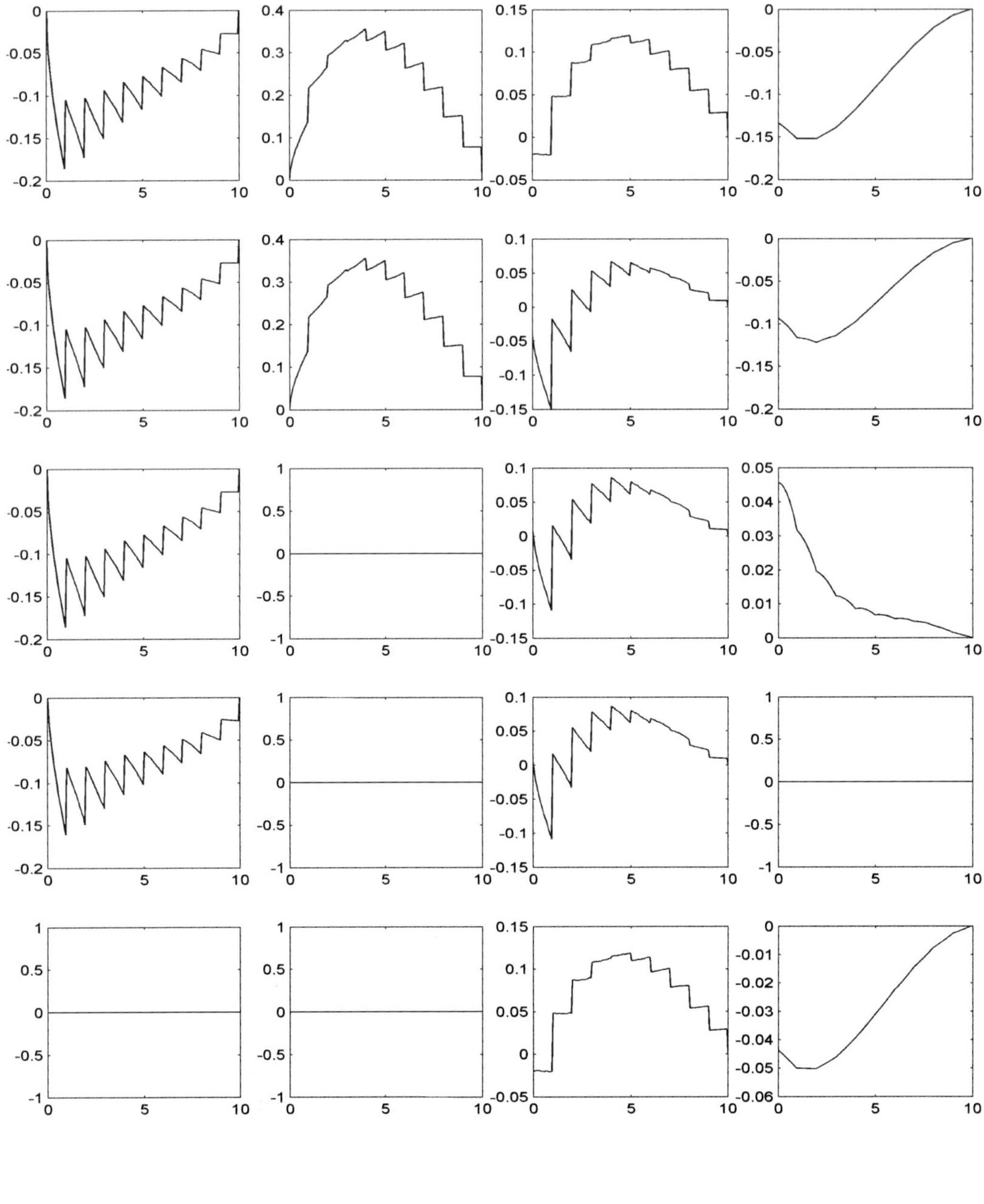

FIGURE 6.6: Expected exposures of the TVA components of a receiver swap in the Lévy model. *Columns (from left to right)*: exposures related to the CVA, the DVA, the LVA and the RC components of the TVA process. *Rows (from top to bottom)*: results corresponding to the five CSA specifications listed in (6.57).

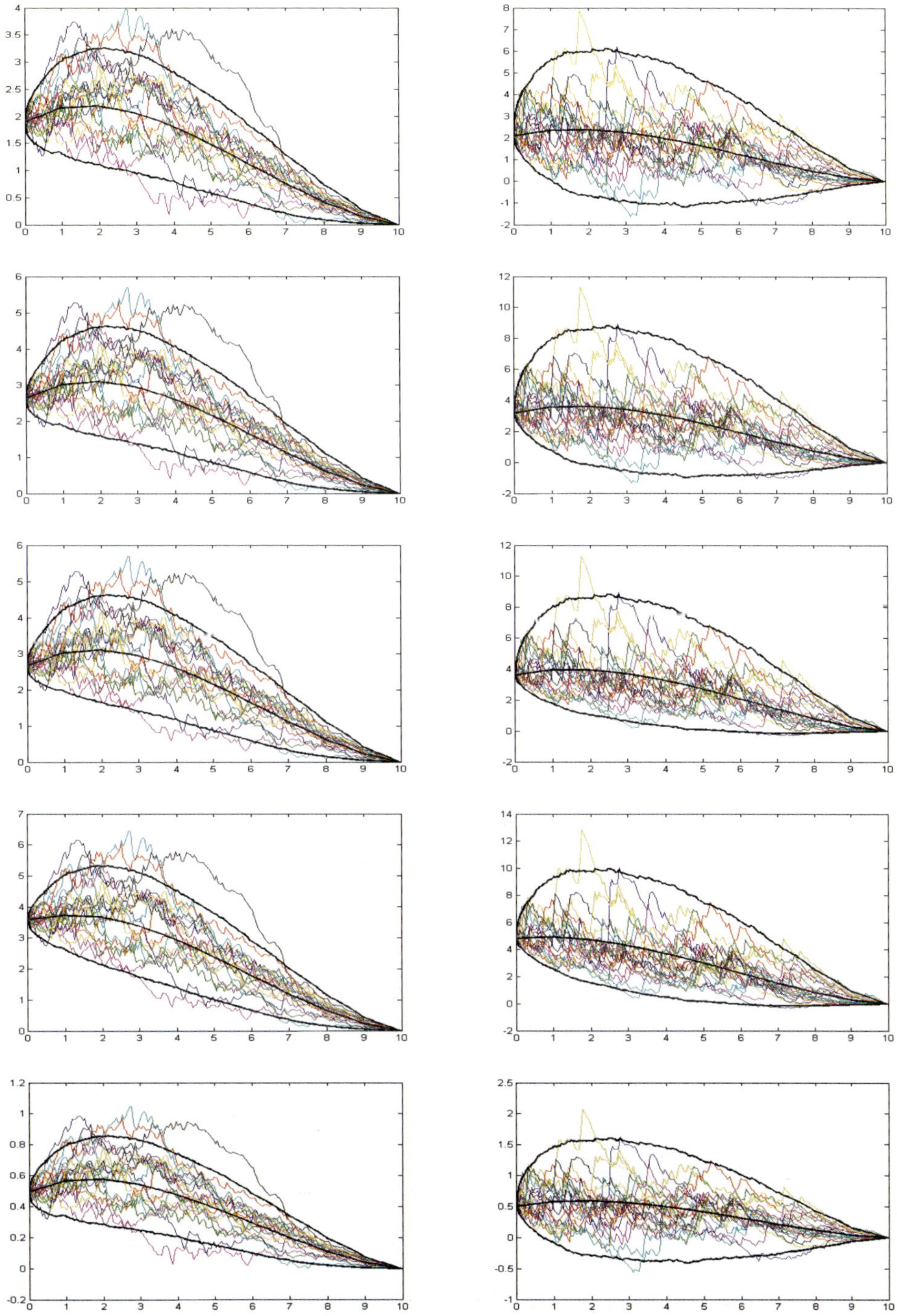

FIGURE 6.7: TVA process $\widetilde{\Theta}_t$ of a payer swap. *Left panels*: results in the Vasicek model. *Right panels*: results in the Lévy model. *Top to bottom panels*: results corresponding to the five CSA specifications listed in (6.57). Each panel shows twenty paths of the TVA process at two hundred time points, along with the process mean and 2.5/97.5-percentiles as a function of time computed using 10^4 simulated paths.

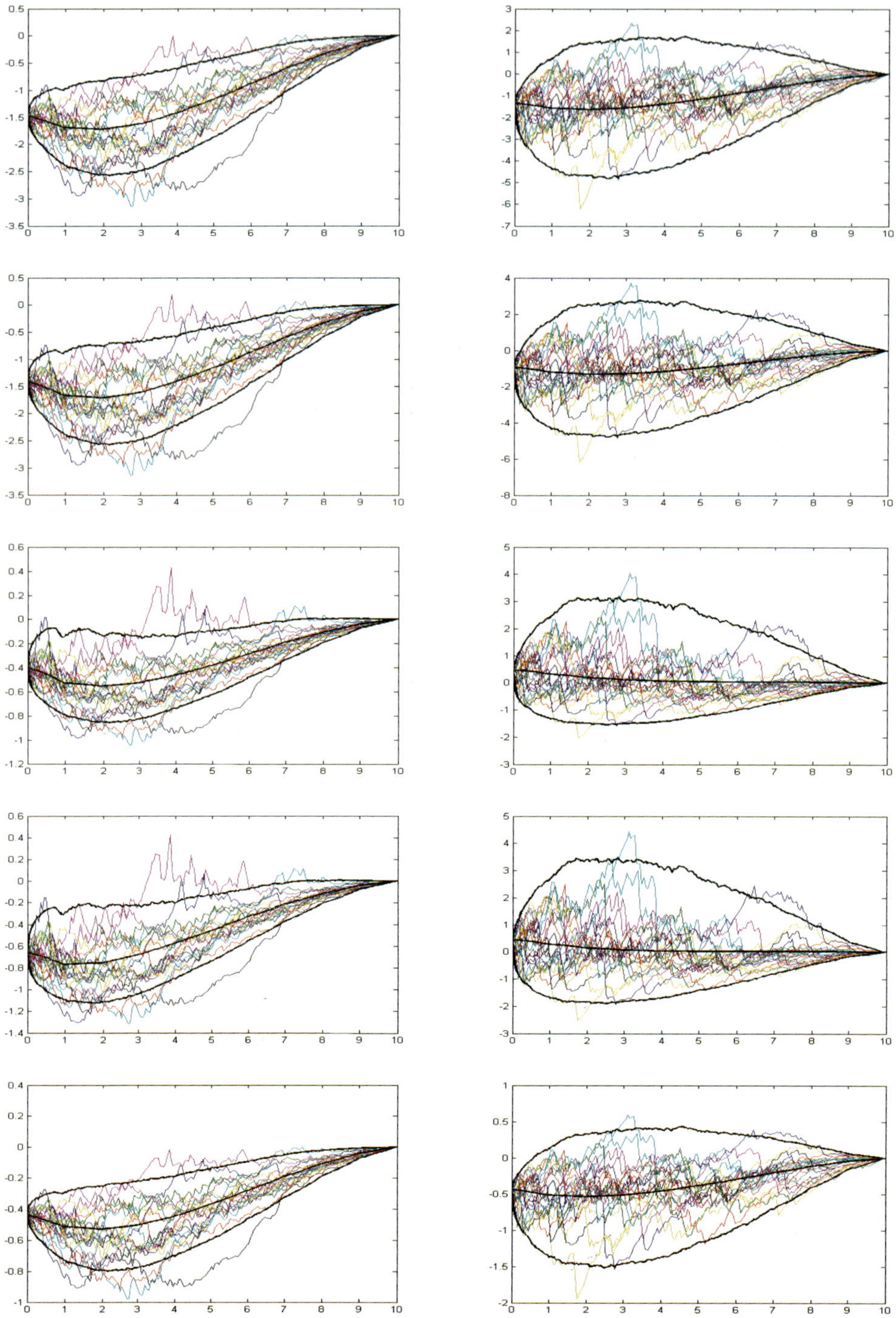

FIGURE 6.8: TVA process $\widetilde{\Theta}_t$ of a receiver swap. *Left panels*: results in the Vasicek model. *Right panels*: results in the Lévy model. *Top to bottom panels*: results corresponding to the five CSA specifications listed in (6.57). Each panel shows twenty paths of the TVA process at two hundred time points, along with the process mean and 2.5/97.5-percentiles as a function of time computed using 10^4 simulated paths.

Conclusion

In this chapter, which is a numerical companion to Chapter 5, we have shown, using the example of an interest-rate swap, how CVA, DVA, LVA and RC, the four "wings" (or pillars) of the TVA, can be computed for various CSA and model specifications. Positive terms such as the CVA (respectively negative terms such as the DVA) are "deal adverse" (respectively "deal friendly"), as they increase (respectively decrease) the TVA (cost of the hedge) and therefore decrease (respectively increase) what the bank can pay for the contract. Beliefs regarding the tangibility of a benefit-at-own-default, which in reality depends on the extent to which own-default is effectively hedged and can therefore be monetized before it happens, are controlled by the choice of the "own-recovery-rate" parameters R_b and $\bar{R}_b$. Larger R_b and $\bar{R}_b$ mean more (less negative) DVA and more LVA and therefore larger TVA, hence less profitability from the deal. In order to emphasize the model-free feature of the TVA computations through nonlinear regression BSDE schemes, we have illustrated this numerically in two alternative short rate models. The results show that the TVA model risk is under reasonable control in the case of two co-calibrated models (models calibrated to the same initial term structure and the same level of he cap's implied volatility). However, the latter observation only applies in the basic immersion setup studied in this part of the book, without dominant wrong-way risk.

Part IV

Dynamic Copula Models

In this part, we focus on counterparty risk embedded within credit derivatives (sometimes referring to this as counterparty credit risk). In the case of bilateral counterparty risk, combined with the related funding issue, the TVA equation is nonlinear (see Chapters 4 through 6). Moreover, in case of portfolio credit risk, the problem becomes highly dimensional due, in part, to the presence of multiple default indicator processes. As is well known (see e.g. Crépey (2013) or Guyon and Henry-Labordère (2012)), high-dimensional nonlinear problems are always a computational challenge. In this part, we focus on the study of unilateral counterparty risk, i.e. $\tau_b = +\infty, \tau = \tau_c$, in the case when funding is provided at the risk-free rate (cf. Sect. 3.2.4.2). This allows us to avoid the nonlinearities associated with the general funding issue.

We consider a contract referencing various credit names between the bank and the counterparty. We label by $i = 1, \dots, n$ the reference credit names and we use "0" to represent the counterparty. We denote by τ_i the default time of name i, by $H^i_t = \mathbb{1}_{\tau_i \le t}$ (respectively $J^i_t = \mathbb{1}_{\tau_i > t}$) the related default (respectively survival) indicator process with natural filtration $\mathbb{H}^i_t = (\mathcal{H}^i_t)$, where $\mathcal{H}^i_t = \sigma(\tau_i \wedge t) \vee \sigma(\tau_i > t)$, and by R_i the corresponding recovery rate (a constant unless explicitly stated), and we write $\mathbf{H}_t = (H^0_t, H^1_t, H^2_t, \dots, H^n_t)$. In particular, $\tau_0 = \tau_c$, $R_0 = R_c$ and $J^0 = J$. The cumulative discounted clean cash flow process of the contract is represented as

$$\beta_t D_t = \int_{[0,t]} \beta_s \varphi(\mathbf{H}_s)\nu(ds) + \int_{[0,t]} \beta_s d\phi(\mathbf{H}_s),$$

where ϕ and φ correspond to the default leg and the fee (sometimes also called premium) legs, respectively, and where ν is a nonnegative measure on the half-line (introduced to cover continuous-time and discrete-time premium payments in a unified setting).

In the case of a payer CDO tranche on names 1 to n with contractual spread S and attachment/detachment point a/b, we have:

- for the default leg:

$$\phi(\mathbf{k}) = \big(L(\mathbf{k}) - a\big)^+ \wedge (b - a) =: L_{a,b}(\mathbf{k}),$$

  where $L(\mathbf{k}) = \frac{1}{n}\sum_{i=1}^n (1 - R_i)k_i$ is the portfolio loss process;

- for the premium leg paid in continuous time, respectively at times t_j of a grid with time step h:

$$\varphi(\mathbf{k})\nu(dt) = -S\big(b - a - L_{a,b}(\mathbf{k})\big)dt,$$

  respectively

$$\varphi(\mathbf{k})\nu(dt) = -\mathrm{h}S\big(b - a - L_{a,b}(\mathbf{k})\big)\sum_i \boldsymbol{\delta}_{\{t_j\}}(dt).$$

In the case of a single-name payer CDS with contractual spread S_i on name i, we have:

- for the default leg:

$$\phi(\mathbf{k}) = (1 - R_i)k_i,$$

- for the fee leg paid in continuous, respectively in discrete time:

$$\varphi(\mathbf{k})\nu(dt) = -S_i(1 - k_i)dt,$$

respectively,

$$\varphi(\mathbf{k})\nu(dt) = -\mathrm{h}S_i(1 - k_i)\sum_i \boldsymbol{\delta}_{\{t_j\}}(dt).$$

All the CDS and CDO notionals are taken equal to one. Unless otherwise stated, all the contractual spreads (S_i or S) are assumed to be setup at par at the contract's inception.

We fix a clean CSA closeout valuation scheme, i.e. $Q = P$, throughout this part. In the unilateral counterparty risk setup of Sect. 3.2.4.2, with risk-free funding rates and $Q = P$, the TVA Θ reduces to the (unilateral) CVA given on $[0, \bar{\tau}]$ by:

$$\beta_t \Theta_t = \mathbb{E}\big[\mathbb{1}_{\tau<T}\beta_\tau \xi \,\big|\, \mathcal{G}_t\big],$$

for an exposure ξ given as

$$\xi = (1 - R_0)\chi^+,$$

where R_0 is the recovery rate of the counterparty and χ represents the algebraic "debt" χ of the counterparty to the bank at time τ, i.e. (for $Q = P$):

$$\chi = P_\tau + \Delta_\tau - \Gamma_\tau,$$

where $\Delta_\tau = \phi(\mathbf{H}_\tau) - \phi(\mathbf{H}_{\tau-})$.

For every real vector $\boldsymbol{\theta}$ representing the state of a credit portfolio model, $\mathrm{supp}(\boldsymbol{\theta}) = \{l : \theta_l \neq 0\}$ (respectively $\mathrm{supp}^c(\boldsymbol{\theta}) = \{l : \theta_l = 0\}$) is the set of survived (respectively defaulted) obligors, and for a process θ_t, where $\boldsymbol{\theta}_t$ represents the state of a credit portfolio model at time t, we denote $\mathcal{J}_t = \mathrm{supp}(\boldsymbol{\theta}_t)$ (respectively $\mathfrak{J}_t = \mathrm{supp}^c(\boldsymbol{\theta}_t)$).

Chapter 7

Dynamic Gaussian Copula Model

In this chapter, we "dynamize" the static Gaussian copula model of portfolio credit risk in order to make it suitable for counterparty risk computations. Toward this aim, we introduce a model filtration made of a reference Brownian filtration progressively enlarged by the default times. This results a multidimensional density model of default times, where the reference filtration is not immersed into the enlarged filtration. In mathematical terms, this lack of immersion means that martingales in the reference filtration are not martingales in the enlarged filtration. In the financial interpretation, this corresponds to a form of default contagion. Computational tractability is ensured by invariance of multivariate Gaussian distributions through conditioning by some components, the ones corresponding to past defaults. The model is also Markov in an augmented state-space that includes past default times.

7.1 Introduction

This chapter is an attempt, a bit in the spirit of Fermanian and Vigneron (2010, 2013), to dynamize the static Gaussian copula model of portfolio credit risk. Toward this end, we introduce a filtration with respect to which conditional expectations are used to define prices at future times, so that one could talk of informational dynamization. However, whereas Fermanian and Vigneron (2010, 2013) use a Brownian filtration, we use a Brownian filtration progressively enlarged by the default times. Moreover, the construction of the model is different from Fermanian and Vigneron (2010, 2013), where a structural approach is used. In our case we rely on the conditional density approach of El Karoui, Jeanblanc, and Jiao (2009).

Sect. 7.2 presents the model. Sect. 7.3 deals with the clean valuation of CDS and CDO tranches and examines the issue of (clean) hedging a CDO tranche by CDS. Sect. 7.4 proceeds with CVA computations.

7.2 Model

First we present a model of n credit names $\{1, \cdots, n\}$. In Sect. 7.4 the model will also include the counterparty, indexed by 0, to make it amenable to the study of counterparty credit risk.

7.2.1 Gaussian Distributions

Let us introduce some notation:

- I denotes a generic subset of $N = \{1, \ldots, n\}$ with complement set $J = N \setminus I$,
- Φ (respectively ϕ) is the standard Gaussian survival function (respectively density function),
- $\Phi_{\rho,\sigma}((z_j)_{j\in J}) = \mathbb{Q}(Z_j > z_j, j \in J)$, where $(Z_j)_{j\in J}$ follows a $|J|$-dimensional (ρ, σ)-Gaussian distribution, or distribution of a $|J|$-dimensional centered Gaussian vector with homogenous variances σ^2 and pairwise correlations ρ.

Note that $\Phi_{\rho,\sigma}((z_j)_{j\in J})$ is invariant under permutation of $(z_j)_{j\in J}$. The following straightforward result is the key feature of the Gaussian copula model of Li (2000).

Lemma 7.2.1 *The $(Z_j)_{j\in J}$ are (ρ, σ)-Gaussian if and only if*

$$(Z_j, j \in J) \stackrel{\mathcal{L}}{=} (\sigma(\sqrt{\rho}Y + \sqrt{1-\rho}Y_j),\ j \in J),$$

where $(Y, Y_1, \ldots, Y_{|J|})$ are i.i.d. standard Gaussian random variables.

Therefore

$$\begin{aligned}\Phi_{\rho,\sigma}((z_j)_{j\in J}) &= \int_{\mathbb{R}} \prod_{j\in J} \Phi\left(\frac{z_j - \sigma\sqrt{\rho}\,y}{\sigma\sqrt{1-\rho}}\right)\phi(y)dy \\ \partial_{z_k}\Phi_{\rho,\sigma}((z_j)_{j\in J}) &= \int_{\mathbb{R}} \frac{-1}{\sigma\sqrt{1-\rho}}\phi\left(\frac{z_k - \sigma\sqrt{\rho}\,y}{\sigma\sqrt{1-\rho}}\right) \prod_{j\in J\setminus\{k\}} \Phi\left(\frac{z_j - \sigma\sqrt{\rho}\,y}{\sigma\sqrt{1-\rho}}\right)\phi(y)dy.\end{aligned} \tag{7.1}$$

The following result belongs to the folklore of Gaussian distributions.

Lemma 7.2.2 *If $(X_l)_{l\in N}$ are $(\varrho, 1)$-Gaussian, then*

$$\left((X_j)_{j\in J} \,\middle|\, (X_i)_{i\in I}\right) \stackrel{\mathcal{L}}{=} (\mu + Z_j)_{j\in J},$$

where $(Z_j)_{j\in J}$ are (ρ, σ)-Gaussian, with

$$\rho = \frac{\varrho}{|I|\varrho + 1}, \quad \sigma^2 = \frac{(|I|-1)\varrho + 1 - \varrho^2|I|}{(|I|-1)\varrho + 1}, \quad \mu = \frac{\varrho\sum_{i\in I} X_i}{(|I|-1)\varrho + 1}. \tag{7.2}$$

Proof. Let $X_I = (X_i)_{i\in I}$ and $X_J = (X_j)_{j\in J}$, with covariance matrix of $\begin{pmatrix} X_I \\ X_J \end{pmatrix}$ denoted by $\begin{pmatrix} \Gamma_I & \Gamma_{J,I}^{\mathrm{T}} \\ \Gamma_{J,I} & \Gamma_J \end{pmatrix}$. The conditional distribution of X_J given X_I is well known to be $\mathcal{N}\left(\Gamma_{J,I}\Gamma_I^{-1}X_I, \Gamma_J - \Gamma_{J,I}\Gamma_I^{-1}\Gamma_{J,I}^T\right)$. We denote by $\mathbf{1}_l$ (respectively $\mathbf{1}_{l\times k}$) the l-dimensional vector (respectively $l \times k$-matrix) whose entries are all equal to 1, and by $\Gamma_l(x, y)$ the $l \times l$-matrix

$$\begin{pmatrix} x & y & \cdots & y \\ y & \ddots & \ddots & \vdots \\ \vdots & \ddots & \ddots & y \\ y & \cdots & y & x \end{pmatrix}.$$

We have for the mean:

$$\begin{aligned} \Gamma_{J,I}\Gamma_I^{-1}X_I &= \varrho\mathbf{1}_{|J|\times|I|}\Gamma_{|I|}^{-1}(1,\varrho)(X_i)_{i\in I} \\ &= \varrho\mathbf{1}_{|J|\times|I|}\Gamma_{|I|}(a,b)(X_i)_{i\in I} \\ &= \mu\mathbf{1}_{|J|}, \end{aligned}$$

where $\Gamma_{|I|}^{-1}(1,\varrho) = \Gamma_{|I|}(a,b)$, with

$$a = -\frac{-(|I|-2)\varrho + 1}{(|I|-1)\varrho^2 + (2-|I|)\varrho - 1},\ b = \frac{\varrho}{(|I|-1)\varrho^2 - (|I|-2)\varrho - 1},$$

and where

$$\mu = \frac{(\varrho^2 - \varrho)\sum_{i\in I} X_i}{(|I|-1)\varrho^2 + (2-|I|)\varrho - 1} = \frac{\varrho\sum_{i\in I} X_i}{(|I|-1)\varrho + 1}.$$

Likewise, for the variance, we have:

$$\begin{aligned} \Gamma_J - \Gamma_{J,I}\Gamma_I^{-1}\Gamma_{J,I}^T &= \Gamma_{|J|}(1,\varrho) - \varrho\mathbf{1}_{|J|\times|I|}\Gamma_{|I|}^{-1}(1,\varrho)\varrho\mathbf{1}_{|I|\times|J|} \\ &= \Gamma_{|J|}(1,\varrho) - \frac{|I|\varrho^2}{(|I|-1)\varrho+1}\mathbf{1}_{|J|\times|J|} \\ &= \sigma^2\Gamma_{|J|}(1,\rho), \end{aligned}$$

where $\sigma^2 = \frac{(|I|-1)\varrho+1-\varrho^2|I|}{(|I|-1)\varrho+1}$ and $\rho = \frac{-\varrho^3+2\varrho^2-\varrho}{(|I|-1)\varrho^2+(2-|I|)\varrho-1} = \frac{\varrho}{|I|\varrho+1}$.

□

7.2.2 Model of Default Times

We consider a stochastic basis $(\Omega, \mathcal{B}_\infty, \mathbb{B}, \mathbb{Q})$, where $\mathbb{B} = (\mathcal{B}_t)_{t\geq 0}$ is the completed filtration of an n-dimensional Brownian motion $\mathbf{B} = (B^1, B^2, \ldots, B^n)$ and $\mathbb{Q}$ represents a pricing measure chosen by the market. The components of $\mathbf{B}$ are mutually correlated with a constant ϱ, i.e. $d\langle B^l, B^k\rangle_t = \varrho dt$ for any two different indices l and k. For any $l \in$

N, let h_l be a differentiable and increasing function on $(0, \infty)$ with $\lim_{0+} h_l(s) = -\infty$ and $\lim_{+\infty} h_l(s) = +\infty$. We write $h(0) = -\infty$.

Given a real valued, square integrable function of time $\varsigma(\cdot)$ such that $\int_0^\infty \varsigma^2(t)dt = 1$, we define n random times (positive random variables) on $(\Omega, \mathcal{B}_\infty, \mathbb{Q})$ by, for $l \in N$,

$$\tau_l = h_l^{-1}\Big(\int_0^{+\infty} \varsigma(v)dB_v^l\Big). \tag{7.3}$$

We assume that $\varsigma(\cdot)$ is normalized so that $\int_0^{+\infty} \varsigma^2(v)dv = 1$. Given the above construction, the random times τ_l satisfy the properties of the static Gaussian copula model of Li (2000), with correlation parameter ϱ and marginal survival function $\Phi \circ h_l$ of τ_l.

We write $\nu^2(t) = \int_t^{+\infty} \varsigma^2(v)dv$, assumed positive for all t; in particular, $\nu(0) = 1$. Letting $\mathbf{m}_t = (m_t^l)_{l\in N}$, where $m_t^l = \int_0^t \varsigma(v)dB_v^l$, we introduce for fixed t a $(\varrho, 1)$-Gaussian vector $\mathbf{X}_t = (X_t^l)_{l\in N}$ with

$$X_t^l = \frac{1}{\nu(t)} \int_t^{+\infty} \varsigma(v)dB_v^l = \frac{h_l(\tau_l) - m_t^l}{\nu(t)}.$$

Hence, for $l \in N$ and $t_l \geq 0$, we have:

$$\{\tau_l > t_l\} = \Big\{X_t^l > \frac{h_l(t_l) - m_t^l}{\nu(t)}\Big\}. \tag{7.4}$$

Note that $\mathbf{X}_t$ is independent of $\mathcal{B}_t$. This allows one to derive the following formula for the joint survival probability given $\mathcal{B}_t$. At time 0 this reduces to the well-known formula in the Gaussian copula model.

Lemma 7.2.3 *For every nonnegative t and $(t_l)_{l\in N}$,*

$$\mathbb{Q}(\tau_l > t_l,\ l \in N \,|\, \mathcal{B}_t) = \int_{\mathbb{R}} \prod_{l=1}^n \Phi\Big(\frac{\Phi^{-1}(\mathbb{Q}(\tau_l > t_l \,|\, \mathcal{B}_t)) - \sqrt{\varrho}y}{\sqrt{1-\varrho}}\Big)\,\phi(y)dy > 0.$$

Proof. It follows from (7.4) that

$$\mathbb{Q}(\tau_l > t_l \,|\, \mathcal{B}_t) = \Phi\Big(\frac{h_l(t_l) - m_t^l}{\nu(t)}\Big) > 0,$$

of which the positivity results from $\nu(t) > 0$.

Recall that $\mathbf{X}_t$ is $(\varrho, 1)$-Gaussian and independent of $\mathcal{B}_t$. Thus, in view of (7.4), application of (7.1) yields:

$$\mathbb{Q}(\tau_l > t_l,\ l \in N \,|\, \mathcal{B}_t) = \int_{\mathbb{R}} \phi(y) \prod_{l=1}^n \Phi\Big(\frac{h_l(t_l) - m_t^l - \nu(t)\sqrt{\varrho}y}{\nu(t)\sqrt{1-\varrho}}\Big)dy.$$

□

For $I \subseteq N$, we define the filtration $\mathbb{G}^I = (\mathcal{G}_t^I, t \geq 0)$ as the initial enlargement of $\mathbb{B}$ by the τ_i for $i \in I$, so (using the same notation for a random variable and the σ-field it generates, i.e. τ_i for $\sigma(\tau_i)$ here)

$$\mathcal{G}_t^I = \mathcal{B}_t \vee \bigvee_{i \in I} \tau_i .$$

Note that this filtration can be shown to be right-continuous by a straightforward multi-default extension of the results of Amendinger (1999), and it is also complete since $\mathbb{B}$ is the completed filtration of $\mathbf{B}$ and the τ_is are in $\mathcal{B}_\infty$.

Let $\boldsymbol{\tau}(I) = (\tau_l(I))_{l \in N}$, with $\tau_l(I) = \tau_l \mathbb{1}_{\{l \in I\}}$.

Lemma 7.2.4 *For every bounded function φ from $\mathbb{R}^{|J|}$ to $\mathbb{R}$, we have (recall $J = N \setminus I$):*

$$\mathbb{E}\Big[\varphi\big(h_j(\tau_j), j \in J\big) \,\Big|\, \mathcal{G}_t^I\Big] = \Gamma_\varphi\big(t, \mathbf{m}_t,\ \boldsymbol{\tau}(I)\big),$$

in which, for $t \in \mathbb{R}_+$, $\mathbf{m} = (m_l)_{l \in N}$ in $\mathbb{R}^n$ and $\boldsymbol{\theta} = (\theta_l)_{l \in N}$ in $\mathbb{R}_+^n$:

$$\Gamma_\varphi(t, \mathbf{m}, \boldsymbol{\theta}) = \mathbb{E}\Big[\varphi\Big(m_j + \nu(t)(\mu + Z_j), j \notin \mathit{supp}(\boldsymbol{\theta})\Big)\Big],$$

where $(Z_j)_{j \notin \mathit{supp}(\boldsymbol{\theta})}$ are (ρ, σ)-Gaussian. In these expressions, ρ, σ and μ are given as in (7.2), with I and X_i respectively replaced by $\mathit{supp}(\boldsymbol{\theta})$ and $\frac{h_i(\theta_i) - m_i}{\nu(t)}$. In particular,

$$\mathbb{Q}(\tau_j > t, j \in J \,|\, \mathcal{G}_t^I) = \Phi_{\rho,\sigma}\left(\frac{h_j(t) - m_t^j}{\nu(t)} - \mu, j \notin \mathit{supp}(\boldsymbol{\tau}(I)))\right). \tag{7.5}$$

Proof. By $\mathcal{B}_t$-measurability of $\mathbf{m}_t$, we have:

$$\begin{aligned}
&\mathbb{E}\Big[\varphi\Big(h_j(\tau_j),\ j \in J\Big)\Big|\mathcal{G}_t^I\Big] \\
&= \mathbb{E}\Big[\varphi\Big(\int_0^{+\infty} \varsigma(u) dB_u^j,\ j \in J\Big)\Big|\mathcal{B}_t \vee \bigvee_{i \in I} \tau_i\Big] \\
&= \mathbb{E}\Big[\varphi\Big(m_j + \int_t^{+\infty} \varsigma(u) dB_u^j,\ j \in J\Big)\Big|\mathcal{B}_t \vee \bigvee_{i \in I}\Big(\frac{h_i(\tau_i) - m_t^i}{\nu(t)}\Big)\Big]_{|m_j = m_t^j, j \in J} \\
&= \mathbb{E}\Big[\varphi\Big(m_j + \nu(t) X_t^j,\ j \in J\Big)\Big|\mathcal{B}_t \vee \bigvee_{i \in I} X_t^i\Big]_{|m_j = m_t^j, j \in J},
\end{aligned}$$

which, by independence of the X_t^l from $\mathcal{B}_t$, reduces to

$$\mathbb{E}[\varphi(m_j + \nu(t) X_t^j\ , j \in J) \mid X_t^i,\, i \in I]_{|m_j = m_t^j, j \in J}.$$

The result then follows by an application of Lemma 7.2.2. As a special case we get (7.5).
□

The natural filtration $\mathbb{B} = (\mathcal{B}_t)_{t\geq 0}$ of $\mathbf{B} = (B^l)_{l\in N}$ is used as a reference filtration. The full model filtration $\mathbb{G} = (\mathcal{G}_t)_{t\geq 0}$ is then defined as the progressive enlargement of $\mathbb{B}$ by the τ_l, i.e.

$$\mathcal{G}_t = \mathcal{B}_t \vee \bigvee_{l\in N} \mathcal{H}_t^l.$$

This filtration can be shown to be right-continuous by a combination of the arguments of Amendinger (1999) and of the appendix[1] of Bélanger, Shreve, and Wong (2001).

7.2.2.1 Conditional Survival Distribution

We recall that

$$\mathfrak{J}_t = \{i \in N \,|\, \tau_i \leq t\}, \text{ respectively } \mathcal{J}_t = N \setminus \mathfrak{J}_t,$$

denote the random set of the indices of the obligor in default, respectively alive, at time t, i.e. for any $I \subseteq N$:

$$\{\mathfrak{J}_t = I\} = \{\tau_i \leq t,\ i \in I;\ \tau_j > t,\ j \in J\}.$$

We also write, for any random function f_t of I:

$$\mathbb{E}(f_t(\mathfrak{J}_t) \,|\, \mathcal{G}_t^{\mathfrak{J}_t}) = \sum_{I\subseteq N} \mathbb{1}_{\{\mathfrak{J}_t = I\}}\, \mathbb{E}(f_t(I) \,|\, \mathcal{G}_t^I).$$

In the following proposition, we compute the conditional survival probability of $(\tau_l)_{l\in N}$, which is given, for all nonnegative t, t_l, by:

$$G_t(t_1, t_2, \ldots, t_n) := \mathbb{Q}(\tau_1 > t_1, \tau_2 > t_2, \ldots, \tau_n > t_n \,|\, \mathcal{G}_t),$$

and the individual pre-default conditional survival probability $G_t^l(t_l)$ such that, for $t_l \geq t$:

$$\mathbb{Q}(\tau_l > t_l \,|\, \mathcal{G}_t) = \mathbb{1}_{\{\tau_l > t\}} G_t^l(t_l).$$

In order to emphasize that ρ, σ and μ, appearing in Lemma 7.2.4, are functions of $(t, \mathbf{m}, \boldsymbol{\theta})$, we write:

$$\rho = \rho(t, \mathbf{m}, \boldsymbol{\theta}),\ \ \sigma = \sigma(t, \mathbf{m}, \boldsymbol{\theta}),\ \ \mu = \mu(t, \mathbf{m}, \boldsymbol{\theta}). \tag{7.6}$$

Accordingly, we define:

$$\rho_t, \sigma_t, \mu_t = \rho, \sigma, \mu\,(t, \mathbf{m}_t, \boldsymbol{\theta}_t),$$

where $\boldsymbol{\theta}_t = (\theta_t^l)_{l\in N}$ with $\theta_t^l = \tau_l \mathbb{1}_{\{\tau_l \leq t\}}$. We also write

$$Z_t^j(t_l) = \frac{h_j(t_l) - m_t^j}{\nu(t)} - \mu_t,\ \ D_t = \mathbb{Q}(\tau_j > t, j \in \mathcal{J}_t \,|\, \mathcal{G}_t^{\mathfrak{J}_t}) = \Phi_{\rho_t, \sigma_t}\Big(Z_t^j(t), j \in \mathcal{J}_t\Big),$$

where the last equality is a consequence of (7.5).

[1] Available online, not present in the *Mathematical Finance* published version of the chapter.

Proposition 7.2.5 *For every nonnegative t and $(t_l)_{l\in N}$, G_t satisfies:*

$$D_t\, G_t(t_1, t_2, \ldots, t_n) \quad = \quad \mathbb{1}_{\{\tau_i > t_i,\ i\in\mathfrak{J}_t\}}\, \Phi_{\rho_t,\sigma_t}\Big(Z_t^j(t \vee t_j),\ j \in \mathcal{J}_t\Big). \tag{7.7}$$

For every l and $t_l \geq t$, G_t^l satisfies:

$$D_t\, G_t^l(t_l) = \Phi_{\rho_t,\sigma_t}\Big(Z_t^l(t_l),\ \big(Z_t^j(t)\big)_{j\in\mathcal{J}_t\setminus\{l\}}\Big), \tag{7.8}$$

with the abuse of notation that an argument $\big(z_l,\ (z_j)_{j\in\mathcal{J}_t\setminus\{l\}}\big)$ of Φ is to be understood as a $|\mathcal{J}_t|$-dimensional vector (note that the ordering of the components doesn't matter).

Proof. By Lemma 13.7.6 (the multi-name version of the "key lemma" of credit risk), we have:

$$G_t(t_1, t_2, \ldots, t_n) = \sum_{I\subseteq N} \mathbb{1}_{\{\mathfrak{J}_t = I\}} \frac{\mathbb{Q}(\tau_i > t_i,\ i \in I;\ \tau_j > t \vee t_j,\ j \in J \,|\, \mathcal{G}_t^I)}{\mathbb{Q}(\tau_j > t,\ j \in J \,|\, \mathcal{G}_t^I)}$$

$$= \sum_{I\subseteq N} \mathbb{1}_{\{\mathfrak{J}_t = I\}} \mathbb{1}_{\{\tau_i > t_i,\ i\in I\}} \frac{\mathbb{Q}(\tau_j > t \vee t_j,\ j \in J \,|\, \mathcal{G}_t^I)}{\mathbb{Q}(\tau_j > t,\ j \in J \,|\, \mathcal{G}_t^I)},$$

where, by Lemma 7.2.4, the numerator $\mathbb{Q}(\tau_j > t \vee t_j,\ j \in J \,|\, \mathcal{G}_t^I)$ equals $\Phi_{\rho_t,\sigma_t}\big(Z_t^j(t \vee t_j), j \in J\big)$ and the denominator $\mathbb{Q}(\tau_j > t,\ j \in J \,|\, \mathcal{G}_t^I)$ equals $\Phi_{\rho_t,\sigma_t}\big(Z_t^j(t), j \in J\big)$, i.e. D_t. This proves (7.7), from which (7.8) follows by an application of (7.7) with $t_j = 0$ for $j \neq l$.

□

Observe that the "effective" Gaussian copula parameter ρ_t at time t depends on past defaults.

7.2.3 Fundamental Martingales

We now discuss the structure of martingales in the dynamic model.

7.2.3.1 Univariate Case

We start by discussing the case of one default time, i.e. $n = 1$, removing the index $l = 1$ from the notation. In view of (7.4), the $\mathbb{B}$-conditional survival probability of τ is given by:

$$\mathbb{Q}(\tau > v \,|\, \mathcal{B}_t) = \Phi\Big(\frac{h(v) - m_t}{\nu(t)}\Big).$$

Hence, τ admits a $\mathbb{B}$-conditional Lebesgue-density given, for t and v in $\mathbb{R}_+$, by:

$$a_t(v) = \frac{\mathbb{Q}(\tau \in dv \,|\, \mathcal{B}_t)}{dv} = \phi\Big(\frac{h(v) - m_t}{\nu(t)}\Big)\frac{h'(v)}{\nu(t)}. \tag{7.9}$$

In particular, this is a Lebesgue-density model in the sense of El Karoui, Jeanblanc, and Jiao (2009). Note that the filtration $\mathbb{B}$ is not immersed into the full model filtration $\mathbb{G}$ (see

Sect. 13.7). Otherwise, the Azéma supermartingale $A_t = \mathbb{Q}(\tau > t \,|\, \mathcal{B}_t)$ of τ would be nonincreasing in t, which it isn't in view of the right-hand side in (7.10).

Lemma 7.2.6 *The dynamics of $a_t(v)$ and A_t are given by:*

$$da_t(v) = a_t(v)\alpha_t(v)dB_t, \; dA_t = -a_t(t)dt + \beta_t dB_t, \tag{7.10}$$

where

$$\alpha_t(v) = -\frac{(m_t - h(v))}{\nu(t)}\frac{\varsigma(t)}{\nu(t)}, \; \beta_t = \phi\left(\frac{m_t - h(t)}{\nu(t)}\right)\frac{\varsigma(t)}{\nu(t)}. \tag{7.11}$$

***Proof*.** Noting that $\phi'(x) = -x\phi(x)$, an application of Itô's formula to the right-hand side of (7.9) yields the left-hand side in (7.10), where the right-hand side follows from the Itô-Ventzell formula[2] $dA_t = d_t A_t(v)_{|t=v} + d_v A_t(v)_{|t=v}$. □

An application of a result of Jeanblanc and Le Cam (2009), specialized to the present case of a density model with a and A continuous, shows that every $\mathbb{B}$-local martingale X is a $\mathbb{G}$-special semimartingale with the following canonical Doob-Meyer decomposition:

$$X_t = Y_t + \int_0^{t\wedge\tau} \frac{d\langle X, A\rangle_u}{A_u} + \left(\int_{t\wedge v}^t \frac{d\langle X, a(v)\rangle_u}{a_u(v)}\right)_{|v=\tau}, \tag{7.12}$$

where Y is a $\mathbb{G}$-local martingale. In particular, the following $\mathbb{G}$-canonical decomposition of the $\mathbb{B}$-Brownian motion B follows from (7.12):

$$\begin{aligned} B_t &= W_t + \int_0^{t\wedge\tau} \frac{d\langle B, A\rangle_u}{A_u} + \left(\int_{t\wedge v}^t \frac{d\langle B, a(v)\rangle_u}{a_u(v)}\right)_{|v=\tau} \\ &= W_t + \int_0^{t\wedge\tau} \frac{\beta_u}{A_u}du + \left(\int_{t\wedge v}^t \alpha_u(v)du\right)_{|v=\tau}, \end{aligned} \tag{7.13}$$

where W is a continuous $\mathbb{G}$-martingale with the same bracket t as the Brownian motion $\mathbf{B}$, hence a $\mathbb{G}$-Brownian motion, and where α and β are defined in (7.11).

By application of the results of Section 4 in El Karoui, Jeanblanc, and Jiao (2009), the $\mathbb{G}$-compensated martingale of the default indicator process $H_t = \mathbb{1}_{\{\tau \le t\}}$ is given by

$$M_t = H_t - \int_0^{t\wedge\tau} \lambda_v dv, \tag{7.14}$$

for the pre-default intensity of τ given by $\lambda_t = \frac{a_t(t)}{A_t}$.

[2] See e.g. page 39 in Jeanblanc, Yor, and Chesney (2010).

7.2.3.2 Portfolio Case

In the portfolio case with n obligors, an immediate multi-default extension of the results of Jeanblanc and Le Cam (2009) shows that, for any l, we have a $\mathbb{G}$-Brownian motion and a compensated jump-to-default $\mathbb{G}$-martingale of the form

$$W_t^l = B_t^l - \int_0^t \gamma_v^l dv,\ M_t^l = H_t^l - \int_0^{t\wedge\tau_l} \lambda_v^l dv,$$

for some processes γ^l and λ^l. Moreover, the collection of the W^l and the M^l has the $\mathbb{G}$-martingale representation property. We denote by $\mathfrak{J}_{t-}$, respectively $\mathcal{J}_{t-}$, the left-limit of $\mathfrak{J}_t$, respectively $\mathcal{J}_t$ (random set of obligors in default, respectively alive, "right before t"). By order set, we mean any subset of the state space $\mathbb{R}_+ \times \mathbb{R}^n \times \mathbb{R}_+^n$ composed of all the triplets $(t, \mathbf{m}, \boldsymbol{\theta})$ corresponding to a given support I of $\boldsymbol{\theta}$ and with the components of $\boldsymbol{\theta}$ in I in a certain order (note there are no joint defaults in a Lebesgue-density model).

Proposition 7.2.7 **(i)** *Writing* $\gamma_t = (\gamma_t^l)_{l\in N}$ *and* $\lambda_t = (\lambda_t^l)_{l\in N}$, *we have:*

$$\gamma_t = \gamma(t, \mathbf{m}_t, \boldsymbol{\theta}_t),\ \lambda_t = \lambda(t, \mathbf{m}_t, \boldsymbol{\theta}_t)$$

for suitable functions γ *and* λ, *differentiable in* $\mathbf{m}$ *on every order set.*
(ii) *The following Itô formula holds for any function* $u = u(t, \mathbf{m}, \boldsymbol{\theta})$ *of class* $\mathcal{C}^{1,2}$ *in* $(t, \mathbf{m})$:

$$\begin{aligned} du(t, \mathbf{m}_t, \boldsymbol{\theta}_t) &= \varsigma(t) \sum_{l\in N} \partial_{m_l} u(t, \mathbf{m}_t, \boldsymbol{\theta}_t) dW_t^l + \sum_{j\in\mathcal{J}_{t-}} \delta_j u(t, \mathbf{m}_t, \boldsymbol{\theta}_{t-}) dM_t^j \\ &\quad + (\partial_t + \mathcal{A})\, u(t, \mathbf{m}_t, \boldsymbol{\theta}_t) dt, \end{aligned} \tag{7.15}$$

where

$$\delta_l u(t, \mathbf{m}, \boldsymbol{\theta}) = u(t, \mathbf{m}, \boldsymbol{\theta}^{l,t}) - u(t, \mathbf{m}, \boldsymbol{\theta}),$$

in which $\boldsymbol{\theta}^{l,t}$ *stands for* $\boldsymbol{\theta}$ *with* θ_l *replaced by t, and where*

$$\mathcal{A}u = \varsigma \sum_{l\in N} \gamma_l \partial_{m_l} u + \frac{\varsigma^2}{2} \left(\sum_{l\in N} \partial^2_{m_l^2} u + \varrho \sum_{l,k\in N,\, l\neq k} \partial^2_{m_l, m_k} u \right) + \sum_{l\in N} \lambda_l \delta_l u. \tag{7.16}$$

(iii) *The process* $(\mathbf{m}_t, \boldsymbol{\theta}_t)$ *is a* $\mathbb{G}$*-Markov process with generator* $\mathcal{A}$ *(under appropriate conditions on the coefficients).*

Proof. Part (i) follows from arguments similar to the ones developed in Cousin, Jeanblanc, and Laurent (2011). The Itô formula (7.15) in part (ii) is a standard Itô formula between the τ_l, amended in the obvious way to account for the jumps of $\boldsymbol{\theta}_t$ at the τ_l (which in this model cannot occur simultaneously). Under appropriate conditions on the coefficients (see Ethier and Kurtz (1986)), the Itô formula of part (ii), where the operator $\mathcal{A}$ is deterministic in view of part (i), implies the Markov property of part (iii). □

In (iii) above, a more rigorous statement would be that the right-hand side in (7.16) defines the time t-generator of the time-inhomogeneous Markov process $(\mathbf{m}_t, \boldsymbol{\theta}_t)$ (and one would typically write $\mathcal{A}_t u$ on the left-hand side). But technicalities of this kind are immaterial in this book, where we only need the corresponding "Itô-Markov formulas", which are established by direct SDE computations without reference to operator theory.

Remark 7.2.8 As opposed to the common shock model of the next chapter, the dynamic Gaussian copula model of this chapter is not a Markov copula model in the sense of Chapter 14. Indeed, in relation, in particular, with the lack of immersion in the DGC model, the individual processes (m_t^l, θ_t^l) are not Markov. Nevertheless, as our numerical results indicate, the DGC model is already suitable for counterparty risk computations (with the reservation, typical for Gaussian copula models, that the dependence structure is not rich enough to allow for calibration of the model to multiple CDO tranches).

7.3 Clean Valuation and Hedging of Credit Derivatives

In this section, we address the issues of valuation and hedging of credit derivatives in the above-dynamic Gaussian copula setup. Note that we are still in a counterparty risk-free environment, so this is clean valuation and hedging. For notational simplicity, we assume zero interest rates, i.e. $\beta = 1$. In a zero interest rate environment, the (ex-dividend, clean) price process of an asset is simply given by the risk-neutral conditional expectation of future cash flows promised by the asset; the cumulative price is the sum of the price process and of the cumulative dividend process (cf. the definition 3.2.2(i)), which also corresponds to the martingale component of the (ex-dividend) price process. Note that for hedging purposes what matters is the dynamics of this cumulative price process (rather than of the price itself).

The extension of the theoretical results to deterministic interest rates is, of course, straightforward. Deterministic interest rates $r(t)$ will be used in the numerics.

7.3.1 Pricing of a CDS

The cumulative cash flow of a credit default swap (CDS) on firm l, with maturity T and recovery R_l, is (for continuously paid fees):

$$\int_0^T (1 - R_l) dH_v^l - \int_0^{T \wedge \tau_l} S_l dv,$$

where the constant S_l represents the contractual spread.

Proposition 7.3.1 **(i)** *The price process of a CDS on firm l, say C^l, is given as $C_t^l = \mathbb{1}_{\{\tau_l > t\}} \widetilde{C}_t^l$, where the pre-default price $\widetilde{C}^l$ is defined as*

$$\widetilde{C}_t^l = C_l(t, \mathbf{m}_t, \boldsymbol{\theta}_t) = (1 - R_l)\big(1 - G_t^l(T)\big) - S_l \int_t^T G_t^l(v) dv, \tag{7.17}$$

in which $G^l_t(\cdot) = G_l(\cdot; t, \mathbf{m}_t, \boldsymbol{\theta}_t)$ *for some function* $G_l(\cdot; t, \mathbf{m}, \boldsymbol{\theta})$ *(see (7.8)).*

(ii) *The dynamics of the cumulative price of a CDS (martingale component of the price process) on firm* l *are given by*

$$\begin{aligned} d\widehat{C}^l_t &= \mathbb{1}_{\{\tau_l \geq t\}}\big[\varsigma(t) \sum_{k \in N} \partial_{m_k} C_l(t, \mathbf{m}_t, \boldsymbol{\theta}_t) dW^k_t + (1 - R_l - C_l(t, \mathbf{m}_t, \boldsymbol{\theta}_{t-}))\, dM^l_t \\ &\quad + \sum_{j \in \mathcal{J}_{t-} \setminus \{l\}} \delta_j C_l(t, \mathbf{m}_t, \boldsymbol{\theta}_{t-}) dM^j_t\big]. \end{aligned} \tag{7.18}$$

Proof. We have

$$\begin{aligned} \mathbb{1}_{\{\tau_l > t\}} C^l_t &= \mathbb{E}\big[\int_t^T (1 - R_l) dH^l_v - \int_t^{T \wedge \tau_l} S_l dv \,|\, \mathcal{G}_t\big] \\ &= (1 - R_l)\mathbb{Q}(t < \tau_l < T \,|\, \mathcal{G}_t) - S_l \int_t^T \mathbb{Q}(\tau_l > v \,|\, \mathcal{G}_t) dv \\ &= \mathbb{1}_{\{\tau_l > t\}}\big[(1 - R_l)\big(1 - G^l_t(T)\big) - S_l \int_t^T G^l_t(v) dv\big], \end{aligned}$$

where the third equality follows from (7.8) and the fact that $G^l_t(t) = 1$. This proves (i). Since the cumulative price is the martingale component of the price process, (ii) immediately follows from (i) by an application of the Itô-Markov formula (7.15) to obtain the martingale part of the process $C_l(t, \mathbf{m}_t, \boldsymbol{\theta}_t)$. □

7.3.2 Pricing of a CDO

We write $N_v = \sum_{l=1}^n \mathbb{1}_{\{\tau_l \leq v\}} = |\mathfrak{J}_v|$, the number of defaults by time v. The conditional distribution

$$\Gamma^k_t(v) = \mathbb{Q}(N_v = k \,|\, \mathcal{G}_t)$$

is of key importance in pricing of CDO tranches. It is given by the next lemma, in which the c_k can be efficiently computed by standard recursive procedures (see e.g. Andersen and Sidenius (2004) or Sect. 5.3.2.1 in Crépey (2013)).

Lemma 7.3.2 *For every* $v \geq t$ *and* $N_t \leq k \leq n$, Γ_t *satisfies:*

$$D_t {\Gamma_t}^k(v) = \int_{\mathbb{R}} c^y_k(v; t, \mathbf{m}_t, \boldsymbol{\theta}_t)\, \phi(y) dy =: \Gamma_k(v; t, \mathbf{m}_t, \boldsymbol{\theta}_t), \tag{7.19}$$

where $c^y_k(v; t, \mathbf{m}, \boldsymbol{\theta})$ *is the order* $(k - N_t)$*-coefficient of the following polynomial* P^y *in* x*, parameterized by a real* y *(and its other arguments* v *and* $t, \mathbf{m}, \boldsymbol{\theta}$*):*

$$P^y(x, v; t, \mathbf{m}, \boldsymbol{\theta}) = \prod_{j \notin supp(\boldsymbol{\theta})} (p^y_j(v) x + q^y_j(v)).$$

Here $p^y_j(v) = p^y_j(v; t, \mathbf{m}, \boldsymbol{\theta})$ and $q^y_j(v) = q^y_j(v; t, \mathbf{m}, \boldsymbol{\theta})$ are shorthand notation for

$$p^y_j(v) = \Phi\Big(\frac{h_j(t) - m_j - \nu(t)\mu - \nu(t)\sigma\sqrt{\rho}y}{\nu(t)\sigma\sqrt{1-\rho}}\Big) - \Phi\Big(\frac{h_j(v) - m_j - \nu(t)\mu - \nu(t)\sigma\sqrt{\rho}y}{\nu(t)\sigma\sqrt{1-\rho}}\Big),$$
$$q^y_j(v) = \Phi\Big(\frac{h_j(v) - m_j - \nu(t)\mu - \nu(t)\sigma\sqrt{\rho}y}{\nu(t)\sigma\sqrt{1-\rho}}\Big), \tag{7.20}$$

where ρ, σ and μ are as in Lemma 7.2.4.

Proof. An application of the key lemma 13.7.6 yields

$$\mathbb{P}(N_v = k \,|\, \mathcal{G}_t) = \sum_{I \subseteq N} \mathbb{1}_{\{\mathfrak{J}_t = I\}} \frac{\mathbb{Q}\big(N_v = k;\ \ \tau_j > t,\ j \in J \,|\, \mathcal{G}^I_t\big)}{\mathbb{Q}\big(\tau_j > t, j \in J \,|\, \mathcal{G}^I_t\big)},$$

in which the denominator $\mathbb{Q}(\tau_j > t, j \in J \,|\, \mathcal{G}^I_t)$ is given by (7.7). For the numerator, setting $N^J_v = \sum_{j \in J} \mathbb{1}_{\{\tau_j \le v\}}$, we have, on $\{\mathfrak{J}_t = I\}$:

$$\mathbb{Q}(N_v = k;\ \tau_j > t,\ j \in J \,|\, \mathcal{G}^I_t) = \mathbb{Q}(N^J_v = k - |I|;\ \tau_j > t,\ j \in J \,|\, \mathcal{G}^I_t).$$

Recalling (7.4), we can apply Lemma 7.2.4 to the function φ such that

$$\varphi\big(h_j(\tau_j),\ j \in J\big) = \mathbb{1}_{\{N^J_v = k - |I|;\ \tau_j > t,\ j \in J\}}.$$

We obtain:

$$\mathbb{Q}(N^J_v = k - |I|;\ \tau_j > t,\ j \in J \,|\, \mathcal{G}^I_t) = \Gamma_\varphi(v; t, \mathbf{m}_t, \boldsymbol{\theta}_t),$$

where

$$\begin{aligned} &\Gamma_\varphi(v; t, \mathbf{m}, \boldsymbol{\theta}) = \mathbb{E}\big[\varphi\big(m_j + \nu(t)Z_j, j \in J\big)\big] \\ =\ &\mathbb{Q}\Big(\sum_{j \in J} \mathbb{1}_{\{h_j(t) < m_j + \nu(t)\mu + \nu(t)Z_j \le h_j(v)\}} = k - |I|;\ h_j(t) < m_j + \nu(t)\mu + \nu(t)Z_j,\ j \in J\Big), \end{aligned}$$

for some (ρ, σ)-Gaussian $(Z_j)_{j \in J}$. We apply Lemma 7.2.1 to represent $(Z_j)_{j \in J}$ in terms of an independent standard Gaussian vector $(Y, (Y_j)_{j \in J})$ which yields, with X^Y_j as a shorthand for $m_j + \nu(t)\mu + \nu(t)\sigma(\sqrt{\rho}Y + \sqrt{1-\rho}Y_j)$,

$$\begin{aligned} \Gamma_\varphi(v; t, \mathbf{m}, \boldsymbol{\theta}) &= \mathbb{Q}\Big(\sum_{j \in J} \mathbb{1}_{\{h_j(t) < X^Y_j \le h_j(v)\}} = k - |I|;\ h_j(t) < X^Y_j, j \in J\Big) \\ &= \int_{\mathbb{R}} \mathbb{Q}\Big(\sum_{j \in J} \varepsilon^y_j = k - |I|\Big)\phi(y)dy, \end{aligned}$$

where the random variables ε^y_j are defined, for y real, by:

$$\varepsilon^y_j = \begin{cases} \infty, & X^y_j \le h_j(t) \\ 1, & h_j(t) < X^y_j \le h_j(v) \\ 0, & X^y_j > h_j(v). \end{cases}$$

Therefore $\mathbb{Q}(\sum_{j\in J}\varepsilon_j^y = k-|I|) = c_k^y(v;t,\mathbf{m},\boldsymbol{\theta})$, for c_k^y as described in the statement of the lemma. □

The cumulative cash flow of a payer CDO tranche of maturity T, attachment point a, detachment point b and contractual spread S is given by

$$\int_0^T \left[dL_v^{a,b} - S(b-a-L_v^{a,b})dv\right],$$

for a tranche cumulative loss process $L_t^{a,b}$ given as

$$L_t^{a,b} = (L_t-a)^+ - (L_t-b)^+ =: L_{a,b}(N_t),$$

where $L_t = (1-R)\frac{N_t}{n}$ is the portfolio loss process (assuming a common recovery R on the n reference names).

Proposition 7.3.3 **(i)** *The price process of a CDO tranche $[a,b]$ is given by*

$$\begin{aligned} C_t^{a,b} = C_{a,b}(t,\mathbf{m}_t,\boldsymbol{\theta}_t) = \sum_{k=N_t}^{n} L_{a,b}(k)\Big(\Gamma_t^k(T) + S\int_t^T \Gamma_t^k(v)dv\Big) \\ - L_{a,b}(N_t) - S(b-a)(T-t), \end{aligned} \tag{7.21}$$

where $\Gamma_t^k(v) = \Gamma_k(v;t,\mathbf{m}_t,\boldsymbol{\theta}_t)$, in which $\Gamma_k(v;t,\mathbf{m}_t,\boldsymbol{\theta}_t)$ is given in (7.19).

(ii) *The dynamics of the cumulative price of a CDO tranche $[a,b]$ (martingale component of the price process) is given by*[3]*:*

$$\begin{aligned} d\widehat{C}_t^{a,b} =& \varsigma(t)\sum_{k\in N}\partial_{m_k}C_{a,b}(t,\mathbf{m}_t,\boldsymbol{\theta}_t)dW_t^k \\ &+ \sum_{j\in\mathcal{J}_{t-}}(\delta_j C_{a,b}(t,\mathbf{m}_t,\boldsymbol{\theta}_{t-}) + L_{a,b}(N_{t-}+1) - L_{a,b}(N_{t-}))dM_t^j. \end{aligned} \tag{7.22}$$

Proof. We have

$$\begin{aligned} C_t^{a,b} &= \mathbb{E}\Big[\int_t^T [dL_v^{a,b} - S(b-a-L_u^{a,b})dv]\Big|\mathcal{G}_t\Big] \\ &= \mathbb{E}[L_T^{a,b}\,|\,\mathcal{G}_t] - L_t^{a,b} - S(b-a)(T-t) + S\int_t^T \mathbb{E}[L_v^{a,b}\,|\,\mathcal{G}_t]dv, \end{aligned}$$

where

$$\mathbb{E}[L_v^{a,b}\,|\,\mathcal{G}_t] = \mathbb{E}[L_{a,b}(N_v)\,|\,\mathcal{G}_t] = \sum_{k=N_t}^{n} L_{a,b}(k)\Gamma_k(v;t,\mathbf{m}_t,\boldsymbol{\theta}_t).$$

This proves (i), from which (ii) follows as in the proof of Proposition 7.3.1. □

[3] It can be verified from the previous formulas, (7.19) in particular, that the function $C_{a,b}$ is regular with respect to $(s,\mathbf{m})$.

7.3.3 Hedging CDO with CDS

Here our goal is to hedge counterparty risk on credit derivatives by using single name CDS contracts. Note, however, that an already initiated CDS contract (of a given contractual spread in particular) cannot be traded dynamically in the market, in the way equity is traded. Indeed, only freshly emitted CDS contracts can be entered into, at no cost and at the related fair market spread, at any given time. What one can effectively use for hedging is a rolling CDS (see Bielecki, Jeanblanc, and Rutkowski (2008) and Bielecki, Crépey, Jeanblanc, and Rutkowski (2011)), i.e. a self-financing trading strategy in market CDS contracts. So, much like with futures contracts, the value of a rolling CDS is null at any point in time, yet due to the trading gains of the strategy the related cumulative value process is not zero. This being said (as we will see in detail in the context of a different model in Sect. 8.3.2, assuming that CDS are contracted at their fair spread), at time $t = 0$ rolling CDS dynamics are the same as the dynamics of the corresponding (standard) CDS, which implies that, at time 0, the relevant rolling CDS deltas coincide with the ones which would be based on standard CDS.

In the dynamic Gaussian copula setup, we thus consider the issue of dynamic hedging of a CDO tranche by individual CDS contracts (being understood from the above that the resulting time-0 hedging ratios are also valid for a more practical hedge by rolling CDS contracts) in various ways. In particular, mimicking the pre-crisis market practice of hedging only the spread risk of a CDO tranche (i.e. neglecting the jump-to-default risk), we can hedge the dW^k-exposures in (7.22) through suitable dynamic positions in individual CDS contracts.

In view of (7.18) and (7.22), until the first default in the portfolio this objective is achieved by the following row-vector ζ_t^{spd} of dynamic positions in the CDS on the reference names of the tranche:

$$\zeta_t^{spd} = \left(\partial_{\mathbf{m}} C_{a,b} (\partial_{\mathbf{m}} \mathbf{C})^{-1}\right)(t, \mathbf{m}_t, \boldsymbol{\theta}_{t-}), \tag{7.23}$$

where $\mathbf{C} = (C_l)_{l \in N}$.

Remark 7.3.4 After the first default time in the portfolio, say τ_k, the matrix $\partial_{\mathbf{m}}\mathbf{C}$ becomes degenerate as the corresponding C_t^k and its sensitivities vanish (whereas $d\widehat{C}_t^{a,b}$ in (7.22) still depends on dW_t^k). Thus, another nonredundant instrument (e.g. nonredundant CDS on one of the surviving names) must be substituted into the CDS for the defaulted name in order to sustain a perfect hedge of the spread risk of the tranche.

In general, numerical computation of the above deltas, ζ_t^{spd}, involves the numerical solution to a linear system. However, at $t = 0$ the system is diagonal and, as we will see below, the corresponding deltas ζ_0^{spd} are found numerically very close to standard "static" Gaussian copula bump-sensitivities (see also Fermanian and Vigneron (2010, 2013) for similar considerations).

As an alternative to the above ζ_t^{spd}, it is possible to compute, in a dynamic Gaussian copula setup, the min-variance deltas, ζ_t^{va}, which minimize the risk-neutral variance of the hedging error (minimization of spread risk and jump-to-default risk altogether, as opposed to focus on spread risk only with ζ_t^{spd}; see Sect. 3.2.5.1). For comparison we will also compute min-variance deltas in the dynamic common-shock model constructed in the next chapter by informational dynamization of a Marshall-Olkin copula setup. Henceforth DGC and are used as acronyms for dynamic Gaussian copula (the model of this chapter) and dynamic Marshall-Olkin (the model of the next chapters). Note that a DGC model can only be fitted to one tranche quote at a time (as it has a unique correlation parameter ϱ), whereas a DMO model has a richer dependence structure that can be jointly fitted to several tranches. The DGC model of this chapter is sufficient to deal with counterparty risk on CDS contracts, but a DMO setup is required for counterparty risk on CDO tranches.

As a common data set for all deltas, we use the North American CDX 17 December 2007 data, a set of credit data on 125 underlying credit names including [0-3%], [3-7%], [7-10%], [10-15%] and [15-30%] CDO tranche market quotes. We refer the reader to Brigo, Pallavicini, and Torresetti (2010) for the notions of compound correlation and base correlation of a CDO tranche. The base correlation is an alternative to compound correlation for cases in which the latter is undefined, as happens in our data set with the junior-mezzanine tranche 3%-7% (see Fig. 7.1).

We are ready to compare the following notions of deltas of a CDO tranche with respect to individual CDS contracts on all or part of the credit names underlying a CDO tranche, all these deltas being calibrated (in the sense implied by their respective definitions) to the same data set of CDX 17 December 2007:

- *market compound (respectively base) spread deltas*: static Gaussian copula bump-sensitivities for a correlation parameter ϱ set equal to the compound (respectively base) correlation of the tranche;

- *DGC compound spread deltas*: time-0 values of the dynamic deltas ζ_t^{spd} in (7.23), for a correlation parameter ϱ set equal to the compound correlation of the tranche;

- *DGC compound min-variance deltas*: time-0 values of the dynamic deltas that minimize the risk-neutral variance of the hedging error in a DGC model, for a correlation parameter ϱ set equal to the compound correlation of the tranche;

- *DMO min-variance deltas*: time-0 values of the dynamic deltas that minimize the risk-neutral variance of the hedging error in a DMO model jointly calibrated to all the tranches and to all the CDS contracts used as hedging instruments.

These deltas are computed using semi-explicit formulas for the $\partial_{m_k} C_l(t, \mathbf{m}, \boldsymbol{\theta})$ and the $\partial_{m_k} C_{a,b}(t, \mathbf{m}, \boldsymbol{\theta})$ that can be derived from Propositions 7.3.1(i) and 7.3.3(i) (see Dong Li (2013)).

The left panel of Fig. 7.2 displays the DGC and the market compound spread deltas for all the individual CDS contracts (represented by decreasing spread on the x-axis) and all the CDO tranches (y-axis), except for the junior mezzanine tranche 3%-7% with undefined compound correlation on these data (see Fig. 7.1). For every tranche, the DGC and the market compound spread deltas are very close (the two delta curves are essentially superposed for each tranche). In particular, this means that the practical notion of market compound spread deltas can be related to a sound dynamic definition, namely the DGC compound spread deltas (see also Fermanian and Vigneron (2010, 2013)).

For comparison, the right panel of Fig. 7.2 also displays the market base spread deltas for all tranches. As opposed to the previous notions of deltas, these are only ad-hoc bump-sensitivities that cannot be related to a sound dynamic approach. But they still provide a possible hedge for the junior-mezzanine tranche 3%-7% with undefined compound correlation. Except for the equity tranche, note the significant difference between the DGC or market compound spread deltas (left panel) and the base spread deltas (right panel).

Fig. 7.3 displays the DGC compound min-variance deltas for the various tranches (other than junior-mezzanine) and for portfolios of hedging CDS contracts comprising respectively all the 125 underlying names, the 61 riskiest names and the 64 safest names in the portfolio (safest and riskiest in the sense of the corresponding CDS spreads at time 0). In Chapter 8 the 61 riskiest names are used for technical reasons that are explained there. This is why we also display here, for comparison purposes, the deltas computed with the 61 riskiest names as well as the ones computed with the complement set of the 64 safest names. The corresponding (DGC or market) compound spread deltas are also displayed on the same graphs. Not surprisingly, for all the tranches there are significant numerical differences between these different notions of deltas.

In the case of the junior-mezzanine tranche 3%-7% with undefined compound correlation, there is no consistently calibrated notion of DGC deltas on our data. Instead, Fig. 7.4 displays the corresponding static market base correlation spread deltas, as well as the dynamic DMO min-variance deltas based on the portfolio of the 61 riskiest names at time 0, where the DMO model is jointly calibrated to these 61 names and to all the five-year CDO tranches at time 0 (see Chapter 8). Note the quite different patterns of these two notions of deltas.

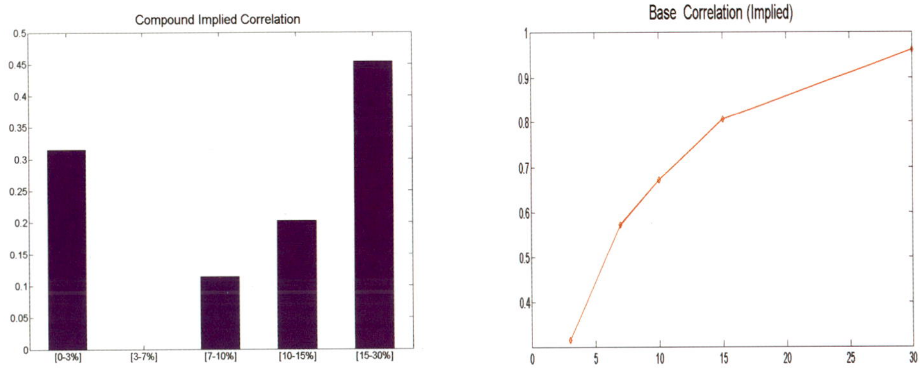

FIGURE 7.1: CDX 17 December 2007 – *Left*: Compound correlation (undefined on these data for the junior mezzanine tranche 3%-7%). *Right*: Base correlation (all tranches).

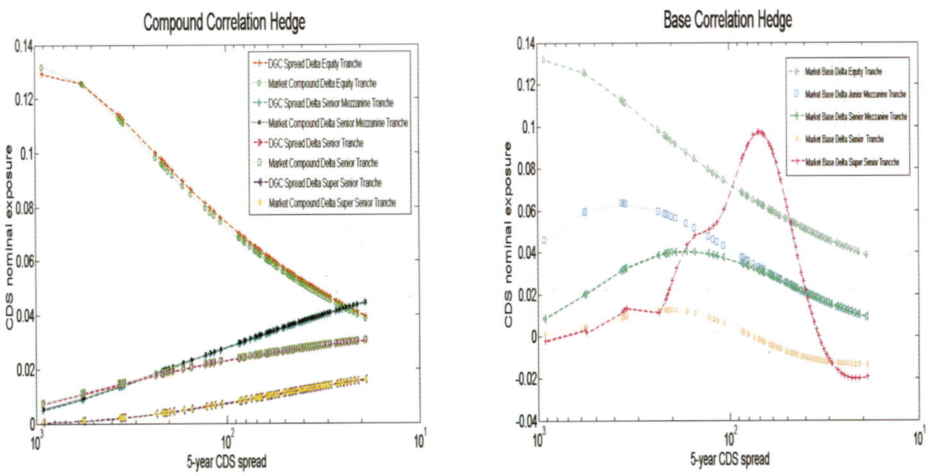

FIGURE 7.2: CDX 17 December 2007 – *Left*: DGC compound spread deltas versus market compound deltas (undefined on these data for the junior mezzanine tranche 3%-7%). *Right*: Market base deltas (all tranches).

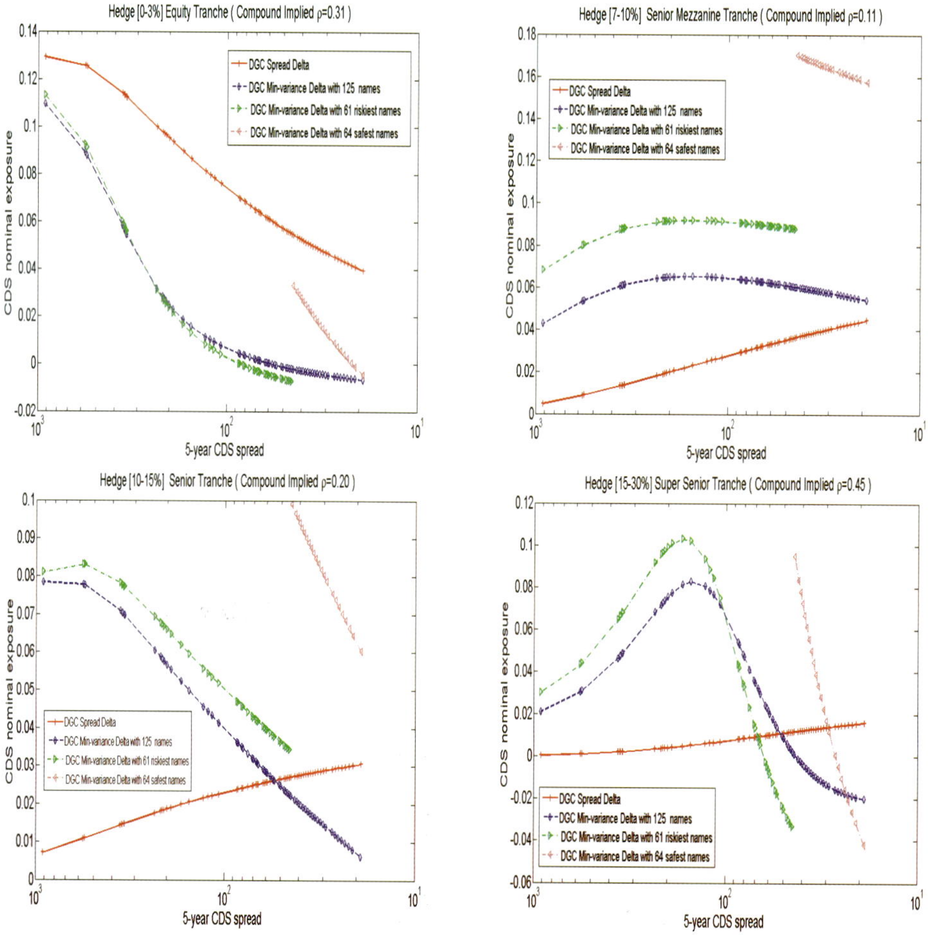

FIGURE 7.3: CDX 17 December 2007 – DGC compound min-variance deltas versus DGC (very close to market) compound spread deltas (all tranches except junior mezzanine for which the compound correlation is undefined on this data set).

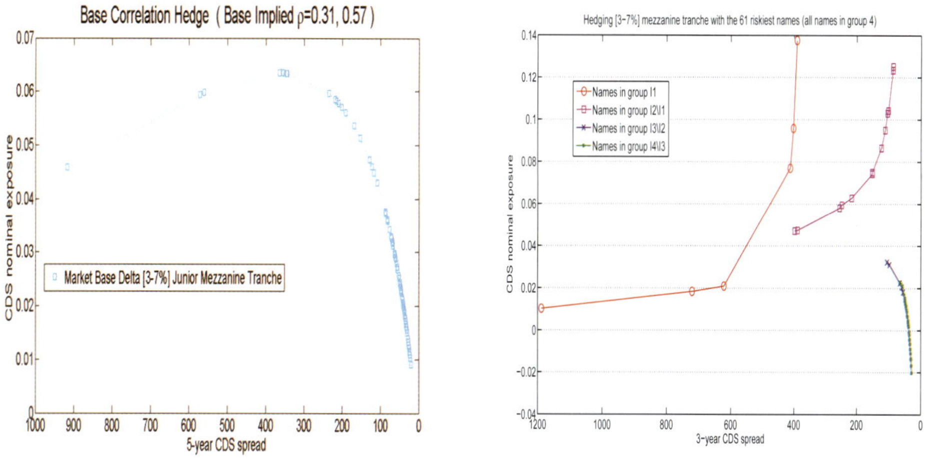

FIGURE 7.4: CDX 17 December 2007 – Junior mezzanine deltas.

7.4 Counterparty Risk

We now consider a credit derivative with maturity T between the default-free bank and a counterparty with default time τ. Our next goal is to assess the corresponding CVA, in the unilateral counterparty credit risk setup of Sect. 3.2.4.2. We still assume $r = 0$, except in the numerics, and we work with the clean CSA closeout valuation scheme $Q = P$. As recalled in the introduction to this part of the book, the unilateral CVA is an option on the clean value P_τ of the contract at time τ. In order to price it, we need a dynamic and tractable model for P_t. Towards this end, we use a dynamic Gaussian copula (DGC) model of the default times $(\tau_0 = \tau, \tau_1, \ldots, \tau_n) = (\tau_l)_{l \in N}$ of the counterparty and of the reference names underlying the credit derivative, where N now represents the set $\{-1, 0, \cdots, n\}$ (which includes the counterparty). Since there are no joint defaults in this model (this will no longer be true in the DMO model of the next chapter), we can assume that the contract promises no cash flow at τ, i.e. $\Delta_\tau = 0$, and that the exposure at default ξ reduces to

$$\xi = (1 - R_0)(P_\tau - \Gamma_\tau)^+.$$

The $\mathbb{G}$-Markov properties of $(\mathbf{m}_t, \boldsymbol{\theta}_t)$ with generator $\mathcal{A}$ implies the following pricing equations for

$$\Theta_t = \mathbb{E}\big[\mathbb{1}_{\tau < T}\xi \,\big|\, \mathcal{G}_t\big] = \mathbb{E}\big[\mathbb{1}_{\theta^0_T \neq 0}\xi \,\big|\, \mathcal{G}_t\big], \ 0 \le t \le \bar{\tau}.$$

Proposition 7.4.1 *If $P_t = P(t, \mathbf{m}_t, \boldsymbol{\theta}_t)$ and $\Gamma_t = \Gamma(t, \mathbf{m}_t, \boldsymbol{\theta}_t)$, then $\Theta_t = \Theta(t, \mathbf{m}_t, \boldsymbol{\theta}_t)$, where Θ_t satisfies the following linear CVA BSDE (assuming the needed smoothness):*

$\Theta_{\bar{\tau}} = \mathbb{1}_{\tau<T}\xi$ *and, for* $t \in [0, \bar{\tau}]$,

$$d\Theta_t = \varsigma(t) \sum_{l \in N} \partial_{m_l} \Theta(t, \mathbf{m}_t, \boldsymbol{\theta}_t) dW_t^l + \sum_{j \in \mathcal{J}_{t-}} \delta_j \Theta(t, \mathbf{m}_t, \boldsymbol{\theta}_{t-}) dM_t^j. \quad (7.24)$$

An equivalent linear CVA PDE, stated in terms of the generator $\mathcal{A}$*, is satisfied by the CVA pricing function* $\Theta = \Theta(t, \mathbf{m}, \boldsymbol{\theta})$.

From the results of Sect. 7.3, the assumption $P_t = P(t, \mathbf{m}_t, \boldsymbol{\theta}_t)$ in this proposition is met for any CDS contract, CDO tranche or (by linearity) any portfolio of them. We also have $\Gamma_t = \Gamma(t, \mathbf{m}_t, \boldsymbol{\theta}_t)$ in the "extreme" cases of no ($\Gamma = 0$) or continuous ($\Gamma = P$) collateralization. More realistic path-dependent ISDA schemes such as the ones of Sect. 3.3.2 can be dealt with by augmentation of the state space, treating the collateral process Γ as an additional factor (see Crépey and Song (2014)).

The equation (7.24) implies that the exposure $\mathbb{1}_{\tau<T}\xi$ at $\bar{\tau}$ can be dynamically replicated by using $2n + 2$ nonredundant hedging instruments, plus the funding riskless (constant) asset. Of course in practice one would more realistically hedge a selection of risk factors. Any specific hedging scheme can be implemented on the basis of the linear BSDE (7.24) or of the equivalent linear PDE. For instance, a rolling CDS referencing the counterparty can be used by the bank for hedging its CVA exposure at time τ. However, for large n, as is the case with a CDO tranche, the CVA BSDE/PDE are untractable numerically due to the curse of dimensionality (unless, perhaps, appropriate particle schemes are considered; see Sect. 12). Even the data in these equations become quite involved due to the combinatorial structure of the coefficients (particularly γ) in the generator $\mathcal{A}$. As a consequence, a Monte-Carlo computation of the CVA, based on the "CCDS formula" (3.23) (for $t = 0$), is preferred.

7.4.1 Numerics

To conclude this chapter, we provide some numerical results regarding the CVA on a CDS computed in the DGC model. These results are derived for τ_0 and τ_1 given as exponential random variables with constant parameters $\lambda_0^\star$ and $\lambda_1^\star$, which in the real life could be calibrated to the related 5yrs CDS market spreads. We use a function $\varsigma(\cdot)$ in (7.3) that is constant before T. We will see that the number

$$\sqrt{\int_0^T \varsigma^2(u)du} = \varsigma(0)\sqrt{T} \in [0, 1]$$

can can be interpreted as a volatility parameter (also depending on T, i.e. "the proportion of the volatility of $\int_0^{+\infty} \varsigma(t)dB_t^1$ before T"), which is denoted by $\%(T)$.

For comparison purposes, we will also show results obtained in the DMO model that will be developed in the next chapter.

7.4.1.1 Spread Volatilities

Since the CVA on a CDS is an option on the clean value of the CDS, an important driver of this CVA is the volatility of CDS spreads. Our next goal is to assess this volatility in terms of CDS option implied volatilities. A CDS (call) option with maturity T_a on name 1 gives the investor the right to enter at T_a a payer CDS on name 1 with contractual spread S_1 and termination time $T_b > T_a$. As explained in Brigo (2005), the corresponding price process is given by:

$$O_t^1 = \mathbb{E}\left[\mathbb{1}_{\{\tau_1 > T_a\}}\big(C^1_{T_a}\big)^- \,\big|\, \mathcal{G}_t\right], \tag{7.25}$$

in which C_t^1 stands for the time-t pre-default value of the CDS. In the market, this option is quoted in terms of its Black-implied volatility, Σ_t, which is defined (at time 0) through the following identity, in which F_0^1 denotes the forward (T_a, T_b)-CDS fair spread on name 1:

$$O_0^1 = \left(\int_{T_a}^{T_b} \mathbb{Q}(\tau_1 > u)du\right)\left(F_0^1\Phi(d_+) - S_1\Phi(d_-)\right), \tag{7.26}$$

where

$$d_\pm = \frac{\ln(F_0^1/K)}{\Sigma_0\sqrt{T_a}} \pm \frac{\Sigma_0\sqrt{T_a}}{2} \tag{7.27}$$

(see Brigo (2005) for more details). In the DGC model with two names 0 and 1, we compute the price O_0^1 of the option by Monte Carlo simulation based on the formula (7.25) at time 0 in which, by (7.17):

$$C^1_{T_a} = (1 - R_1)\big(1 - G^1_{T_a}(T_b)\big) - S_1\int_{T_a}^{T_b} G^1_{T_a}(u)du,$$

where G^1 is given by the formula (7.8) (here with two names).

We set $T_a = 3$ years, $T_b = 10$ years, $R_1 = 40\%, S_1 = \lambda_1^\star(1 - R_1)$ and we use a constant $r = 5\%$ (with the obvious amendments to all the above formulas in case of a nonnull but constant r). For a model parameter $\varrho = 40\%$, Figure 7.5 shows the prices (left panel) and the corresponding implied volatilities (right panel) of the CDS option as $\%(T_a)$ varies from 0 to 1, for four values $0.0083, 0.0125, 0.0167$ and 0.0250 of the default intensity $\lambda_1^\star$, respectively corresponding, for the chosen recovery of 40%, to clean credit spreads of 50, 75, 150 and 200 basis points. The prices O_0^1 are computed by Monte Carlo simulation based on 5000 scenarios and the implied volatilities are deduced from the prices by numerical solution of (7.26) in Σ_0. The profile of the implied volatility (see the right panel of Figure 7.5) justifies the interpretation of the quantity $\%(T_a) = \varsigma(0)\sqrt{T_a}$ as a volatility parameter. The range of implied volatilities obtained as $\%(T_a)$ varies from 0 to 1 is very wide, from a few percent to more than 200% of implied volatility. The implied volatility is slightly decreasing in $\lambda_1^\star$, a feature that will also be observed in the DMO setup (see Sect. 9.6.4). As is clear from the equation (7.18), due to absence of immersion in a DGC model, the dynamics for the price of the CDS depend not only on the underlying name 1, but also on name 0, i.e.on the counterparty, present as well in the model. However, intuitively, the impact of name 0 should be rather limited quantitatively. This is confirmed in Fig. 7.6, which shows the prices (left panel) and the corresponding implied volatilities (right panel) of the CDS option as $\%(T_a)$ varies from 0 to 1, for $\lambda_1^\star$ fixed to 0.167 (corresponding to a credit spread $S_1^\star$ of 100 basis points) and for three values $10\%, 40\%$ and 70% of the correlation parameter ϱ. As visible in the figure, the three curves corresponding to the three different values of ϱ are quite close to each other.

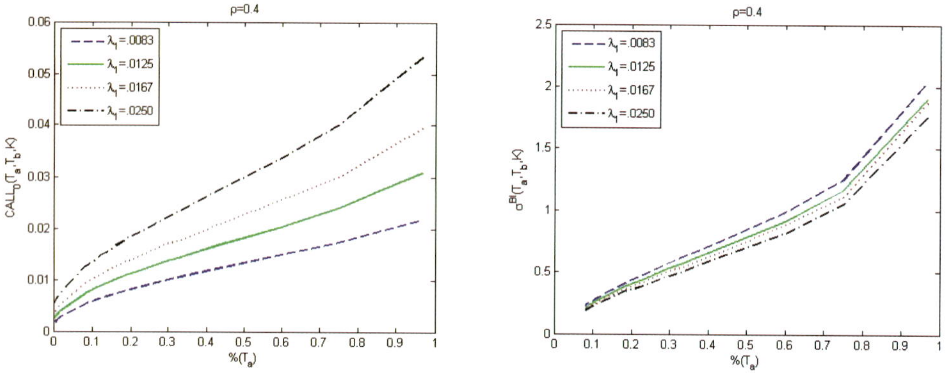

FIGURE 7.5: DGC prices computed by Monte Carlo with 5000 scenarios (*left*) and corresponding implied volatilities *(right)* of CDS options referencing name 1, for four values of $\lambda_1^\star$, as $\%(T_a)$ varies from 0 to 1 (contractual spread of the underlying CDS $S_1^\star = (1 - R_1)\lambda_1^\star$ with $R_1 = 40\%$, correlation parameter $\varrho = 40\%$).

7.4.1.2 CVA

Having checked that the model is adequately responsive in terms of volatility of CDS spreads, we show on the left panel of Fig. 7.7 the time-0 CVA on the CDS, Θ_0, computed by Monte Carlo simulation based on the "CCDS formula" (3.23), using the same parameters as before and 10^5 scenarios. This time-0 CVA is shown for a level of the Gaussian correlation parameter ϱ increasing from 0 to 1, for $\lambda_1^\star = 0.0140$ (i.e. $S_1^\star = 84$ bps). The right panel of Fig. 7.7 shows the values of Θ_0 in the DMO setup calibrated to the same data (see Sect. 9.3.1.2 for the related calibration issues). In a DMO setup, the dependence between names mostly stems from the possibility of joint defaults. In the context of counterparty risk on credit derivatives, joint defaults between the counterparty and the reference names can be a factor of strong wrong-way risk (adverse dependence between the exposure ξ and the default time of a party). Here, for instance, a joint default of the counterparty and the reference name of the CDS impacts the bank at a huge level of exposure, i.e. the protection payment not paid by the counterparty, corresponding to a nonzero and in fact very large Δ_τ-term in (3.23)-(3.24). Consistent with these observations, the DMO CVAs of Fig. 7.7 are significantly larger than the DGC CVAs (at least for high ϱ, for which the CVA is important). This illustrates the dynamic and optional flavor of the CVA and gives an idea of the related model risk (cf. also Sect. 6.4.4) in presence of wrong-way risk.

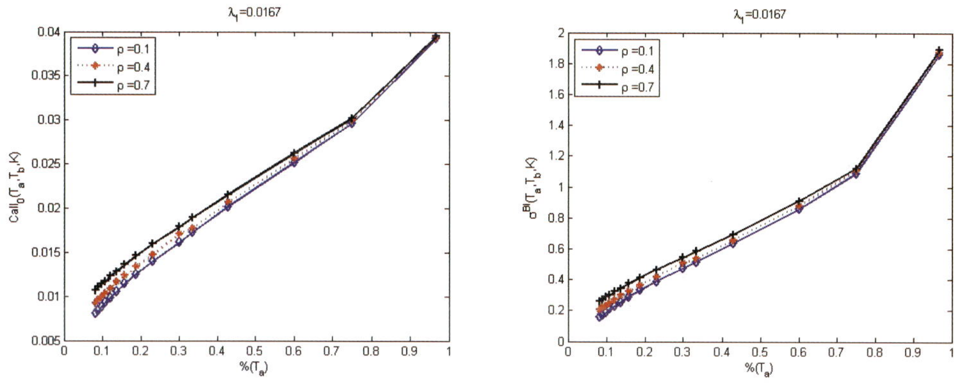

FIGURE 7.6: DGC prices computed by Monte Carlo with 5000 scenarios (*left*) and corresponding implied volatilities *(right)* of CDS options on name one for a contractual spread of the underlying CDS for three values of ϱ as $\%T_a$ varies from 0 to 1 (contractual spread of the underlying CDS $S_1^\star = (1 - R_1)\lambda_1^\star = (1 - 40\%) \times 0.167 = 100$ basis points, correlation parameter $\varrho = 40\%$).

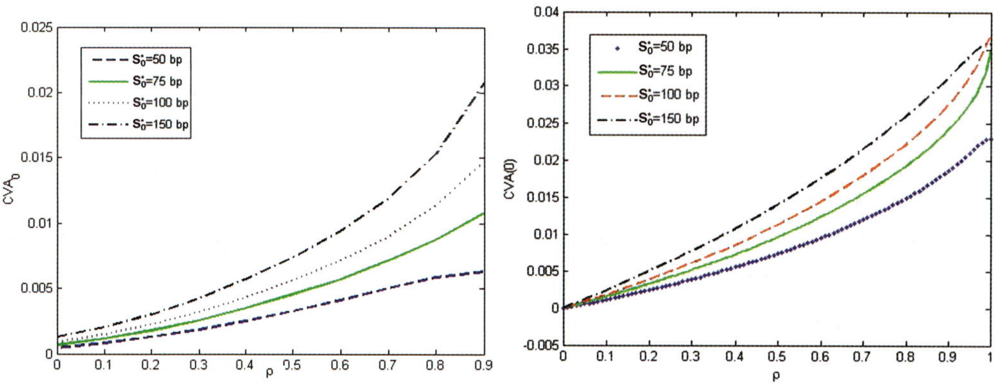

FIGURE 7.7: CVA_0 versus ϱ for $\lambda_1^\star = 0.0140$ (credit spread $S_1^\star = 84$ basis points) in the DGC *(left)* and in the DMO *(right)* model.

Chapter 8

Common-Shock Model

8.1 Introduction

The CDO market has been deeply and adversely impacted by the last financial crises. In particular, CDO issuances have become quite rare. Nevertheless, there are huge notionals of CDO contracts outstanding and market participants continue to be confronted with the task of hedging and collateralizing their positions in these contracts up to maturity date. Moreover, according to the current regulation (see Basel Committee on Banking Supervision (2011b)), tranches on standard indices and their associated liquid hedging positions continue to be charged as hedge sets under the internal VaR-based method, which also contributes to making the issue of hedging still important for standardized CDO tranches. For studies of this issue we refer the reader to, among others, Cont and Kan (2011), Cousin, Crépey, and Kan (2012), Laurent, Cousin, and Fermanian (2011) or Frey and Backhaus (2010).

The dynamic Gaussian copula CVA model of Chapter 7 may be sufficient to deal with counterparty risk on CDS contracts. However, a Gaussian copula dependence structure, parameterized by a single parameter ϱ, is not sufficiently rich for application to counterparty risk on, say, CDO tranches. In this chapter we introduce a Markov copula common-shock model of portfolio credit risk with a richer dependence structure, in which wrong-way risk is represented by the possibility of simultaneous defaults.

As we did in Chapter 7, we first present a model with n credit names $\{1, \cdots, n\}$. In Sect. 8.5 we also include the counterparty, indexed by 0, to make it amenable to the study of CVA.

8.2 Model of Default Times

In this section, we construct a bottom-up Markov model consisting of a factor process $\mathbf{X}$ and a vector $\mathbf{H}$ representing the default indicator processes of n credit names. More specifically, $\mathbf{H}_t$ is a vector in $\{0,1\}^n$ where the i-th entry, H_t^i, is the indicator function for the event of a default of obligor i up to time t. The purpose of the factor process $\mathbf{X}$ is to more realistically model diffusive randomness of credit spreads.

In this model defaults are the consequence of some "shocks" associated with groups of obligors. Denoting by $\mathcal{Z}$ the set of all nonempty subsets of names, we define the following pre-specified set of groups

$$\mathcal{Y} = \{\{1\}, \ldots, \{n\}, I_1, \ldots, I_m\},$$

where $I_1, \ldots, I_m$ are elements of $\mathcal{Z}$ and where each group I_j contains at least two obligors. The shocks are divided in two categories: the ones associated with singletons $\{1\}, \ldots, \{n\}$ can only trigger the default of names $1, \ldots, n$ individually, while the others associated with multi-name groups $I_1, \ldots, I_m$ may simultaneously trigger the default of all names in these groups. Note that several shocks may affect the same particular name, so that only the one occurring first effectively triggers the default of that name. As a result, when a shock associated with a specific group occurs at time t, it only triggers the default of names that are still alive in the group at that time. In the following, the elements Y of $\mathcal{Y}$ will be used to designate shocks, and we let $\mathcal{I} = (I_l)_{1 \le l \le m}$ denote the pre-specified collection of multi-name groups of obligors. Shock intensities of the form $\lambda_Y(t, \mathbf{X}_t)$ will be specified later in terms of a Markov factor process $\mathbf{X}_t = (X_t^Y)_{Y \in \mathcal{Y}}$. Letting $\Lambda_t^Y = \int_0^t \lambda_Y(s, \mathbf{X}_s) ds$, we define

$$\eta_Y = \inf\{t > 0 \, : \, \Lambda_t^Y > E_Y\}, \tag{8.1}$$

where the random variables E_Y are i.i.d. and exponentially distributed with parameter 1. Note that for $Y \neq Y'$, we have:

$$\mathbb{Q}(\eta_Y = \eta_{Y'}) = 0. \tag{8.2}$$

For every obligor i we let

$$\tau_i = \min_{\{Y \in \mathcal{Y};\, i \in Y\}} \eta_Y, \tag{8.3}$$

which defines the default time of obligor i in the common-shock model. We also introduce the indicator processes $K_t^Y = \mathbb{1}_{\{\eta_Y \le t\}}$ and $H_t^i = \mathbb{1}_{\{\tau_i \le t\}}$. In particular, $\mathbf{H} = (H_t^i)_{i \in N}$. Finally, we assume that processes $\mathbf{X}$ and $\mathbf{H}$ do not jump simultaneously, so that their square bracket $[\mathbf{X}, \mathbf{H}]$ is zero. The model filtration is given as $\mathbb{G} = \mathbb{X} \vee \mathbb{H}$, the filtration generated by the factor process $\mathbf{X}$ and the point process $\mathbf{H} = (H_t^i)_{i \in N}$.

This model is a doubly stochastic (via the stochastic intensities Λ^Y) and dynamic (via introduction of the filtration $\mathbb{G}$) generalization of the model of MarshallOlkinMarshall and Olkin (1967). Note that in Bielecki, Cousin, Crépey, and Herbertsson (2013a) we construct the model the other way round, i.e. we first construct a suitable Markov process $(\mathbf{X}_t, \mathbf{H}_t)$ and then define the τ_i as the jump times of the H^i.

Example 8.2.1 Fig. 8.1 shows one possible default path in our model with $n = 5$ and $\mathcal{Y} = \{\{1\}, \{2\}, \{3\}, \{4\}, \{5\}, \{4, 5\}, \{2, 3, 4\}, \{1, 2\}\}$. The inner oval shows which shocks happened and caused the observed default scenarios at successive default times. At the first shock, default of name 2 is observed as the consequence of triggering-event $\{2\}$. At the second shock, names 4 and 5 have defaulted simultaneously as a consequence of triggering-event $\{4, 5\}$. At the third shock, the shock $\{2, 3, 4\}$ triggers the default of name 3 alone as name 2 and 4 have already defaulted. At the fourth shock, default of name 1 alone is observed as the consequence of shock $\{1, 2\}$. Note that the information of the arrival of the shocks cannot be deduced from the mere observation of the sequence of states followed by $\mathbf{H}_t$.

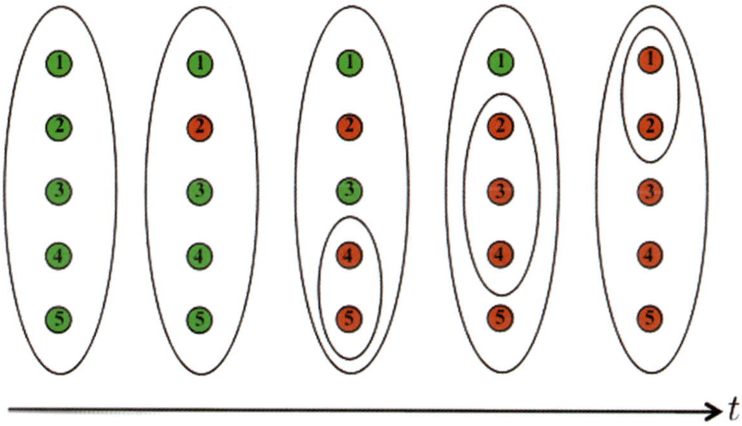

FIGURE 8.1: One possible default path in a model with $n = 5$ and $\mathcal{Y} = \{\{1\}, \{2\}, \{3\}, \{4\}, \{5\}, \{4, 5\}, \{2, 3, 4\}, \{1, 2\}\}$.

For any $Z \in \mathcal{Z}$, let the symbol $\mathbf{k}^Z$ represent the vector obtained from $\mathbf{k} = (k_1, \dots, k_n)$ by replacing each component k_i, $i \in Z$, by the number one (whenever k_i is not equal to one already). For $\mathbf{l} \neq \mathbf{k}$ in $\{0, 1\}^n$, we denote by $\tau_{\mathbf{k},\mathbf{l}}$ the jump of $\mathbf{H}$ from $\mathbf{H}_{t-} = \mathbf{k}$ to $\mathbf{H}_t = \mathbf{l}$ and by $H_t^{\mathbf{k},\mathbf{l}}$ the related indicator processes. Thus,

$$\tau_{\mathbf{k},\mathbf{l}} = \min_{\{Y \in \mathcal{Y};\, \mathbf{k}^Y = \mathbf{l}\}} \eta_Y, \quad H_t^{\mathbf{k},\mathbf{l}} = \mathbb{1}_{\{\tau_{\mathbf{k},\mathbf{l}} \le t\}}. \tag{8.4}$$

Lemma 8.2.2 *The* $(\mathbb{X}, \mathbb{G})$*-martingale intensity* $\lambda_{\mathbf{k},\mathbf{l}}(t, \mathbf{X}_t)$ *of* $H_t^{\mathbf{k},\mathbf{l}}$ *is equal to the sum of the intensities* $\lambda_Y(t, \mathbf{X}_t)$ *of the shocks* $Y \in \mathcal{Y}$ *such that, if the joint default of the survivors in group* Y *occurred at time t, then the state of* $\mathbf{H}$ *would move from* $\mathbf{k}$ *to* $\mathbf{l}$*. I.e.*

$$\lambda_{\mathbf{k},\mathbf{l}}(t, \mathbf{X}_t) = \sum_{\{Y \in \mathcal{Y};\, \mathbf{k}^Y = \mathbf{l}\}} \lambda_Y(t, \mathbf{X}_t). \tag{8.5}$$

In particular, the pre-default intensity of default of name i *is given, for* $t \in [0, T]$*, by:*

$$\lambda_t^i := \sum_{\{Y \in \mathcal{Y};\, i \in Y\}} \lambda_Y(t, \mathbf{X}_t), \tag{8.6}$$

where the sum in this expression is taken over all pre-specified groups that contain name

i. In other words, the processes $M^{\mathbf{k},\mathbf{l}}$ and M^i defined by

$$M_t^{\mathbf{k},\mathbf{l}} = H_t^{\mathbf{k},\mathbf{l}} - \int_0^{t\wedge\tau_{\mathbf{k},\mathbf{l}}} \lambda_{\mathbf{k},\mathbf{l}}(t, \mathbf{X}_t)dt, \;\; M_t^i = H_t^i - \int_0^{t\wedge\tau_i} \lambda_s^i ds \tag{8.7}$$

are $\mathbb{G}$-martingales.

Proof. Let $\mathbb{Y}$ represent the filtration generated by $\mathbf{X}$ and the collection of the K^Y. In view of (8.2) and (8.4) it follows from Lemma 7.1.2 in Bielecki and Rutkowski (2001) that the $(\mathbb{Y}, \mathbb{G})$–martingale intensity of $H_t^{\mathbf{k},\mathbf{l}}$ is $\lambda_{\mathbf{k},\mathbf{l}}(t, \mathbf{X}_t)$. Since $H_t^{\mathbf{k},\mathbf{l}}$ is $\mathbb{G}$-adapted, therefore $\lambda_{\mathbf{k},\mathbf{l}}(t, \mathbf{X}_t)$ is also the $(\mathbb{X}, \mathbb{G})$-martingale intensity of $H_t^{\mathbf{k},\mathbf{l}}$. □

Note that this model excludes direct contagion effects in which intensities of surviving names would be affected by past defaults, as opposed to the bottom-up contagion models of e.g. Cont and Minca (2013), Herbertsson (2011), or Laurent, Cousin, and Fermanian (2011). To provide some understanding in this regard, we give a simple illustrative example.

Example 8.2.3 Take $n = 3$, so that the state space of $\mathbf{H}$ contains the following 8 elements:

$$(0,0,0), (1,0,0), (0,1,0), (0,0,1), (1,1,0), (1,0,1), (0,1,1), (1,1,1).$$

Let $\mathcal{Y}$ be given as $\mathcal{Y} = \{\{1\}, \{2\}, \{3\}, \{1,2\}, \{1,2,3\}\}$. This is an example of the nested structure of $\mathcal{I}$ that we use in our numerics, here with $I_1 = \{1,2\} \subset I_2 = \{1,2,3\}$. Suppose for simplicity that λ_Y does not depend either on t or on $\mathbf{x}$ (dependence on $t, \mathbf{x}$ will be dealt with in Sect. 8.2.3). The generator of the chain $\mathbf{H}$ is given in matrix-form by

$$A = \begin{bmatrix} \cdot & \lambda_{\{1\}} & \lambda_{\{2\}} & \lambda_{\{3\}} & \lambda_{\{1,2\}} & 0 & 0 & \lambda_{\{1,2,3\}} \\ 0 & \cdot & 0 & 0 & \lambda_{\{2\}} + \lambda_{\{1,2\}} & \lambda_{\{3\}} & 0 & \lambda_{\{1,2,3\}} \\ 0 & 0 & \cdot & 0 & \lambda_{\{1\}} + \lambda_{\{1,2\}} & 0 & \lambda_{\{3\}} & \lambda_{\{1,2,3\}} \\ 0 & 0 & 0 & \cdot & 0 & \lambda_{\{1\}} & \lambda_{\{2\}} & \lambda_{\{1,2,3\}} + \lambda_{\{1,2\}} \\ 0 & 0 & 0 & 0 & \cdot & 0 & 0 & \lambda_{\{3\}} + \lambda_{\{1,2,3\}} \\ 0 & 0 & 0 & 0 & 0 & \cdot & 0 & \lambda_{\{2\}} + \lambda_{\{1,2,3\}} + \lambda_{\{1,2\}} \\ 0 & 0 & 0 & 0 & 0 & 0 & \cdot & \lambda_{\{1\}} + \lambda_{\{1,2,3\}} + \lambda_{\{1,2\}} \\ 0 & 0 & 0 & 0 & 0 & 0 & 0 & 0 \end{bmatrix} \tag{8.8}$$

where '$\cdot$' represents the sum of all other elements in a row multiplied by -1. Now, consider shock $\{1,2,3\}$. Suppose that at some point of time obligor 2 has defaulted, but obligors 1 and 3 are still alive, so that process $\mathbf{H}$ is in state $(0,1,0)$. In this case the two survivors in the group $\{1,2,3\}$ may default simultaneously with intensity $\lambda_{\{1,2,3\}}$. Note that here $\lambda_{\{1,2,3\}}$ cannot be interpreted as intensity of all three defaulting simultaneously, as obligor 2 has already defaulted. In fact, the only state of the model in which $\lambda_{\{1,2,3\}}$ can be interpreted as the intensity of all three defaulting is state $(0,0,0)$. Note that obligor 1 defaults with intensity $\lambda_{\{1\}}+\lambda_{\{1,2,3\}}+\lambda_{\{1,2\}}$, regardless of the state of the pool, as long as company 1 is alive. Similarly, obligor 2 will default with intensity $\lambda_{\{2\}}+\lambda_{\{1,2,3\}}+\lambda_{\{1,2\}}$, regardless of the state of the pool, as long as companies 1 is alive. Also, obligors 1 and 2 will default together with intensity $\lambda_{\{1,2,3\}}+\lambda_{\{1,2\}}$ regardless of the state of the pool, as long as company 1 and 2 still are alive. The above observations are consequences of the fact that the generator matrix A satisfies condition (14.11) of Chapter 14, so that process $\mathbf{H}$ is a Markov copula model satisfying so-called strong Markov consistency with respect to each of its components (i.e. each subvector of $\mathbf{H}$ is Markov in the full model filtration $\mathbb{G}$).

8.2.1 Conditional Joint Survival Function

In this subsection we give an explicit formula for the conditional joint survival function of the τ_i. We define

$$\Lambda^Y_{s,t} := \int_s^t \lambda_Y(u,\mathbf{X}_u)du, \;\; \Lambda^Y_t := \int_0^t \lambda_Y(u,\mathbf{X}_u)du \tag{8.9}$$

and, given nonnegative constants t_i,

$$\theta^Y_t = \max_{i\in Y\cap \mathrm{supp}^c(\mathbf{H}_t)} t_i, \tag{8.10}$$

where we use the convention that $\max\emptyset=-\infty$. Note that the set $Y\cap\mathrm{supp}^c(\mathbf{H}_t)$ in (8.10) represents the set of survivors in Y at time t.

Proposition 8.2.4 *For any nonnegative constants $t, t_1, \ldots, t_n$ we have:*

$$\begin{aligned}&\mathbb{Q}\left(\tau_1>t_1,\ldots,\tau_n>t_n \mid \mathcal{G}_t\right)\\ &\quad = \mathbb{1}_{\{t_i<\tau_i,\; i\in supp(\mathbf{H}_t)\}}\mathbb{E}\left\{\exp\Big(-\sum_{Y\in\mathcal{Y}}\Lambda^Y_{t,\theta^Y_t}\Big)\,\Big|\,\mathbf{X}_t\right\}.\end{aligned} \tag{8.11}$$

Proof. Let $N=\{1,\cdots,n\}$. For $I\subseteq N$, we define the filtration $\mathbb{X}^I=(\mathcal{X}^I_t)_{t\geq 0}$ as the initial enlargement of $\mathbb{X}$ by the $\tau_i, i\in I$, i.e.

$$\mathcal{X}^I_t=\mathcal{X}_t\vee\bigvee_{i\in I}\tau_i.$$

For every $I\subseteq N$, we write $J=N\setminus I$. An application of the multiname key lemma 13.7.6 yields:

$$\begin{aligned}&\mathbb{Q}\left(\tau_1>t_1,\ldots,\tau_n>t_n \mid \mathcal{G}_t\right)\\ &\quad =\sum_{I\subsetneq N}\mathbb{1}_{\{\mathfrak{J}_t=I\}}\mathbb{1}_{\{\tau_i>t_i,\; i\in I\}}\frac{\mathbb{Q}(\tau_j>t\vee t_j,\; j\in J\,|\,\mathcal{X}^I_t)}{\mathbb{Q}(\tau_j>t,\; j\in J\,|\,\mathcal{X}^I_t)}.\end{aligned} \tag{8.12}$$

Let $\mathfrak{J}_t$ denote the random set of the indices of the obligor in default at time t. In the common-shock model, writing

$$\begin{aligned}
&\mathcal{Y}_J = \{Y \in \mathcal{Y};\, Y \cap J \neq \emptyset\},\ \ \bar{\mathcal{Y}}_J = \mathcal{Y} \setminus \mathcal{Y}_J,\\
&\tau_i^J = \min_{\{Y \in \bar{\mathcal{Y}}_J;\, i \in Y\}} \eta_Y,\ \ \bar{\mathcal{X}}_t^I = \mathcal{X}_t \vee \bigvee_{i \in I} \tau_i^J,\\
&\bar{t}_Y = \max_{j \in Y \cap J} (t \vee t_j),\ \ \bar{t} = \max_{Y \in \mathcal{Y}} \bar{t}_Y = \max_{Y \in \mathcal{Y}_J} \bar{t}_Y,
\end{aligned}$$

we have on $\{\mathfrak{J}_t = I\}$ (and therefore $\{\tau_i = \tau_i^J,\, i \in I\}$):

$$\begin{aligned}
\mathbb{Q}(\tau_j > & t \vee t_j,\ j \in J \,|\, \mathcal{X}_t^I) = \mathbb{Q}(\tau_j > t \vee t_j,\ j \in J \,|\, \bar{\mathcal{X}}_t^I)\\
&= \mathbb{Q}(\eta_Y > \bar{t}_Y,\ Y \in \mathcal{Y}_J \,|\, \bar{\mathcal{X}}_t^I) = \mathbb{Q}\left(E_Y > \Lambda_{\bar{t}_Y},\ Y \in \mathcal{Y}_J \,|\, \bar{\mathcal{X}}_t^I\right)\\
&= \mathbb{E}\left(\mathbb{Q}\left(E_Y > \Lambda_{\bar{t}_Y},\ Y \in \mathcal{Y}_J \,|\, \bar{\mathcal{X}}_t^I\right) \,|\, \mathcal{X}_t^I\right) = \mathbb{E}\left\{\exp\left(-\sum_{Y \in \mathcal{Y}_J} \Lambda_{\bar{t}_Y}^Y\right) \Big|\, \bar{\mathcal{X}}_t^I\right\},
\end{aligned}$$

where $\mathbb{Q}\left(E_Y > \Lambda_{\bar{t}_Y},\ Y \in \mathcal{Y}_J \,|\, \bar{\mathcal{X}}_t^I\right) = \exp\left(-\sum_{Y \in \mathcal{Y}_J} \Lambda_{\bar{t}_Y}^Y\right)$ in the last identity holds by independence of $(E_Y)_{Y \in \mathcal{Y}_J}$ from $\bar{\mathcal{X}}_t^I$ and by $\bar{\mathcal{X}}_{\bar{t}}^I$- (in fact, $\mathcal{X}_{\bar{t}}$-) measurability of $(\Lambda_{\bar{t}_Y}^Y)_{Y \in \mathcal{Y}_J}$.

Note moreover that we have on $\{\mathfrak{J}_t = I\}$:

$$\mathcal{X}_t \subseteq \bar{\mathcal{X}}_t^I \subseteq \mathcal{X}_t \vee \bigvee_{Y \in \bar{\mathcal{Y}}_J} E_Y,$$

where the E_Y, for $Y \in \bar{\mathcal{Y}}_J$, are independent from $\mathbb{X}$. Therefore, on $\{\mathfrak{J}_t = I\}$:

$$\begin{aligned}
\mathbb{E}\left\{\exp\left(-\sum_{Y \in \mathcal{Y}_J} \Lambda_{\bar{t}_Y}^Y\right) \Big|\, \mathcal{X}_t\right\} &= \mathbb{E}\left\{\exp\left(-\sum_{Y \in \mathcal{Y}_J} \Lambda_{\bar{t}_Y}^Y\right) \Big|\, \mathcal{X}_t \vee \bigvee_{Y \in \bar{\mathcal{Y}}_J} E_Y\right\}\\
&= \mathbb{E}\left\{\exp\left(-\sum_{Y \in \mathcal{Y}_J} \Lambda_{\bar{t}_Y}^Y\right) \Big|\, \bar{\mathcal{X}}_t^I\right\} = \mathbb{Q}\left(\tau_j > t \vee t_j,\ j \in J \,|\, \mathcal{X}_t^I\right).
\end{aligned}$$

The Markov property of $\mathbf{X}_t$ finally yields that

$$\begin{aligned}
\mathbb{Q}(\tau_j > t \vee t_j, j \in J \,|\, \mathcal{X}_t^I) &= \exp\left(-\sum_{Y \in \mathcal{Y}_J} \Lambda_t^Y\right) \mathbb{E}\left\{\exp\left(-\sum_{Y \in \mathcal{Y}_J} \Lambda_{t,\bar{t}_Y}^Y\right) \Big|\, \mathbf{X}_t\right\}\\
&= \mathbb{Q}(\tau_j > t,\ j \in J \,|\, \mathcal{X}_t^I) \mathbb{E}\left\{\exp\left(-\sum_{Y \in \mathcal{Y}_J} \Lambda_{t,\bar{t}_Y}^Y\right) \Big|\, \mathbf{X}_t\right\}.
\end{aligned}$$

Substituting this into (8.12) yields that

$$\mathbb{Q}\left(\tau_1 > t_1, \ldots, \tau_n > t_n \mid \mathcal{G}_t\right) = \sum_{I \subsetneq N} \mathbb{1}_{\{\mathfrak{J}_t = I\}} \mathbb{1}_{\{\tau_i > t_i,\, i \in I\}} \mathbb{E}\left\{\exp\left(-\sum_{Y \in \mathcal{Y}} \Lambda_{t,\bar{t}_Y}^Y\right) \Big|\, \mathbf{X}_t\right\},$$

which is (8.11). □

We will illustrate the above proposition using the following example.

Example 8.2.5 In case of two obligors with $\mathcal{Y} = \{\{1\},\{2\},\{1,2\}\}$, the formula (8.11) reduces to

$$\begin{aligned}\mathbb{Q}\left(\tau_1 > t_1, \tau_2 > t_2 \mid \mathcal{G}_t\right) &= \mathbb{1}_{\{\tau_1>t\}}\mathbb{1}_{\{\tau_2>t\}}\mathbb{E}\left\{\exp\big(-\sum_{Y\in\mathcal{Y}}\int_t^{t_1\vee t_2}\lambda_Y(s,\mathbf{X}_s)ds\big)\,\Big|\,\mathbf{X}_t\right\}\\ &+\mathbb{1}_{\{t_2<\tau_2\le t\}}\mathbb{1}_{\{\tau_1>t\}}\mathbb{E}\left\{\exp\big(-\int_t^{t_1}\lambda^1_s\,ds\big)\,\Big|\,\mathbf{X}_t\right\}\\ &+\mathbb{1}_{\{t_1<\tau_1\le t\}}\mathbb{1}_{\{\tau_2>t\}}\mathbb{E}\left\{\exp\big(-\int_t^{t_2}\lambda^2_s\,ds\big)\,\Big|\,\mathbf{X}_t\right\}\\ &+\mathbb{1}_{\{t_1<\tau_1\le t\}}\mathbb{1}_{\{t_2<\tau_2\le t\}},\end{aligned}$$

where the individual intensities λ^1 and λ^2 are defined in (8.6).

8.2.1.1 Conditional Common-Shock Model

Conditionally on $\mathcal{G}_t$, it is possible to define a "common-shock model of the surviving names at time t" such that the law of the default times in this conditional common-shock model is the same as the corresponding conditional distribution in the original model. This will be useful for CVA computations Towards this end, we introduce a family of so-called conditional common-shock copula models, parameterized by the current time t, as follows. For every $Y \in \mathcal{Y}$, we define

$$\eta_Y(t) := \inf\{\theta > t;\ \Lambda^Y_\theta > \Lambda^Y_t + E_Y(t)\},$$

where the random variables $E_Y(t)$ are i.i.d. exponentially distributed with parameter 1. For every obligor i, we let

$$\tau_i(t) = \min_{\{Y\in\mathcal{Y};\, i\in Y\}} \eta_Y(t),$$

which defines the default time of obligor i in the conditional common-shock model starting at time t. Accordingly, we define the indicator processes

$$K^Y_\theta(t) = \mathbb{1}_{\{\eta_Y(t)\le\theta\}},\quad H^i_\theta(t) = \mathbb{1}_{\{\tau_i(t)\le\theta\}}, \tag{8.13}$$

for $\theta \ge t$. Let $Z \in \mathcal{Z}$ be a set of obligors. We will prove that on the event $\{\mathrm{supp}^c(\mathbf{H}_t) = Z\}$, the $\mathcal{G}_t$-conditional law of $(\tau_i)_{i\in\mathrm{supp}^c(\mathbf{H}_t)}$ is equal to the $\mathbf{X}_t$-conditional of $(\tau_i(t))_{i\in Z}$. Let N_θ be $\sum_{1\le i\le n} H^i_\theta$, the cumulative number of defaulted obligors in the original model up to time θ, and let $N_\theta(t,Z)$ be $n - |Z| + \sum_{i\in Z} H^i_\theta(t)$, the cumulative number of defaulted obligors in the conditional model up to time θ.

Proposition 8.2.6 *Let $Z \in \mathcal{Z}$ and $t \geq 0$.*
(i) *For every $t_1, \ldots, t_n \geq t$, we have*

$$\mathbb{1}_{\{supp^c(\mathbf{H}_t)=Z\}}\mathbb{Q}\left(\tau_i > t_i,\, i \in supp^c(\mathbf{H}_t) \,\middle|\, \mathcal{G}_t\right) = \qquad (8.14)$$
$$\mathbb{1}_{\{supp^c(\mathbf{H}_t)=Z\}}\mathbb{Q}\left(\tau_i(t) > t_i,\, i \in Z \,\middle|\, \mathbf{X}_t\right);$$

(ii) *For $\theta \geq t$ and $k = n - |Z|, \ldots, n$, we have that*

$$\mathbb{1}_{\{supp^c(\mathbf{H}_t)=Z\}}\mathbb{Q}\left(N_\theta = k \,\middle|\, \mathcal{G}_t\right) = \mathbb{1}_{\{supp^c(\mathbf{H}_t)=Z\}}\mathbb{Q}\left(N_\theta(t,Z) = k \,\middle|\, \mathbf{X}_t\right).$$

Proof. Part (ii) readily follows from part (i), so we only prove (i). For every obligor i, let $\widetilde{t}_i = \mathbb{1}_{i \in \mathrm{supp}^c(\mathbf{H}_t)} t_i$. For $Y \in \mathcal{Y}$, we have:

$$\max_{i \in Y \cap \mathrm{supp}^c(\mathbf{H}_t)} \widetilde{t}_i = \max_{i \in Y \cap \mathrm{supp}^c(\mathbf{H}_t)} t_i = \theta_t^Y.$$

Therefore, an application of Proposition 8.2.4 to the sequence of times $(\widetilde{t}_i)_{1 \leq i \leq n}$ yields:

$$\begin{aligned}
&\mathbb{1}_{\{\mathrm{supp}^c(\mathbf{H}_t)=Z\}}\mathbb{Q}\left(\tau_i > t_i,\, i \in \mathrm{supp}^c(\mathbf{H}_t) \,\middle|\, \mathcal{G}_t\right) \\
&= \mathbb{1}_{\{\mathrm{supp}^c(\mathbf{H}_t)=Z\}}\mathbb{Q}\left(\left(\tau_i > t_i,\, i \in Z\right),\, \left(\tau_i > 0,\, i \in Z^c\right) \,\middle|\, \mathcal{G}_t\right) \\
&= \mathbb{1}_{\{\mathrm{supp}^c(\mathbf{H}_t)=Z\}}\mathbb{E}\left\{\exp\left(-\sum_{Y \in \mathcal{Y}} \Lambda^Y_{t,\theta_t^Y}\right) \,\middle|\, \mathbf{X}_t\right\} \\
&= \mathbb{1}_{\{\mathrm{supp}^c(\mathbf{H}_t)=Z\}}\mathbb{E}\left\{\exp\left(-\sum_{Y \in \mathcal{Y}} \Lambda^Y_{t,\max_{i \in Y \cap Z} t_i}\right) \,\middle|\, \mathbf{X}_t\right\} \qquad (8.15)
\end{aligned}$$

where, by application of Proposition 8.2.4 in the $\mathcal{G}_t$-conditional common-shock model:

$$\mathbb{E}\left\{\exp\left(-\sum_{Y \in \mathcal{Y}} \Lambda^Y_{t,\max_{i \in Y \cap Z} t_i}\right) \,\middle|\, \mathbf{X}_t\right\} = \mathbb{Q}(\tau_i(t) > t_i,\, i \in Z \,|\, \mathcal{G}_t).$$

□

8.2.2 Itô-Markov Formula

In this subsection we establish an Itô formula for functions of $(\mathbf{X}_t, \mathbf{H}_t)$. In particular, this formula also implies the $\mathbb{G}$-Markov property of $(\mathbf{X}_t, \mathbf{H}_t)$.

For any set $Z \in \mathcal{Z}$, let the set-event indicator process H^Z denote the indicator process of a joint default of the names in Z, but only in Z. The following lemma provides the structure of the so-called compensated set-event martingales M^Z, which we will use later as fundamental martingales to represent the pure jump martingale components of the various price processes involved. For Y in $\mathcal{Y}$, we write

$$Y_t = Y \cap \mathrm{supp}^c(\mathbf{H}_{t-}), \qquad (8.16)$$

the set-valued process representing the survived obligors in Y right before time t.

Lemma 8.2.7 *For every set $Z \in \mathcal{Z}$, the intensity of H^Z is given by $\ell_Z(t, \mathbf{X}_t, \mathbf{H}_t)$, so that*

$$dM_t^Z = dH_t^Z - \ell_Z(t, \mathbf{X}_t, \mathbf{H}_t)dt \tag{8.17}$$

is a martingale, for a set-event intensity function $\ell_Z(t, \mathbf{x}, \mathbf{k})$ defined as

$$\ell_Z(t, \mathbf{x}, \mathbf{k}) = \sum_{Y \in \mathcal{Y};\, Y \cap supp^c(\mathbf{k}) = Z} \lambda_Y(t, \mathbf{x}). \tag{8.18}$$

Proof. By definition of the set-event indicator process H^Z, we have

$$dH_t^Z = \sum_{\{\mathbf{k},\mathbf{l} \in \{0,1\}^n\,;\, \mathrm{supp}(\mathbf{l}) \setminus \mathrm{supp}(\mathbf{k}) = Z\}} dH_t^{\mathbf{k},\mathbf{l}}.$$

So, by (8.5),

$$\begin{aligned}
\ell_t^Z &= \sum_{\{\mathbf{k},\mathbf{l} \in \{0,1\}^n\,;\, \mathrm{supp}(\mathbf{l}) \setminus \mathrm{supp}(\mathbf{k}) = Z\}} \mathbb{1}_{\{\mathbf{H}_{t-} = \mathbf{k}\}} \sum_{\{Y \in \mathcal{Y};\, \mathbf{k}^Y = \mathbf{l}\}} \lambda_Y(t, \mathbf{X}_t) \\
&= \sum_{\{\mathbf{l} \in \{0,1\}^n\,;\, \mathrm{supp}(\mathbf{l}) \setminus \mathrm{supp}(\mathbf{H}_{t-}) = Z\}} \sum_{\{Y \in \mathcal{Y};\, \mathbf{H}_{t-}^Y = \mathbf{l}\}} \lambda_Y(t, \mathbf{X}_t) \\
&= \sum_{\{Y \in \mathcal{Y};\, \mathrm{supp}(\mathbf{H}_{t-}^Y) \setminus \mathrm{supp}(\mathbf{H}_{t-}) = Z\}} \lambda_Y(t, \mathbf{X}_t) \\
&= \sum_{\{Y \in \mathcal{Y};\, Y_t = Z\}} \lambda_Y(t, \mathbf{X}_t).
\end{aligned}$$

□

From now on we assume that the process $\mathbf{X}$ is a jump-diffusion process (see Sect. 13.3) driven by a vector martingale $\mathbf{M}$ composed of Brownian motions and/or a compensated jump measure (see the remark 13.3.1) and with generator denoted here by $\mathcal{A}_{\mathbf{x}}$ (cf. (13.38)). Consequently, there exists a suitable operator $\mathcal{B}_{\mathbf{x}}$ (see the remark 13.3.1) such that for a sufficiently regular function $u = u(t, \mathbf{x})$, we have (cf. (13.37)):

$$du(t, \mathbf{X}_t) = \Big(\partial_t + \mathcal{A}_{\mathbf{x}}\Big)u(t, \mathbf{X}_t)dt + \mathcal{B}_{\mathbf{x}}u(t, \mathbf{X}_{t-}) \cdot d\mathbf{M}_t, \tag{8.19}$$

where $\mathcal{B}_{\mathbf{x}}u(t, \mathbf{X}_{t-}) \cdot d\mathbf{M}_t$ represents the relevant stochastic integral.

We now derive the Itô formula relevant for the full model $(\mathbf{X}_t, \mathbf{H}_t)$. Towards this end, we denote

$$\delta_Z u(t, \mathbf{x}, \mathbf{k}) = u(t, \mathbf{x}, \mathbf{k}^Z) - u(t, \mathbf{x}, \mathbf{k})$$

and we define

$$\mathcal{Z}_t = \{Z \in \mathcal{Z};\, Z = Y_t \text{ for at least one } Y \in \mathcal{Y}\} \setminus \emptyset$$

to be the collection of all nonempty sets of survivors of sets Y in $\mathcal{Y}$ right before time t.

Proposition 8.2.8 *For a sufficiently regular function $u = u(t, \mathbf{x}, \mathbf{k})$, we have*

$$du(t, \mathbf{X}_t, \mathbf{H}_t) = \Big(\partial_t + \mathcal{A}\Big)u(t, \mathbf{X}_t, \mathbf{H}_t)dt \;+\; \mathcal{B}_{\mathbf{x}}u(t, \mathbf{X}_{t-}, \mathbf{H}_{t-}) \cdot d\mathbf{M}_t \\ \quad + \sum_{Z \in \mathcal{Z}_t} \delta_Z u(t, \mathbf{X}_t, \mathbf{H}_{t-})dM^Z_t, \tag{8.20}$$

where

$$\mathcal{A}u(t, \mathbf{x}, \mathbf{k}) = \mathcal{A}_{\mathbf{x}}u(t, \mathbf{x}, \mathbf{k}) + \sum_{Y \in \mathcal{Y}} \lambda_Y(t, \mathbf{x})\delta_Y u(t, \mathbf{x}, \mathbf{k}). \tag{8.21}$$

Hence, $(\mathbf{X}, \mathbf{H})$ is a (time-inhomogeneous) Markov process with infinitesimal generator $\mathcal{A}$ given by (8.21) *(under appropriate conditions on the coefficients).*

Proof. Observe that $[M^Y, M^Z] = 0$ for $Y \neq Z$. Also, by assumption, $[\mathbf{X}, \mathbf{H}] = 0$. We thus have the following Itô formula (cf. (13.37)):

$$du(t, \mathbf{X}_t, \mathbf{H}_t) = \Big(\partial_t + \mathcal{A}_{\mathbf{x}}\Big)u(t, \mathbf{X}_t, \mathbf{H}_t)dt \quad + \mathcal{B}_{\mathbf{x}}u(t, \mathbf{X}_{t-}, \mathbf{H}_{t-}) \cdot d\mathbf{M}_t \\ + \sum_{Z \in \mathcal{Z}} \delta_Z u(t, \mathbf{X}_t, \mathbf{H}_{t-})dH^Z_t. \tag{8.22}$$

Moreover, the structure (8.18) of the event-set intensities implies that

$$\sum_{Z \in \mathcal{Z}} \delta_Z u(t, \mathbf{X}_t, \mathbf{H}_{t-})dH^Z_t = \sum_{Z \in \mathcal{Z}_t} \delta_Z u(t, \mathbf{X}_t, \mathbf{H}_{t-})dH^Z_t,$$

which we may further rewrite as

$$\sum_{Z \in \mathcal{Z}_t} \Big(\delta_Z u(t, \mathbf{X}_t, \mathbf{H}_{t-})dH^Z_t - \ell_Z(t, \mathbf{X}_t, \mathbf{H}_t)\delta_Z u(t, \mathbf{X}_t, \mathbf{H}_t)dt\Big) \\ + \sum_{Z \in \mathcal{Z}_t} \ell_Z(t, \mathbf{X}_t, \mathbf{H}_{t-})\delta_Z u(t, \mathbf{X}_t, \mathbf{H}_{t-})dt.$$

Here the first term is $\sum_{Z \in \mathcal{Z}_t} \delta_Z u(t, \mathbf{X}_t, \mathbf{H}_{t-})dM^Z_t$ and, by (8.18), we have for the second term:

$$\sum_{Z \in \mathcal{Z}_t} \ell_Z(t, \mathbf{X}_t, \mathbf{H}_{t-})\delta_Z u(t, \mathbf{X}_t, \mathbf{H}_{t-}) \\ = \sum_{Z \in \mathcal{Z}_t} \sum_{Y \in \mathcal{Y}; Y_t = Z} \lambda_Y(t, \mathbf{X}_t)\delta_Z u(t, \mathbf{X}_t, \mathbf{H}_{t-}) \\ = \sum_{Y \in \mathcal{Y}} \lambda_Y(t, \mathbf{X}_t)\delta_Y u(t, \mathbf{X}_t, \mathbf{H}_{t-}),$$

the last identity resulting from

$$\delta_Z u(t, \mathbf{X}_t, \mathbf{H}_{t-}) = \delta_Y u(t, \mathbf{X}_t, \mathbf{H}_{t-}),$$

for all Y and Z such that $Y_t = Z$. Thus (8.22) can be rewritten as (8.20). As in the proof of Proposition 7.2.7, the the Markov property of $(\mathbf{X}, \mathbf{H})$ follows from an appropriate version of the local martingale characterization of Markov processes (under appropriate conditions on the coefficients that can be found in Ethier and Kurtz (1986)). □

The martingale dimension[1] of the model is equal to the sum of the number of martingales M^Z and the number of martingales constituting $\mathbf{M}$ (i.e., in the terminology of Sect. 13.1.2, "the number of marks e in $\mathbf{M}(dt, de)$"). In our model, there are $(2^n - 1)$ fundamental martingales M^Z so that, due to this combinatorially exploding martingale dimension, the model might be computationally intractable for n greater than a few units. However, due to the specific structure of the intensities in the model, the sum in (8.21) only runs over the set of shocks $\mathcal{Y}$, which is of cardinality $n + m$.

Remark 8.2.9 By construction, the Markov process $(\mathbf{X}, \mathbf{H})$ satisfies the Markov consistency condition with respect to each pair $(\mathbf{X}, H^i)$, $i = 1, 2, \ldots, d$ (see Chapter 14). It thus can serve as a Markov copula for processes Y^i such that $Y^i \stackrel{\mathcal{L}}{=} (\mathbf{X}, H^i)$, $i = 1, 2, \ldots, d$. (see also the end of the example 8.2.3 regarding the simpler situation without factor process $\mathbf{X}$).

8.2.3 Intensity Structure

We say that a Markov process is affine if its conditional Laplace transform is given in exponential form (see Filipović (2005) for the general theory and Sect. 13.4.1 for a specific example).

Assumption 8.2.10 **(i)** The jump-diffusion $\mathbf{X}_t$ and the shock intensity functions $\lambda_Y(t, \mathbf{x})$ are such that the shock intensities $\lambda_Y(t, \mathbf{X}_t)$ are $\mathbb{X}$-affine. Moreover, for any subset $\widetilde{\mathcal{Y}}$ of $\mathcal{Y}$ containing at most one singleton $\{i\}$, the process $\sum_{Y \in \widetilde{\mathcal{Y}}} \lambda_Y(t, \mathbf{X}_t)$ is $\mathbb{X}$-affine.
(ii) For every name i and time $\theta > t$, the random variables $\Lambda^{\{i\}}_{t,\theta}$ and $(\Lambda^Y_{t,\theta})_{Y \in \mathcal{Y} \setminus \{i\}}$ are conditionally independent given $\mathbf{X}_t$, i.e.

$$\mathbb{E}\left(\varphi(\Lambda^{\{i\}}_{t,\theta})\psi((\Lambda^Y_{t,\theta})_{Y \in \mathcal{Y}\setminus\{i\}})\Big|\mathcal{G}_t\right) = \mathbb{E}\left(\varphi(\Lambda^{\{i\}}_{t,\theta})\Big|\mathcal{G}_t\right)\mathbb{E}\left(\psi((\Lambda^Y_{t,\theta})_{Y \in \mathcal{Y}\setminus\{i\}})\Big|\mathcal{G}_t\right),$$

for any Borel functions φ, ψ.

In the following lemma, part (i), respectively (ii), immediately follows from the second part of the assumption 8.2.10(i), respectively (ii).

Lemma 8.2.11 **(i)** *The single-name default intensities* λ^i_t *of* (8.6) *are* $\mathbb{X}$*-affine.*
(ii) $K^{\{i\}}_\theta(t)$ *and* $(K^Y_\theta(t))_{Y \in \mathcal{Y}\setminus\{i\}}$ *(defined in* (8.13)*) are conditionally independent given* $\mathbf{X}_t$.

[1] Minimal number of martingales which can be used as integrators to represent all the martingales in the model.

Observe that as a consequence of part (i), every pair (λ^i_t, H^i_t) (and not only $(\mathbf{X}_t, H^i_t)$, as seen in the remark 8.2.9) is Markov in the big model filtration $\mathbb{G}$.

We now provide three concrete specifications. Note that the second one provides a fully stochastic specification of the λ_Y (including the "systemic"[2] intensities λ_I).

Example 8.2.12 (i) (Deterministic group intensities). The idiosyncratic intensities $\lambda_{\{i\}}(t, \mathbf{X}_t)$ are $\mathbb{X}$-affine and the systemic intensities $\lambda_Y(t, \mathbf{X}_t)$ are Borel functions of time, i.e. the functions $\lambda_I(t, \mathbf{x})$ do not depend on $\mathbf{x}$, for $I \in \mathcal{I}$.
(ii) (Extended CIR intensities). For every $Y \in \mathcal{Y}$, $\lambda_Y(t, \mathbf{X}_t) = X^Y_t$, where

$$dX^Y_t = a(b_Y(t) - X^Y_t)dt + c\sqrt{X^Y_t}\,dW^Y_t, \tag{8.23}$$

for nonnegative constants a, c (independent of Y) and for a nonnegative function $b_Y(t)$, where the W^Y, $Y \in \mathcal{Y}$, are independent standard Brownian motions.
(iii) (Lévy Hull-White intensities). For every $Y \in \mathcal{Y}$,

$$dX^Y_t = a(b_Y(t) - X^Y_t)dt + dZ^Y_t, \tag{8.24}$$

where the Z^Y are independent Lévy subordinators[3].

In the second specification, the last part of the assumption 8.2.11(i) holds from the fact that the SDEs for the factors X^Y have the same coefficients except for the $b_Y(t)$. Thus $X^i := \sum_{\{Y\in\mathcal{Y};\, i\in Y\}} X^Y$ satisfies the following extended CIR SDE:

$$dX^i_t = a(b_i(t) - X^i_t)dt + c\sqrt{X^i_t}dW^i_t, \tag{8.25}$$

for the function $b_i(t) = \sum_{\{Y\in\mathcal{Y};\, i\in Y\}} b_Y(t)$ and the standard Brownian motion

$$dW^i_t = \sum_{i\in Y} \frac{\sqrt{X^Y_t}}{\sqrt{\sum_{i\in Y} X^Y_t}} dW^Y_t.$$

In the third specification, stable affinity is due to the additive stability of the class of subordinators (see e.g. Crépey, Grbac, and Nguyen (2012)).

8.3 Clean Pricing, Calibration and Hedging

This section briefly discusses the clean pricing, calibration and hedging issues in the common-shock model. In Sect. 8.3.1 we derive the price dynamics for CDS, for

[2]The choice of the word systemic is made here for convenience and has no direct connection with the concept of systemic risk and the related literature.

[3]I.e. nondecreasing Lévy subordinators (see e.g. Jeanblanc, Yor, and Chesney (2010)).

the wealth of so called rolling CDS strategies (see the beginning of Sect. 7.3.3) and for CDO tranches. In Sect. 8.3.2 these dynamics are used for deriving min-variance hedging strategies. In Sect. 8.3.3 we derive fast, semi-explicit pricing schemes for CDO tranches. Note that semi-explicit exponential-affine formulas are readily available for CDS contracts given the affine intensity structure postulated in the assumption 8.2.10(i).

As we did in the previous chapter (see sect. 7.3), we assume zero interest rates in the theoretical part, for notational simplicity. Time-dependent deterministic interest rates will be used in the numerical part.

8.3.1 Pricing Equations

In this subsection we derive price dynamics for CDS contracts and CDO tranches in the common-shock model. All prices are considered from the perspective of the protection buyers. These dynamics will be useful for deriving the min-variance hedging strategies in Sect. 8.3.2.

Let S_i denote the T-year contractual CDS spread on name i, with deterministic recovery rate $R_i \in [0, 1)$. Let S denote the T-year contractual CDO tranche spread for the tranche $[a, b]$, with tranche loss function $L_{a,b}$ (see the introduction to the current part of the book). Below, the notation is the same as in the Itô formula (8.20).

Proposition 8.3.1 **(i)** *The clean price P^i and cumulative dividend clean price $\widehat{P}^i$ of the single-name CDS on name i are given, for $t \in [0, T]$, by:*

$$\begin{aligned} P^i_t &= v_i(t, \mathbf{X}_t, \mathbf{H}_t), \\ d\widehat{P}^i_t &= \mathcal{B}_{\mathbf{X}} v_i(t, \mathbf{X}_{t-}, \mathbf{H}_{t-}) \cdot d\mathbf{M}_t + \sum_{Z \in \mathcal{Z}_t} \Delta^i_Z v_i(t, \mathbf{X}_t, \mathbf{H}_{t-}) dM^Z_t, \end{aligned} \tag{8.26}$$

where

$$\Delta^i_Z v_i(t, \mathbf{x}, \mathbf{k}) = \mathbb{1}_{i \in Z} \left((1 - R_i) - v_i(t, \mathbf{x}, \mathbf{k}) \right) \tag{8.27}$$

for a suitable pricing function v_i.
(ii) *The clean price process P and the clean cumulative dividend price price process $\widehat{P}$ of a CDO tranche $[a, b]$ are given, for $t \in [0, T]$, by:*

$$\begin{aligned} P_t &= u(t, \mathbf{X}_t, \mathbf{H}_t), \\ d\widehat{P}_t &= \mathcal{B}_{\mathbf{X}} u(t, \mathbf{X}_{t-}, \mathbf{H}_{t-}) \cdot d\mathbf{M}_t + \sum_{Z \in \mathcal{Z}_t} \Delta_Z u(t, \mathbf{X}_t, \mathbf{H}_{t-}) dM^Z_t, \end{aligned} \tag{8.28}$$

where

$$\Delta_Z u(t, \mathbf{x}, \mathbf{k}) = \delta_Z u(t, \mathbf{x}, \mathbf{k}) + L_{a,b}(\mathbf{k}^Z) - L_{a,b}(\mathbf{k}) \tag{8.29}$$

for a suitable pricing function u.

Proof. The first lines in (8.26) and (8.28) follow from the Markov property of the model $(\mathbf{X}_t, \mathbf{H}_t)$. The second lines readily follow using the Itô formula (8.20) and the martingale properties of cumulative prices (for $r = 0$). □

The pricing functions v_i and u satisfy related Kolmogorov equations. We don't state these since for n greater than a few units they are numerically intractable. We will only use the structure of the pricing equations provided by the above proposition. For numerical purposes we will rely on the affine properties of our intensity setup regarding CDS contracts, and on the common-shock semi-explicit pricing schemes of Sect. 8.3.3 for CDO tranches.

8.3.1.1 Rolling CDS

As explained at the beginning of Sect. 7.3.3, an already initiated CDS contract (of a given contractual spread in particular) cannot be traded dynamically in the market, What one effectively uses for hedging is a rolling CDS, i.e. a self-financing trading strategy in market CDS contracts. The following result is a version of Lemma 2.4 in Bielecki, Jeanblanc, and Rutkowski (2008) adapted to the present set-up (see also Lemma 2.2 in Bielecki, Crépey, Jeanblanc, and Rutkowski (2011)).

Proposition 8.3.2 *The clean price Q^i and the cumulative clean price dynamics $d\widehat{Q}^i$ of a rolling CDS on name i are such that, for $t \in [0, \tau_i \wedge T] : Q^i_t = 0$ and*

$$d\widehat{Q}^i_t = \big(\mathcal{B}_{\mathbf{x}} p_i(t, \mathbf{X}_{t-}) - S_i(t, \mathbf{X}_{t-}) \mathcal{B}_{\mathbf{x}} f_i(t, \mathbf{X}_{t-})\big) \cdot d\mathbf{M}_t + (1 - R_i) \sum_{Z \in \mathcal{Z}_t; i \in Z} dM^Z_t, \tag{8.30}$$

where p_i and f_i are given as

$$f_i(t, \mathbf{X}_t) = \mathbb{E}\left(\int_t^T e^{-\int_t^s \lambda^i_\zeta d\zeta} ds \,\Big|\, \mathbf{X}_t \right),$$
$$p_i(t, \mathbf{X}_t) = (1 - R_i) \mathbb{E}\left(\int_t^T e^{-\int_t^s \lambda^i_\zeta d\zeta} \lambda^i_s ds \,\Big|\, \mathbf{X}_t \right)$$

and where $S_i(t, \mathbf{x}) = p_i(t, \mathbf{x}) / f_i(t, \mathbf{x})$ is the corresponding CDS fair spread function.

Note that for a given contracted spread, say S_i, the pricing function v_i used in the statement of Proposition 8.3.1 is in fact given by

$$v_i(t, \mathbf{x}, \mathbf{k}) = p_i(t, \mathbf{x}, \mathbf{k}) - S_i f_i(t, \mathbf{x}, \mathbf{k}) =: v_i^{S_i}(t, \mathbf{x}, \mathbf{k}). \tag{8.31}$$

Remark 8.3.3 CDS dynamics are different from the dynamics of the corresponding rolling CDS, unless the market CDS spread S_i is constant over time (and thus equal to the contracted spread), in which case $v_i = 0$. However, under our assumptions in this book the probability of a default at any constant time is zero, so that, in particular, $dM^Z_0 = 0$ for all Z. Therefore, assuming that a CDS is contracted at its fair spread at time zero, we see that its dynamics are the same as the dynamics of the corresponding rolling CDS at time $t = 0$.

8.3.2 Min-Variance Hedging

In this subsection we use the price dynamics from Sect. 8.3.1 to derive min-variance hedging strategies of a CDO tranche using rolling CDS. As explained in the beginning of Sect. 7.3.3, dynamic hedging using CDS contracts is not feasible in practice, unless CDS contracts can be traded in an exchange, where existing contracts would be bought or sold without marking to market, so that the spread is constant. But dynamic hedging using rolling CDS is a proper mathematical model for dynamic hedging that would be done by means of terminating existing CDS contracts and entering into new ones as either protection seller or protection buyer. However, the remark 8.3.3 implies that at time 0 the min-variance hedge in rolling CDS coincides with the min-variance hedge which would be based on standard (non-rolling) CDS issued at par at inception.

8.3.2.1 Hedging of a CDO Tranche Using Rolling CDS Contracts

Consider a CDO tranche with price process $P_t = u(t, \mathbf{X}_t, \mathbf{H}_t)$ to be hedged by the first d rolling CDS contracts with cumulative dividend price processes $\widehat{Q}^i_t$, $i = 1, \dots, d$, as given in Proposition 8.3.2. We write $\widehat{\mathbf{Q}} = (\widehat{Q}^i)_{1 \le i \le d}$ and, for any real vector $\mathbf{S} = (S_i)_{1 \le i \le d}$,

$$\mathbf{v}^{\mathbf{S}}(t, \mathbf{x}, \mathbf{k}) = (p_i(t, \mathbf{x}, \mathbf{k}) - S_i f_i(t, \mathbf{x}, \mathbf{k}))_{1 \le i \le d} = \left(v_i^{S_i}(t, \mathbf{x}, \mathbf{k})\right)_{1 \le i \le d},$$
$$\boldsymbol{\Delta}_Z \mathbf{v}^{\mathbf{S}}(t, \mathbf{x}, \mathbf{k}) = \left(\mathbb{1}_{i \in Z}\left((1 - R_i) - (p_i(t, \mathbf{x}, \mathbf{k}) - S_i f_i(t, \mathbf{x}, \mathbf{k}))\right)\right)_{1 \le i \le d}.$$

Letting also $\mathbf{S}_t = \mathbf{S}(t, \mathbf{X}_{t-}, \mathbf{H}_{t-})$, in view of (8.30) we have:

$$d\widehat{\mathbf{Q}}_t = \left(\mathcal{B}_{\mathbf{x}} \mathbf{v}^{\mathbf{S}}(t, \mathbf{X}_{t-}, \mathbf{H}_{t-})\right)_{|\mathbf{S}=\mathbf{S}_t} \cdot d\mathbf{M}_t + \sum_{Z \in \mathcal{Z}_t} \left(\boldsymbol{\Delta}_Z \mathbf{v}^{\mathbf{S}}(t, \mathbf{X}_t, \mathbf{H}_{t-})\right)_{|\mathbf{S}=\mathbf{S}_t} dM_t^Z. \tag{8.32}$$

Let ζ be a d-dimensional row-vector process, representing the number of units held in the first d rolling CDS contracts which are used in a self-financing hedging strategy for the CDO tranche (using the constant asset as savings account). Given (8.28) and (8.32), the tracking error (e_t) of the hedged portfolio satisfies $e_0 = 0$ and, for $t \in [0, T]$ (cf. (3.28) with $\beta = 1$ here),

$$\begin{aligned} de_t &= d\widehat{P}_t - \zeta_t d\widehat{\mathbf{Q}}_t \\ &= \left(\mathcal{B}_{\mathbf{x}} u(t, \mathbf{X}_{t-}, \mathbf{H}_{t-}) - \zeta_t \left(\mathcal{B}_{\mathbf{x}} \mathbf{v}^{\mathbf{S}}(t, \mathbf{X}_{t-}, \mathbf{H}_{t-})\right)_{|\mathbf{S}=\mathbf{S}_t}\right) \cdot d\mathbf{M}_t \\ &\quad + \sum_{Z \in \mathcal{Z}_t} \left(\Delta_Z u(t, \mathbf{X}_t, \mathbf{H}_{t-}) - \zeta_t \left(\boldsymbol{\Delta}_Z \mathbf{v}^{\mathbf{S}}(t, \mathbf{X}_t, \mathbf{H}_{t-})\right)_{|\mathbf{S}=\mathbf{S}_t}\right) dM_t^Z. \end{aligned} \tag{8.33}$$

Since the martingale dimension of the model is exponential in n, replication is typically out of reach[4] in the common-shock model. However, in view of (8.33), we can find min-variance hedging formulas.

[4]See the comments following Proposition 8.2.8.

Proposition 8.3.4 *The min-variance hedging strategy* ζ^{va} *is given as*

$$\zeta_t^{va} = \frac{d\langle \widehat{P}, \widehat{\mathbf{Q}}\rangle_t}{dt}\left(\frac{d\langle \widehat{\mathbf{Q}}\rangle_t}{dt}\right)^{-1} = \zeta(t, \mathbf{X}_t, \mathbf{H}_{t-}) \tag{8.34}$$

with

$$\zeta_{va}(t, \mathbf{x}, \mathbf{k}) = \left(\left(\mathcal{C}(u, \mathbf{v}^{\mathbf{S}})\mathcal{C}(\mathbf{v}^{\mathbf{S}}, \mathbf{v}^{\mathbf{S}})^{-1}\right)(t, \mathbf{x}, \mathbf{k})\right)_{|\mathbf{S}=\mathbf{S}(t,\mathbf{x},\mathbf{k})},$$

where for any $\mathbf{S}$:

$$\begin{aligned} \mathcal{C}(u, \mathbf{v}^{\mathbf{S}}) &= \mathcal{C}_{\mathbf{x}}(u, \mathbf{v}^{\mathbf{S}}) + \sum_{Y\in\mathcal{Y}} \lambda_Y \Delta_Y u (\boldsymbol{\Delta}_Y \mathbf{v}^{\mathbf{S}})^{\mathsf{T}}, \\ \mathcal{C}(\mathbf{v}^{\mathbf{S}}, \mathbf{v}^{\mathbf{S}}) &= \mathcal{C}_{\mathbf{x}}(\mathbf{v}^{\mathbf{S}}, \mathbf{v}^{\mathbf{S}}) + \sum_{Y\in\mathcal{Y}} \lambda_Y \boldsymbol{\Delta}_Y \mathbf{v}^{\mathbf{S}} (\boldsymbol{\Delta}_Y \mathbf{v}^{\mathbf{S}})^{\mathsf{T}}, \end{aligned} \tag{8.35}$$

in which the row-vector $\mathcal{C}_{\mathbf{x}}(u, \mathbf{v}^{\mathbf{S}})$ *and the square matrix* $\mathcal{C}_{\mathbf{x}}(\mathbf{v}^{\mathbf{S}}, \mathbf{v}^{\mathbf{S}})$ *are defined componentwise based on the following expression for the* carré du champ *of two real functions* f *and* g *of* $\mathbf{x}$ *(see* (13.40)-(13.41)*):*

$$\mathcal{C}_{\mathbf{x}}(f, g) = \mathcal{A}_{\mathbf{x}}(fg) - f\mathcal{A}_{\mathbf{x}}g - g\mathcal{A}_{\mathbf{x}}f. \tag{8.36}$$

In particular, ζ_0^{va} *coincides with the time-0 min-variance hedge that would be based on standard (non-rolling) CDS contracts issued at par at inception.*

Proof. The first identity in (8.34) is a classical risk-neutral min-variance hedging regression formula (see Sect. 3.2.5.1). Moreover, in view of (8.28) and (8.32) and assuming for notational simplicity that only one rolling CDS, say the i-th one, is used for hedging so that $\widehat{\mathbf{Q}} = \widehat{Q}^i$ etc., we have:

$$
\begin{aligned}
d\langle \widehat{P}, \widehat{Q}^i \rangle_t &= d\left\langle \int_0^{\cdot} \mathcal{B}_{\mathbf{x}} u(t, \mathbf{X}_{t-}, \mathbf{H}_{t-}) \cdot d\mathbf{M}_t + \sum_{Z \in \mathcal{Z}_t} \Delta_Z u(t, \mathbf{X}_t, \mathbf{H}_{t-}) dM_t^Z, \right. \\
&\qquad \left. \int_0^{\cdot} \left(\mathcal{B}_{\mathbf{x}} v_i^{S_i}(t, \mathbf{X}_{t-}, \mathbf{H}_{t-})\right)_{|_{S_i = S_t^i}} \cdot d\mathbf{M}_t + \sum_{Z \in \mathcal{Z}_t} \left(\Delta_Z^i v_i^{S_i}(t, \mathbf{X}_t, \mathbf{H}_{t-})\right)_{|_{S_i = S_t^i}} dM_t^Z \right\rangle_t \\
&= d\left\langle \int_0^{\cdot} \mathcal{B}_{\mathbf{x}} u(t, \mathbf{X}_{t-}, \mathbf{H}_{t-}) \cdot d\mathbf{M}_t + \sum_{Z \in \mathcal{Z}_t} \Delta_Z u(t, \mathbf{X}_t, \mathbf{H}_{t-}) dM_t^Z, \right. \\
&\qquad \int_0^{\cdot} \mathcal{B}_{\mathbf{x}}((1 - R_i) - p_i(t, \mathbf{X}_t, \mathbf{H}_{t-})) \cdot d\mathbf{M}_t - S_t^i \mathcal{B}_{\mathbf{x}} f_i(t, \mathbf{X}_{t-}, \mathbf{H}_{t-}) \cdot d\mathbf{M}_t \\
&\qquad \left. + \sum_{Z \in \mathcal{Z}_t; i \in Z} ((1 - R_i) - p_i(t, \mathbf{X}_t, \mathbf{H}_{t-})) dM_t^Z + S_t^i \sum_{Z \in \mathcal{Z}_t; i \in Z} f_i(t, \mathbf{X}_t, \mathbf{H}_{t-}) dM_t^Z \right\rangle_t \\
&= d\left\langle \int_0^{\cdot} \mathcal{B}_{\mathbf{x}} u(t, \mathbf{X}_{t-}, \mathbf{H}_{t-}) \cdot d\mathbf{M}_t + \sum_{Z \in \mathcal{Z}_t} \Delta_Z u(t, \mathbf{X}_t, \mathbf{H}_{t-}) dM_t^Z, \right. \\
&\qquad \left. \int_0^{\cdot} \mathcal{B}_{\mathbf{x}}((1 - R_i) - p_i(t, \mathbf{X}_t, \mathbf{H}_{t-})) \cdot d\mathbf{M}_t + \sum_{Z \in \mathcal{Z}_t; i \in Z} ((1 - R_i) - p_i(t, \mathbf{X}_t, \mathbf{H}_{t-})) dM_t^Z \right\rangle_t \\
&\quad - S_t^i d\left\langle \int_0^{\cdot} \mathcal{B}_{\mathbf{x}} u(t, \mathbf{X}_{t-}, \mathbf{H}_{t-}) \cdot d\mathbf{M}_t + \sum_{Z \in \mathcal{Z}_t} \Delta_Z u(t, \mathbf{X}_t, \mathbf{H}_{t-}) dM_t^Z, \right. \\
&\qquad \left. - \int_0^{\cdot} \mathcal{B}_{\mathbf{x}} f_i(t, \mathbf{X}_{t-}, \mathbf{H}_{t-}) \cdot d\mathbf{M}_t + \sum_{Z \in \mathcal{Z}_t; i \in Z} f_i(t, \mathbf{X}_t, \mathbf{H}_{t-}) dM_t^Z \right\rangle_t \\
&= \mathcal{C}_{\mathbf{x}}(u, (1 - R_i) - p_i)(t, \mathbf{X}_t, \mathbf{H}_t) dt + \sum_{Z \in \mathcal{Z}_t; i \in Z} \lambda_Z \Delta_Z u(t, \mathbf{X}_{t-}, \mathbf{H}_{t-}) p_i(t, \mathbf{X}_{t-}, \mathbf{H}_{t-}) dt \\
&\quad - S_t^i \left(\mathcal{C}_{\mathbf{x}}(u, f_i)(t, \mathbf{X}_t, \mathbf{H}_t) dt + \sum_{Z \in \mathcal{Z}_t; i \in Z} \lambda_Z \Delta_Z u(t, \mathbf{X}_{t-}, \mathbf{H}_{t-}) f_i(t, \mathbf{X}_{t-}, \mathbf{H}_{t-}) \right) dt \\
&= \mathcal{C}_{\mathbf{x}}(u, (1 - R_i) - p_i)(t, \mathbf{X}_t, \mathbf{H}_t) dt \\
&\qquad + \sum_{Z \in \mathcal{Z}_t; i \in Z} \lambda_Z \Delta_Z u(t, \mathbf{X}_{t-}, \mathbf{H}_{t-}) ((1 - R_i) - p_i(t, \mathbf{X}_{t-}, \mathbf{H}_{t-})) dt \\
&\quad - \left(S_i \left(\mathcal{C}_{\mathbf{x}}(u, f_i)(t, \mathbf{X}_t, \mathbf{H}_t) dt + \sum_{Z \in \mathcal{Z}_t; i \in Z} \lambda_Z \Delta_Z u(t, \mathbf{X}_{t-}, \mathbf{H}_{t-}) f_i(t, \mathbf{X}_{t-}, \mathbf{H}_{t-}) \right) \right)_{|_{S_i = S_t^i}} dt \\
&= \left(\mathcal{C}_{\mathbf{x}}(u, v_i^{S_i})(t, \mathbf{X}_t, \mathbf{H}_t) \right)_{|S_i = S_t^i} dt \\
&\qquad + \sum_{Z \in \mathcal{Z}_t} \lambda_Z \Delta_Z u(t, \mathbf{X}_{t-}, \mathbf{H}_{t-}) \left(\Delta_Z^i v_i^{S_i}(t, \mathbf{X}_{t-}, \mathbf{H}_{t-}) \right)_{|S_i = S_t^i} dt \\
&= \left(\mathcal{C}(u, v_i^{S_i})(t, \mathbf{X}_{t-}, \mathbf{H}_{t-}) \right)_{|S_i = S_t^i} dt,
\end{aligned}
$$

where the last identity uses a simplification in the sum over Z similar to that observed in the proof of the Itô formula (8.20). Similarly

$$\frac{d\langle\widehat{\mathbf{Q}}\rangle_t}{dt} = \big(\mathcal{C}(\mathbf{v}^{\mathbf{S}}, \mathbf{v}^{\mathbf{S}})(t, \mathbf{X}_t, \mathbf{H}_t)\big)_{|\mathbf{S}=\mathbf{S}_t}. \quad (8.37)$$

Finally, the statement that ζ_0^{va} coincides with the min-variance hedging strategy that would be based on standard (non-rolling) CDS contracts issued at par is justified by the remark 8.3.3. □

See (13.41) for the explicit expressions of the *carré du champ* (8.36) of two functions f and g under a jump-diffusion specification for X. Then, in (8.35), the u-related terms can be computed by the explicit schemes of Sect. 8.3.3 and 8.3.4; the v_i-related terms are also explicit under our affine intensity structure. We will illustrate in Sect. 8.4.5 the tractability of this approach for computing min-variance hedging deltas. An important point is that, due to the specific structure of the intensities, the sums in (8.35) are over the set $\mathcal{Y}$ of shocks, of cardinality $(n+m)$, rather than over the set $\mathcal{Z}$ of all the set-events Z, of cardinality 2^n.

8.3.3 Convolution Recursion Pricing Schemes

In this subsection we derive exact fast recursive convolution algorithms for the portfolio loss distribution in our common-shock model. In the case where the recovery rate is the same for all names, i.e. $R_i = R$, the aggregate loss L_t is equal to $(1-R)N_t$, where N_t is the total number of defaults that have occurred in the model up to time t. Therefore, the time-t price of CDO tranches is determined by the probabilities $\mathbb{Q}\left[N_\theta = k \,|\, \mathcal{G}_t\right]$ for $k = |\mathbf{H}_t|, \dots, n$ and $\theta \geq t \geq 0$. Using the conditional common-shock model of Sect. 8.2.1, it follows from Proposition 8.2.6(ii) that, on the event $\{\mathrm{supp}^c(\mathbf{H}_t) = Z\}$, we have:

$$\mathbb{Q}\left[N_\theta = k \,|\, \mathcal{G}_t\right] = \mathbb{Q}\left[N_\theta(t, Z) = k \,|\, \mathbf{X}_t\right].$$

We will focus on computation of the latter probabilities, which are derived in formula (8.39) below.

Henceforth we assume a nested structure of the sets I_j, first proposed in Bielecki, Vidozzi, and Vidozzi (2008b) and given by:

$$I_1 \subset \ldots \subset I_m. \quad (8.38)$$

As we will detail in the remark 8.3.5, the nested structure (8.38) yields a particularly tractable expression for the portfolio loss distribution. This nested structure also makes sense financially with regard to the hierarchical structure of risks which is reflected in standard CDO tranches. Of course a dynamic group structure would be preferable from a financial point of view. In the same vein one could criticize the absence of direct contagion effects in this model (even if a stylized form of "instantaneous contagion" is present through joint defaults). However, it should be stressed that we are building a stylized pricing model. With counterparty credit risk applications in mind, efficient time-t pricing schemes, along with efficient time-0 calibration to joint CDS and CDO data, are the main issues.

We set $I_0 = \emptyset$ and $K_\theta^{I_0}(t) = 1$ (cf. (8.13)). In view of (8.38), the events

$$\Omega_\theta^j(t) := \{K_\theta^{I_j}(t) = 1, K_\theta^{I_{j+1}}(t) = 0, \ldots, K_\theta^{I_m}(t) = 0\}, \ 0 \le j \le m$$

form a partition of Ω. Hence[5]:

$$\mathbb{Q}(N_\theta(t, Z) = k \,|\, \mathbf{X}_t) = \sum_{0 \le j \le m} \mathbb{Q}\big(N_\theta(t, Z) = k \,|\, \Omega_\theta^j(t), \mathbf{X}_t\big)\mathbb{Q}\big(\Omega_\theta^j(t) \,|\, \mathbf{X}_t\big), \quad (8.39)$$

where, by construction of the $K_\theta^I(t)$, we have:

$$\begin{aligned}
\mathbb{Q}\big(\Omega_\theta^j(t) \,|\, \mathbf{X}_t\big) &= \mathbb{E}\Big(\mathbb{Q}\big(\eta_{I_j}(t) \le \theta, \min_{j+1 \le l \le m} \eta_{I_l}(t) > \theta \,\big|\, (\mathbf{X}_s)_{t \le s \le \theta}\big)\Big|\mathbf{X}_t\Big) \\
&= \mathbb{E}\Big(\big(1 - e^{-\Lambda_{t,\theta}^{I_j}}\big)e^{-\sum_{j+1 \le l \le m} \Lambda_{t,\theta}^{I_l}}\Big|\mathbf{X}_t\Big) \qquad (8.40) \\
&= \mathbb{E}\Big(e^{-\sum_{j+1 \le l \le m} \Lambda_{t,\theta}^{I_l}} - e^{-\sum_{j \le l \le m} \Lambda_{t,\theta}^{I_l}}\Big|\mathbf{X}_t\Big).
\end{aligned}$$

The last expression is given in explicit form in view of the affine intensity structure postulated in the assumption 8.2.10.

We now turn to the computation of the term

$$\mathbb{Q}\big(N_\theta(t, Z) = k \,|\, \Omega_\theta^j(t), \mathbf{X}_t\big) \quad (8.41)$$

appearing in (8.39). Recall first that $N_\theta(t, Z) = n - |Z| + \sum_{i \in Z} H_\theta^i(t)$. We know that for any group $j = 1, \ldots, m$, given $\Omega_\theta^j(t)$, the marginal default indicators $H_\theta^i(t)$ for $i \in Z$ are such that:

$$H_\theta^i(t) = \begin{cases} 1, & i \in I_j, \\ K_\theta^{\{i\}}(t), & \text{otherwise.} \end{cases} \quad (8.42)$$

Consequently,

$$\mathbb{Q}\big(N_\theta(t, Z) = k \,|\, \Omega_\theta^j(t), \mathbf{X}_t\big) = \mathbb{Q}\big(n - |Z \backslash I_j| + \sum_{i \in Z \backslash I_j} K_\theta^{\{i\}}(t) = k \,|\, \Omega_\theta^j(t), \mathbf{X}_t\big). \quad (8.43)$$

Moreover, by the lemma 8.2.11(ii), the $K_\theta^{\{i\}}(t)$ are conditionally independent of $\Omega_\theta^j(t)$ given $\mathbf{X}_t$, so that

$$\begin{aligned}
&\mathbb{Q}\big(n - |Z \setminus I_j| + \sum_{i \in Z \backslash I_j} K_\theta^{\{i\}}(t) = k \,|\, \Omega_\theta^j(t), \mathbf{X}_t\big) = \\
&\qquad \mathbb{Q}\big(n - |Z \setminus I_j| + \sum_{i \in Z \backslash I_j} K_\theta^{\{i\}}(t) = k \,|\, \mathbf{X}_t\big).
\end{aligned} \quad (8.44)$$

[5]The expression $P(A|B, X)$, where A, B are event-sets and X is a random variable, is to be interpreted as $P(A|\mathbb{1}_B, X)$.

Finally, again by the lemma 8.2.11(ii), for $i \neq l$, $K_\theta^{\{i\}}(t)$ and $K_\theta^{\{l\}}(t)$ are conditionally independent given $\mathbf{X}_t$. Therefore, conditionally on $\Omega_\theta^j(t)$ and $\mathbf{X}_t$, $\mathbf{H}_\theta(t) = (H_\theta^i(t))_{i=1,\cdots,n}$ is a vector of independent Bernoulli random variables with parameter $p = (p_\theta^{i,j}(t))_{i=1,\cdots,n}$, where

$$p_\theta^{i,j}(t) = \begin{cases} 1, & i \in I_j, \\ 1 - \mathbb{E}\left\{\exp\left(-\Lambda_{t,\theta}^{\{i\}}\right) \mid \mathbf{X}_t\right\} & \text{otherwise.} \end{cases} \tag{8.45}$$

Note that in view of the assumption 8.2.10 (i) we have:

$$\mathbb{E}\left\{\exp\left(-\Lambda_{t,\theta}^{\{i\}}\right) \mid \mathbf{X}_t\right\} = \mathbb{E}\left\{\exp\left(-\Lambda_{t,\theta}^{\{i\}}\right) \mid \lambda_t^{\{i\}}\right\}, \tag{8.46}$$

which is explicit in our affine intensities framework. All this allows for exact computation of the probability in (8.41) by a standard recursive convolution procedure (see e.g. Andersen and Sidenius (2004) or Crépey (2013)).

Remark 8.3.5 The fact that there are only m terms in the sum (8.39) is due to the nested structure of the groups I_j in (8.39). A recursive convolution procedure would be possible for an arbitrary structuring of the groups I_j. However, a general structuring of the m groups I_j would imply 2^m terms instead of m in the sum of (8.39), which in practice would only work for very few groups m. Again, we emphasize that the nested structure (8.39) of the I_j, or equivalently the tranched structure of the $I_j \setminus I_{j-1}$, is quite natural from the financial point of view (especially with application to CDO tranches in mind).

8.3.4 Random Recoveries

In this subsection we outline how to generalize the model to the case of random recoveries. The implementation details are omitted and we refer the readers to Bielecki, Cousin, Crépey, and Herbertsson (2013b) for a more comprehensive description.

Let $\mathbf{L} = (L^i)_{1 \le i \le n}$ represent the $[0,1]^n$-valued vector process of the losses given defaults in the pool of names. The process $\mathbf{L}$ is a multivariate process, where each component L_t^i represents the fractional loss that name i may have suffered due to default until time t (in particular, $\mathbf{L}_0 = \mathbf{0}$). Assuming a unit notional for each name, the cumulative loss process for the entire portfolio is given by $L_t = \frac{1}{n}\sum_{i=1}^n L_t^i$. We postulate that random recoveries for all the obligors are i.i.d. and independent from any other randomness in the model. The default times are defined as before, but at every time of jump of $\mathbf{H}$, an independent draw is made for every newly defaulted name i, determining the recovery R_i of name i. In particular, the recovery rates resulting from a joint default are drawn independently for the affected names. By introducing stochastic recoveries, we can no longer use the exact recursive convolution procedures of Sect. 8.3.3 for pricing CDO tranches. Instead, we use an approximate procedure based on exponential approximation of the so called hockey stick function, as presented in Iscoe, Jackson, Kreinin, and Ma (2010, 2013) based on Beylkin and Monzon (2005). We now briefly outline how to use this method for computing the price of a CDO tranche in our model.

Recall that the loss process for the tranche $[a, b]$ is given by $L_t^{a,b} = (L_t - a)^+ - (L_t - b)^+$, which can be rewritten as

$$L_t^{a,b} = b\left(1 - h\left(\frac{L_t}{b}\right)\right) - a\left(1 - h\left(\frac{L_t}{a}\right)\right), \tag{8.47}$$

where $h(x)$ is the so-called hockey stick function given by

$$h(x) = \begin{cases} 1 - x & \text{if } 0 \leq x \leq 1, \\ 0 & \text{if } 1 < x. \end{cases} \tag{8.48}$$

It is shown in Beylkin and Monzon (2005) that, for any fixed $\epsilon > 0$, the function $h(x)$ can be approximated by a function $h_{\text{exp}}^{(q)}(x)$ on $[0, d]$ for any $d > 0$ so that $|h(x) - h_{\text{exp}}^{(q)}(x)| \leq \epsilon$ for all $x \in [0, d]$, where $q = q(\epsilon)$ is positive integer and $h_{\text{exp}}^{(q)}(x)$ is given by

$$h_{\text{exp}}^{(q)}(x) = \sum_{\ell=1}^{q} \omega_\ell \exp\left(\gamma_\ell \frac{x}{d}\right). \tag{8.49}$$

Here $(\omega_\ell)_{\ell=1}^q$ and $(\gamma_\ell)_{\ell=1}^q$ are complex numbers obtained as roots of polynomials whose coefficients can be computed numerically in a straightforward way. Fig. 8.2 visualizes the approximation $h_{\text{exp}}^{(q)}(x)$ of $h(x)$ for $d = 2$ on $x \in [0, 10]$ with $q = 5$ and $q = 50$, as well as the approximation error $|h(x) - h_{\text{exp}}^{(q)}(x)|$ for the same q. As can be seen in Fig. 8.2, the approximation is already fairly good for small values values of q and also works well on the entire interval $[0, 10]$, not just on $[0, d] = [0, 2]$. In the rest of this chapter we use $d = 2$ in (8.49), as in Iscoe, Jackson, Kreinin, and Ma (2010, 2013).

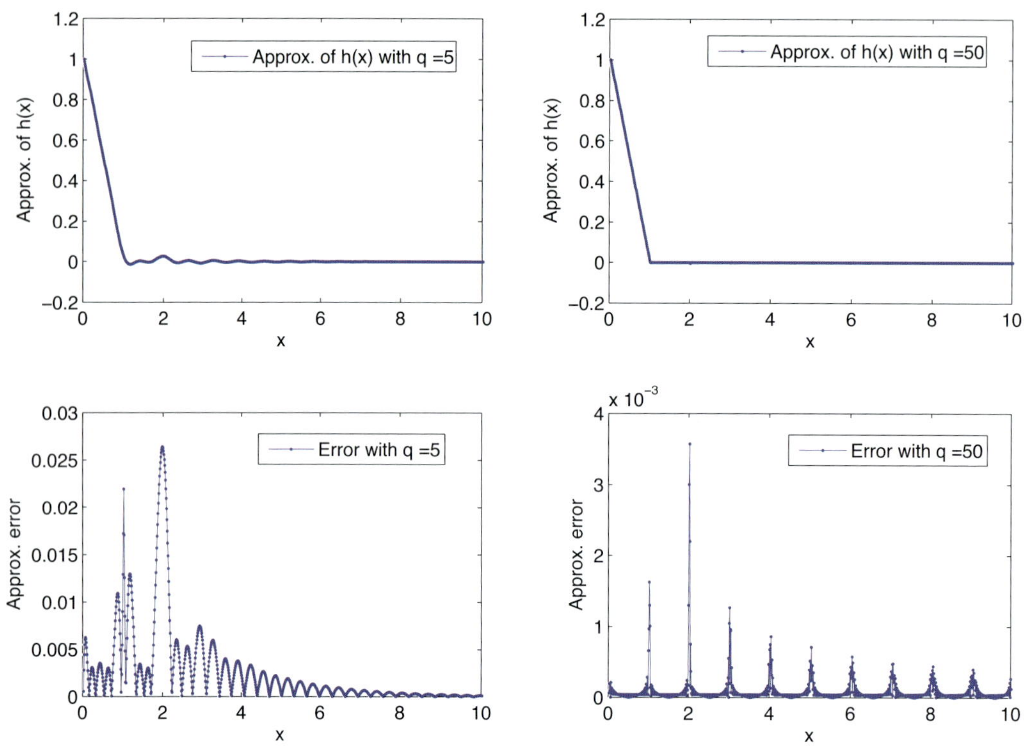

FIGURE 8.2: The function $h_{\text{exp}}^{(q)}(x)$ as approximation of $h(x)$ for $x \in [0, 10]$ with $q = 5$ and $q = 50$ *(top)* and the corresponding approximation errors *(bottom)*.

In view of (8.47)-(8.49), for any two time points $\theta > t$ the time-t pricing of a CDO tranche reduces to computation of conditional expectations of the form

$$\mathbb{E}\Big(e^{\gamma_\ell \frac{L_\theta}{2c}} \,|\, \mathcal{G}_t\Big), \tag{8.50}$$

for $\ell = 1, 2, \ldots, q$ and different attachment points c and time horizons $\theta > t$. Note that the case $t = 0$ is used in the calibration, while the case $t > 0$ is needed for CVA computations. Since the algorithm for computing $\mathbb{E}\Big(e^{\gamma_\ell \frac{L_\theta}{2c}} \,|\, \mathcal{G}_t\Big)$ is the same for each $\ell = 1, 2, \ldots, q$ and any attachment point c, we will henceforth for notational convenience write $\mathbb{E}\big(e^{\gamma L_\theta} \,|\, \mathcal{G}_t\big)$ instead of $\mathbb{E}\Big(e^{\gamma_\ell \frac{L_\theta}{2c}} \,|\, \mathcal{G}_t\Big)$.

Now, using the conditional common-shock model of Sect. 8.3.3, much like in Proposition 8.2.6(ii) we have, for $Z \in \mathcal{Z}$,

$$\mathbb{1}_{\{\text{supp}^c(\mathbf{L}_t)=Z\}}\mathbb{E}\big(e^{\gamma L_\theta} \,|\, \mathcal{G}_t\big) = \mathbb{1}_{\{\text{supp}^c(\mathbf{L}_t)=Z\}}\mathbb{E}\big(e^{\gamma L_\theta(t,Z)} \,|\, \mathbf{X}_t\big), \tag{8.51}$$

where $L_\theta(t, Z) := \sum_{i \notin Z} L_t^i + \sum_{i \in Z}(1 - R_i)H_\theta^i(t)$ and $H_\theta^i(t)$ is defined as in (8.42). Furthermore, R_i is a random recovery with values in $[0, 1]$. Then we have, as in (8.39), that

$$\mathbb{E}\big(e^{\gamma L_\theta(t,Z)} \,|\, \mathbf{X}_t\big) = \sum_{0\le j\le m} \mathbb{E}\big(e^{\gamma L_\theta(t,Z)} \,|\, \Omega^j_\theta(t), \mathbf{X}_t\big)\mathbb{Q}\big(\Omega^j_\theta(t) \,|\, \mathbf{X}_t\big), \tag{8.52}$$

where $\mathbb{Q}\big(\Omega^j_\theta(t) \,|\, \mathbf{X}_t\big)$ is given by (8.40). Moreover, adapting in an obvious way the argument leading to (8.45) and (8.46), we obtain on the set $\{\mathrm{supp}^c(\mathbf{L}_t) = Z\}$ that

$$\begin{aligned}
&\mathbb{E}\big(e^{\gamma L_\theta(t,Z)} \,|\, \Omega^j_\theta(t), \mathbf{X}_t\big)\\
&= e^{\gamma \sum_{i\notin Z} L^i_t}\mathbb{E}\big(e^{\gamma \sum_{i\in Z}(1-R_i)H^i_\theta(t)} \,|\, \Omega^j_\theta(t), \mathbf{X}_t\big)\\
&= e^{\gamma \sum_{i\notin Z} L^i_t}\prod_{i\in Z}\mathbb{E}\big(e^{\gamma(1-R_i)H^i_\theta(t)} \,|\, \Omega^j_\theta(t), \mathbf{X}_t\big)
\end{aligned}$$

where, for any i:

$$\mathbb{E}\big(e^{\gamma(1-R_i)H^i_\theta(t)} \,|\, \Omega^j_\theta(t), \mathbf{X}_t\big) = \begin{cases} \mathbb{E}\big(e^{\gamma(1-R_i)}\big), & i \in I_j,\\ \mathbb{E}\big(e^{\gamma(1-R_i)K^{\{i\}}_\theta(t)} \,|\, X^{\{i\}}_t\big), & \text{otherwise.}\end{cases} \tag{8.53}$$

The independence of R_i and $H^i_\theta(t)$ on one hand, and of R_i and $X^{\{i\}}_t$ on the other hand, implies that

$$\mathbb{E}\big(e^{\gamma(1-R_i)K^{\{i\}}_\theta(t)} \,|\, X^{\{i\}}_t\big) = 1 - p^{i,j}_\theta(t)\left(1 - \mathbb{E}e^{\gamma(1-R_i)}\right), \tag{8.54}$$

where $p^{i,j}_\theta(t)$ was defined in (8.45).

In Sect. 8.4.2 we will give an example of a random recovery distribution that will be used with the above hockey-stick method when calibrating the model against CDO tranches market data. As will be seen in Sect. 8.4.3, using stochastic recoveries will, for some data sets, render significantly better calibration results compared with constant recoveries.

8.4 Numerical Results

In this section we briefly discuss the calibration of the model and a few numerical results regarding the loss-distributions and the min-variance hedging. Sect. 8.4.1 outlines the calibration methodology with piecewise-constant default intensities and with constant recoveries, while Sect. 8.4.2 describes the calibration procedure with stochastic recoveries and with piecewise-constant default intensities. Sect. 8.4.3 then presents the numerical calibration of the model against market data, both with constant recoveries and with stochastic recoveries. We also present the implied loss-distributions in our fitted model (for the case with constant recoveries). In Sect. 8.4.4 we consider that individual and joint defaults are driven by stochastic default intensities, and we describe the calibration methodology and results for a particular model specification. Sect. 8.4.5 discusses min-variance hedging sensitivities in calibrated models (using constant recoveries).

We use the convention that the premium legs of all credit products are payed at $t_1 < t_2 < \ldots < t_p = T$, where $t_j - t_{j-1} = \mathrm{h} = 3$ months.

8.4.1 Calibration Methodology with Piecewise Constant Default Intensities and Constant Recoveries

In this subsection we discuss one of the calibration methodologies that will be used in Sect. 8.4.3 for calibrating the model against CDO tranches on the iTraxx Europe and CDX.NA.IG series. This first calibration methodology will use piecewise-constant default intensities and constant recoveries in the convolution pricing algorithm described in Sect. 8.3.3.

The first step is to calibrate the single-name CDS curve for every obligor. Given the T-year market CDS spread S_i^* for obligor i, we want to find the individual default parameters for this obligor so that $P_0^i(S_i^*) = 0$, i.e.

$$S_i^* = \frac{(1 - R_i)\mathbb{Q}\left(\tau_i < T\right)}{\mathrm{h}\sum_{0<t_j\leq T}\mathbb{Q}\left(\tau_i > t_j\right)} \tag{8.55}$$

(for zero interest rates and a constant recovery R_i). Hence the first step is to extract the implied individual survival function $G_i^\star(t) := \mathbb{Q}\left(\tau_i > t\right)$ from the CDS curve of every obligor i by using a standard bootstrapping procedure based on (8.55).

Given the $G_i^\star(t)$, the law of the total number of defaults N_t at a fixed time t is a function of the joint default intensity functions $\lambda_I(t)$, as described by the recursive algorithm of Sect. 8.3.3. The second step is therefore to calibrate the common-shock intensities $\lambda_I(t)$, so that the model CDO tranche spreads coincide with the corresponding market spreads. This is done by pricing the tranches using the recursive algorithm of Sect. 8.3.3, for the $\lambda_I(t)$ parameterized as nonnegative and piecewise-constant functions of time. Moreover, in view of (8.6), for any obligor i and time t we impose the constraint[6]

$$\sum_{I\in\mathcal{I};\, i\in I} \lambda_I(t) \leq \lambda_i^\star(t), \tag{8.56}$$

where $\lambda_i^\star = -\frac{d\ln(G_i^\star)}{dt}$ (the "hazard intensity function" of name i).

For constant systemic intensities $\lambda_I(t) = \lambda_I$, the constraints (8.56) reduce to

$$\sum_{I\ni i}\lambda_I \leq \underline{\lambda}_i := \inf_{t\in[0,T]} \lambda_i^\star(t) \quad \text{for any obligor } i.$$

Given the nested structure of the groups I_j specified in (8.38), this is equivalent to

$$\sum_{j=l}^{m}\lambda_{I_j} \leq \underline{\lambda}_{I_l} := \min_{i\in I_l\setminus I_{l-1}} \underline{\lambda}_i \quad \text{for any group } l. \tag{8.57}$$

[6]We note that $\lambda_i^\star(t)$ corresponds to λ_t^i in (8.6). Here the model intensities are deterministic, and the present notation reflects this.

For piecewise-constant systemic intensities on a time grid (T_k), the condition (8.57) extends to:

$$\sum_{j=l}^{m} \lambda_{I_j}^{k} \leq \underline{\lambda}_{I_l}^{k} := \min_{i \in I_l \setminus I_{l-1}} \underline{\lambda}_i^{k} \quad \text{for any } l, k, \tag{8.58}$$

where $\underline{\lambda}_i^k := \inf_{t \in [T_{k-1}, T_k]} \lambda_i^\star(t)$. Note that calibrating all CDS names in the portfolio, including the safest ones, implies via (8.57) or (8.58) a quite constrained region for the systemic intensities. This region can be expanded by relaxing the system of constraints for the systemic intensities, excluding the safest CDS from the calibration.

We will use a time grid consisting of two maturities T_1 and T_2. Hence, the single-name CDS contracts constituting the entities in the credit portfolio are bootstrapped from their market spreads for $T = T_1$ and $T = T_2$. This is done by using piecewise-constant individual default intensity functions λ_i on the time intervals $[T_0 = 0, T_1]$ and $[T_1, T_2]$.

Before we leave this subsection, we give a few more details on the calibration of the systemic intensities for the m groups, done in the second calibration step. From now on we assume that the systemic intensities $\{\lambda_{I_j}(t)\}_{j=1}^m$ are piecewise-constant, so that $\lambda_{I_j}(t) = \lambda_{I_j}^{(1)}$ for $t \in [0, T_1]$ and $\lambda_{I_j}(t) = \lambda_{I_j}^{(2)}$ for $t \in [T_1, T_2]$ and for any group j. The systemic intensities $\boldsymbol{\lambda} = (\lambda_{I_j}^{(k)})_{j,k} = \{\lambda_{I_j}^{(k)} : j = 1, \ldots, m \text{ and } k = 1, 2\}$ are then calibrated so that the five-year model spread $S_{a,b}(\boldsymbol{\lambda})$ coincides with the corresponding market spread $S_{a,b}^\star$ for each tranche l. More specifically, the parameters $\boldsymbol{\lambda}^\star = (\lambda_{I_j}^{(k)})_{j,k}^\star$ are obtained according to

$$\boldsymbol{\lambda}^\star = \underset{\boldsymbol{\lambda}}{\operatorname{argmin}} \sum_{(a,b)} \left(\frac{S_{a,b}(\boldsymbol{\lambda}) - S_{a,b}^\star}{S_{a,b}^\star} \right)^2 \tag{8.59}$$

under the constraints that all elements in $\boldsymbol{\lambda}$ are nonnegative and that $\boldsymbol{\lambda}$ satisfies the inequalities (8.58) for any group I_l and in each time interval $[T_{k-1}, T_k]$. In $S_{a,b}(\boldsymbol{\lambda})$ we have emphasized that the model spread for tranche $[a, b]$ is a function of $\boldsymbol{\lambda} = (\lambda_{I_j}^{(k)})_{j,k}$ but we suppressed the dependence on other parameters such as interest rate, premium payment frequency or λ_i, $i = 1, \ldots, n$. In the calibration we used an interest rate of 3% and the quarterly premium payments. For each data set we use a constant recovery of 40%. The objective function (8.59) is minimized by using the Matlab optimization routine `fmincon` together with the constraints given by equations (8.58).

In Sect. 8.4.3 we use the above setting for performing calibrations of our two data sets.

8.4.2 Calibration Methodology with Piecewise Constant Default Intensities and Stochastic Recoveries

In this subsection we discuss the second calibration methodology, is used in Sect. 8.4.3 when calibrating the common-shock model with stochastic recoveries (and piecewise-constant default intensities as above). As was explained in Sect. 8.3.4, stochastic recoveries require a more sophisticated method for computing the tranche loss distribution. The methodology and constraints connected to the piecewise-constant default intensities are the same as in Sect. 8.4.1. Therefore we will only discuss in this subsection the distribution for the individual stochastic recoveries R_i, as well as the accompanying constraints used in the calibration. This distribution will determine the quantity $\mathbb{E}\left(e^{\gamma(1-R_i)}\right)$ in (8.54), which is needed for computing the tranche loss distribution.

We assume that the individual recoveries R_i are i.i.d. binomial mixture random variables of the following form:

$$R_i \sim \frac{1}{q}\text{Bin}\left(l, R^\star(p_0 + (1-\varepsilon)p_1)\right), \quad l = 0, \ldots, q, \tag{8.60}$$

where ε is a Bernoulli with $\mathbb{Q}\left[\varepsilon = 1\right] = p$, where $R^\star$, p, p_0 and p_1 are positive constants and where q is an integer (in the numerics we use $q = 10$). As a result, the distribution of recovery rate is given by

$$\begin{aligned} \mathbb{Q}\left[R_i = \frac{l}{q}\right] = \; & p\binom{q}{l}p_0^l(1-p_0)^{q-l} \\ & + (1-p)\binom{q}{l}\left(p_0 + p_1\right)^l\left(1 - (p_0 + p_1)\right)^{q-l}. \end{aligned} \tag{8.61}$$

Moreover, in order to have a calibration of the single-name CDS that is separate from the calibration of the systemic parameters, we impose the constraint $\mathbb{E}\left[R_i\right] = R^\star$, so that $R^\star$ in (8.60) represents the average recovery for each obligor in the portfolio. The condition $\mathbb{E}\left[R_i\right] = R^\star$ leads to the following constraint on the parameters q (see Bielecki, Cousin, Crépey, and Herbertsson (2013b) for a detailed derivation):

$$p < \min\left(1, \frac{1}{p_0}, \frac{1-R^\star}{1-R^\star p_0}\right). \tag{8.62}$$

Furthermore, the constraint $\mathbb{E}\left[R_i\right] = R^\star$ also implies that $p_1 = \frac{1-p_0}{1-p}$, so that p_1 can be viewed as a function of p and p_0. The constraint (8.62), along with some other constraints for the systemic intensities, will be imposed in our calibration of the CDO tranches. In our calibrations the parameters p_0 and $R^\star$ will be treated as given exogenously, where we set $R^\star = 40\%$, while p_0 can be any positive scalar satisfying $p_0 < \frac{1}{R^\star}$. The scalar p_0 will give us some freedom to fine-tune the calibration. A more detailed description of the constraints for p, p_0 and p_1 is given in Bielecki, Cousin, Crépey, and Herbertsson (2013b).

In this subsection we combine the random recoveries in (8.60) with piecewise-constant default intensities as described in Sect. 8.4.1, so that the parameters to be calibrated are of the form $\boldsymbol{\theta} = (\boldsymbol{\lambda}, p)$, where $\boldsymbol{\lambda}$ is as in Sect. 8.4.1. Consequently, using the same notation as in Sect. 8.4.1, the parameters $\boldsymbol{\theta} = (\boldsymbol{\lambda}, p)$ are obtained according to

$$\boldsymbol{\theta}^\star = \underset{\boldsymbol{\theta}}{\text{argmin}} \sum_{(a,b)} \left(\frac{S_{a,b}(\boldsymbol{\theta}) - S^\star_{a,b}}{S^\star_{a,b}}\right)^2, \tag{8.63}$$

where $\boldsymbol{\lambda}$ must satisfy the same constraints as in Sect. 8.4.1, while p must obey (8.62). In Sect. 8.4.3 we use this setting with stochastic recoveries as an alternative to the one of Sect. 8.4.1 when calibrating the model against two different CDO data sets.

Finally, note that if the i.i.d recoveries R_i would follow other distributions than (8.60), then, in order to proceed with our calibration, we would simply need to modify $\mathbb{E}e^{\gamma(1-R_i)}$ in (8.54) in Sect. 8.3.4, whereas the rest of the computations would parallel computations done for the binomial case. Of course, changing (8.60) would also imply that the constraints in (8.62) would no longer be relevant and some other relevant conditions would need to be used.

8.4.3 Calibration Results with Piecewise Constant Default Intensities

In this subsection we calibrate our model against CDO tranches on the iTraxx Europe and CDX.NA.IG series with maturity of five years. We use the calibration methodology described in Sects. 8.4.1 and 8.4.2.

Accordingly, the default intensities for 125 single-name CDS contracts constituting the entities in these series are bootstrapped from their market spreads for $T_1 = 3$ and $T_2 = 5$, using piecewise-constant individual default intensities on the time intervals $[0, 3]$ and $[3, 5]$. Fig. 8.3 displays the 3 and 5-year market CDS spreads sorted in decreasing order for the 125 obligors used in the single-name bootstrapping, for the two portfolios CDX.NA.IG sampled on December 17, 2007 and the iTraxx Europe series sampled on March 31, 2008.

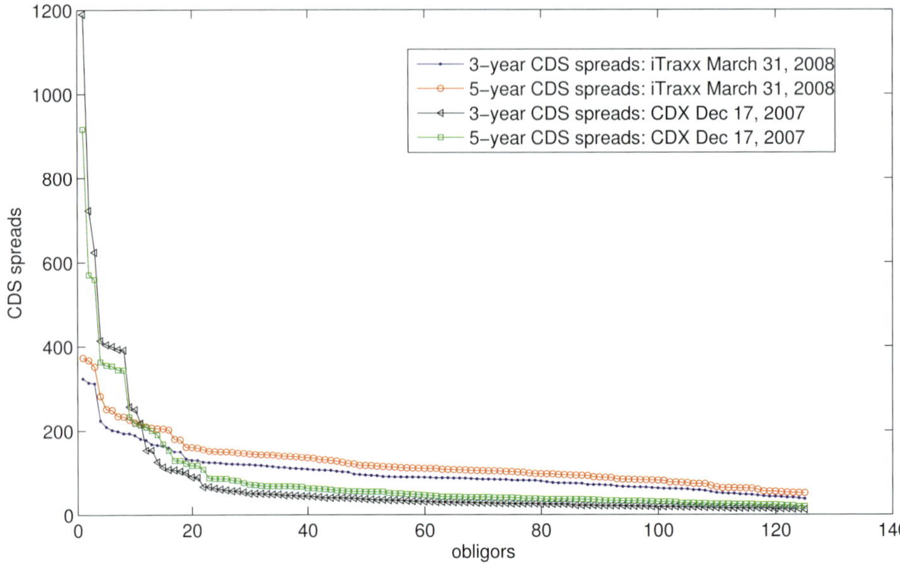

FIGURE 8.3: The 3 and 5-year market CDS spreads for the 125 obligors used in the single-name bootstrapping, for the two portfolios CDX.NA.IG sampled on December 17, 2007 and the iTraxx Europe series sampled on March 31, 2008. The CDS spreads are sorted in decreasing order.

When calibrating the systemic intensities $\boldsymbol{\lambda} = (\lambda_{I_j}^{(k)})_{j,k}$ for the CDX.NA.IG Series 9, December 17, 2007, we used 5 groups $I_1, I_2, \ldots, I_5$, where $I_j = \{1, \ldots, i_j\}$ for $i_j = 6, 19, 25, 61, 125$, after having labelled the obligors by decreasing level of riskiness based on their average 3-year and 5-year CDS spreads. Consequently, obligor 1 has the highest average CDS spread while company 125 has the lowest average CDS spread. Moreover, the obligors in the set $I_5 \setminus I_4$ consisting of the 64 safest companies are assumed to have zero idiosyncratic intensities $\lambda_{\{i\}}(t)$ and the corresponding CDS contracts are excluded from the calibration, which in turn relaxes the constraints for $\boldsymbol{\lambda}$ in (8.58). Hence the obligors in $I_5 \setminus I_4$ can only go bankrupt due to a simultaneous default of the companies in the group $I_5 = \{1, \ldots, 125\}$, i.e. in an Armageddon event. With this structure the calibration of the December 17, 2007 data set with constant recoveries is very good, as can be seen in Table 8.1.

TABLE 8.1: CDX.NA.IG Series 9, December 17, 2007 and iTraxx Europe Series 9, March 31, 2008. The market and model spreads and the corresponding absolute errors, both in bp and in percent of the market spread. The $[0, 3]$ spread is quoted in %. All maturities are for five years.

CDX 2007-12-17: Calibration with constant recovery

Tranche	$[0, 3]$	$[3, 7]$	$[7, 10]$	$[10, 15]$	$[15, 30]$
Market spread	48.07	254.0	124.0	61.00	41.00
Model spread	48.07	254.0	124.0	61.00	38.94
Absolute error in bp	0.010	0.000	0.000	0.000	2.061
Relative error in %	0.0001	0.000	0.000	0.000	5.027

CDX 2007-12-17: Calibration with stochastic recovery

Tranche	$[0, 3]$	$[3, 7]$	$[7, 10]$	$[10, 15]$	$[15, 30]$
Market spread	48.07	254.0	124.0	61.00	41.00
Model spread	48.07	254.0	124.0	61.00	41.00
Absolute error in bp	0.000	0.000	0.000	0.000	0.000
Relative error in %	0.000	0.000	0.000	0.000	0.000

iTraxx Europe 2008-03-31: Calibration with constant recovery

Tranche	$[0, 3]$	$[3, 6]$	$[6, 9]$	$[9, 12]$	$[12, 22]$
Market spread	40.15	479.5	309.5	215.1	109.4
Model spread	41.68	429.7	309.4	215.1	103.7
Absolute error in bp	153.1	49.81	0.0441	0.0331	5.711
Relative error in %	3.812	10.39	0.0142	0.0154	5.218

iTraxx Europe 2008-03-31: Calibration with stochastic recovery

Tranche	$[0, 3]$	$[3, 6]$	$[6, 9]$	$[9, 12]$	$[12, 22]$
Market spread	40.15	479.5	309.5	215.1	109.4
Model spread	40.54	463.6	307.8	215.7	108.3
Absolute error in bp	39.69	15.90	1.676	0.5905	1.153
Relative error in %	0.9886	3.316	0.5414	0.2745	1.053

By using stochastic recoveries specified as in (8.60) and (8.61) we get a perfect fit of the same data set. The calibrated systemic intensities $\boldsymbol{\lambda}$ for the 5 groups in the December 17, 2007 data set, both for constant and stochastic recoveries, are displayed in the left subplot in Fig. 8.4. Note that the systemic intensities $\lambda_{I_j}^{(1)}$ for the first pillar (i.e. on the interval $[0, 3]$) follow the same trends both in the constant and stochastic recovery case, while the systemic intensities $\lambda_{I_j}^{(2)}$ for the second pillar (i.e. on the interval $[3, 5]$) have a less common trend.

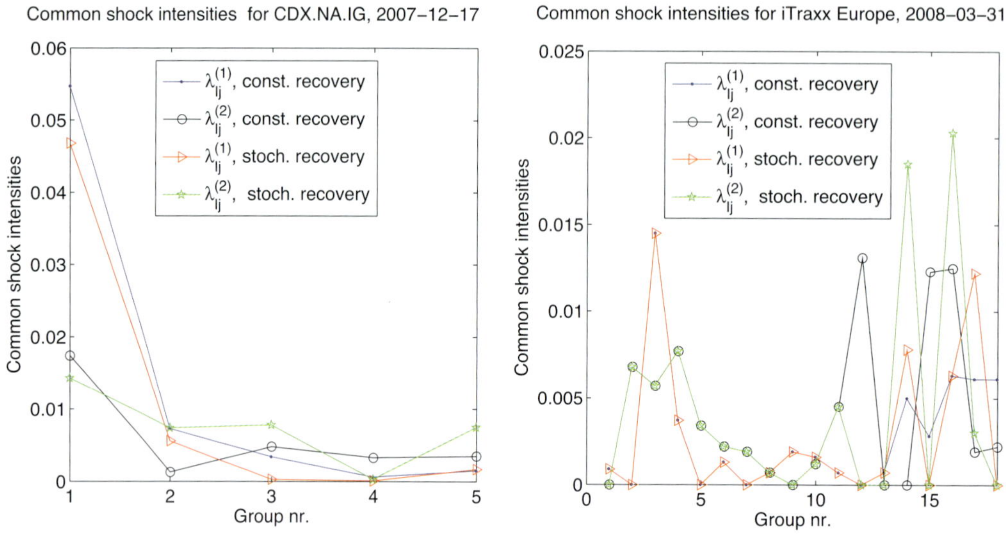

FIGURE 8.4: The calibrated systemic intensities $(\lambda_{I_j}^{(k)})_{j,k}$ both in the constant and stochastic recovery case for the two portfolios CDX.NA.IG sampled on December 17, 2007 *(left)* and the iTraxx Europe series sampled on March 31, 2008 *(right)*.

The calibration of the systemic intensities $\boldsymbol{\lambda} = (\lambda_{I_j}^{(k)})_{j,k}$ for the data sampled at March 31, 2008 is more demanding. We use 18 groups $I_1, I_2, \ldots, I_{18}$ where $I_j = \{1, \ldots, i_j\}$ for $i_j = 1, 2, \ldots, 11, 13, 14, 15, 19, 25, 79, 125$. In order to improve the fit, as in the 2007 case, we relax the constraints for $\boldsymbol{\lambda}$ in (8.58) by excluding from the calibration the CDS corresponding to the obligors in $I_{18} \setminus I_{17}$. We hence assume that the obligors in $I_{18} \setminus I_{17}$ can only go bankrupt due to an Armageddon event, i.e. jointly with all the surviving companies in the group $I_{18} = \{1, \ldots, 125\}$. In this setting, the calibration of the 2008 data set with constant recoveries yields an acceptable fit except for the $[3, 6]$ tranche, as can be seen in Table 8.1. However, by including stochastic recoveries (8.60), (8.61) the fit is substantially improved as seen in Table 8.1. As expected, in both recovery versions, the more groups added the better the fit, which explain why we use as many as 18 groups.

The calibrated systemic intensities $\boldsymbol{\lambda}$ for the 18 groups in the March 2008 data set, both for constant and stochastic recoveries, are displayed in the right subplot in Fig. 8.4. In this subplot we note that for the 13 first groups $I_1, \ldots, I_{13}$, the systemic intensities $\lambda_{I_j}^{(1)}$ for the first pillar are identical in the constant and stochastic recovery case, and then diverge quite a lot on the last five groups $I_{14}, \ldots, I_{18}$, except for group I_{16}. Similarly, in the same subplot we also see that for the 11 first groups $I_1, \ldots, I_{11}$, the systemic intensities $\lambda_{I_j}^{(2)}$ for the second pillar are identical in the constant and stochastic recovery case, and then differ quite a lot on the last seven groups, except for group I_{13}. The optimal parameters p and p_0 used in the stochastic recovery model were given by $p = 0.4405$ and $p_0 = 0.4$ for the 2007 data set and $p = 0.6002$ and $p_0 = 0.4$ for the 2008 case.

Finally, let us discuss the choice of the groupings $I_1 \subset I_2 \subset \ldots \subset I_m$ in our calibrations. First, for the CDX.NA.IG Series 9, December 17, 2007 data set, we used $m = 5$ groups with $i_5 = 125$. For $j = 1, 2$ and 4 the choice of i_j corresponds to the number of defaults needed for the loss process with constant recovery of 40% to reach the j-th attachment points. Hence, using the expression $i_j \cdot \frac{1-R}{n}$, with $R = 40\%$ and $n = 125$, to approximate the attachment points of $3\%, 10\%, 30\%$, we ended up with the following choices $i_1 = 6, i_2 = 19, i_4 = 61$. The choice of $i_3 = 25$ implies a loss of 12% and gave a better fit than choosing i_3 to exactly match 15%. Finally, no group was chosen to match the attachment point of 7% since this actually worsened the calibration for all groupings that we tried. With the above grouping structure we got very good fits in the constant recovery case, and almost perfect fit with stochastic recovery, as is seen in Table 8.1.

Next, we used the above procedure on the market CDO data from the iTraxx Europe series sampled on March 31, 2008. However, as it turned out, we could not achieve good calibration using this procedure. Instead, more groups had to be added and we tried different groupings which led to the optimal choice rendering the calibration presented in Table 8.1. Here, it was of interest to study the sensitivity of the calibrations with respect to the choice of groupings of the form $I_1 \subset I_2 \subset \ldots \subset I_m$, where $I_j = \{1, \ldots, i_j\}$ for $i_j \in \{1, 2, \ldots, m\}$ and $i_1 < \ldots < i_m = 125$, on the March 31 2008 data set. Three such groupings are displayed in Table 8.2 and the corresponding calibration results are showed in Table 8.3.

TABLE 8.2: Three different groupings (denoted A,B and C) consisting of $m = 7, 9, 13$ groups having the structure $I_1 \subset I_2 \subset \ldots \subset I_m$, where $I_j = \{1, \ldots, i_j\}$ for $i_j \in \{1, 2, \ldots, m\}$ and $i_1 < \ldots < i_m = 125$.

Three different groupings

i_j	i_1	i_2	i_3	i_4	i_5	i_6	i_7	i_8	i_9	i_{10}	i_{11}	i_{12}	i_{13}
Grouping A	6	14	15	19	25	79	125						
Grouping B	2	4	6	14	15	19	25	79	125				
Grouping C	2	4	6	8	9	10	11	14	15	19	25	79	125

TABLE 8.3: The relative calibration error in percent of the market spread, for the three different groupings A, B and C in Table 8.2, when calibrated against CDO tranche on iTraxx Europe Series 9, March 31, 2008 (see also in Table 8.1).

Relative calibration error in % (constant recovery)

Tranche	$[0,3]$	$[3,6]$	$[6,9]$	$[9,12]$	$[12,22]$
Error for grouping A	6.875	18.33	0.0606	0.0235	4.8411
Error for grouping B	6.622	16.05	0.0499	0.0206	5.5676
Error for grouping C	4.107	11.76	0.0458	0.0319	3.3076

Relative calibration error in % (stochastic recovery)

Tranche	$[0,3]$	$[3,6]$	$[6,9]$	$[9,12]$	$[12,22]$
Error for grouping A	3.929	9.174	2.902	1.053	2.109
Error for grouping B	2.962	7.381	2.807	1.002	1.982
Error for grouping C	1.439	4.402	0.5094	0.2907	1.235

From Table 8.3 we see that, in the case with constant recovery, the relative calibration error in percent of the market spread decreased monotonically for the first three tranches as the number of groups increased. Furthermore, in the case with stochastic recovery, the relative calibration error decreased monotonically for all five tranches as the number of groups increased in each grouping. The rest of the parameters in the calibration were the same as in the optimal calibration in Table 8.1.

Finally, we remark that the two optimal groupings used in Table 8.1 in the two different data sets CDX.NA.IG Series 9, December 17, 2007 and iTraxx Europe Series 9, March 31, 2008 differ quite a lot. However, the CDX.NA.IG Series is composed of North American obligors while the iTraxx Europe Series is formed by European companies. Thus, there is no model risk or inconsistency created by using different groupings for these two data sets coming from two disjoint markets. If, on the other hand, the same series is calibrated and assessed (e.g. for hedging) at different time points in a short time span, it is of course desirable to use the same grouping in order to avoid model risk.

8.4.3.1 The Implied Loss Distribution

After fitting our model to market spreads, we used the calibrated systemic parameters $\boldsymbol{\lambda} = (\lambda_{I_j}^{(k)})_{j,k}$, together with the calibrated individual default intensities, to study the credit-loss distribution in the portfolio. In this chapter we only focus on a few examples derived from the loss distribution with constant recoveries evaluated at $T = 5$ years.

Allowing for joint defaults of the obligors in the groups I_j, together with imposing the restriction that the safest obligors do not default individually, leads to some interesting effects of the loss distribution, as can be seen in Figs. 8.5 and 8.6. For example, we clearly see that the support of the loss-distributions is limited in practice to a rather small set. Specifically, the upper and lower graphs in Fig. 8.5 indicate that $\mathbb{Q}\left[N_5 = k\right]$ roughly has support on the set $\{1, \ldots, 35\} \cup \{61\} \cup \{125\}$ for the 2007 data set and on $\{1, \ldots, 40\} \cup$

$\{79\} \cup \{125\}$ for the 2008 data set. This becomes clearer in a log-loss distribution, as is seen in the upper and lower graphs in Fig. 8.6.

From the upper graph in Fig. 8.6 we see that the default-distribution is nonzero on $\{36, \ldots, 61\}$ in the 2007 case and nonzero on $\{41, \ldots, 79\}$ for the 2008 sample, but the actual size of the loss-probabilities are in the range 10^{-10} to 10^{-70}. Such low values will obviously be treated as zero in any practically relevant computation. Furthermore, the reasons for the empty gap in the upper graph in Fig. 8.6 on the interval $\{62, \ldots, 124\}$ for the 2007 case is due to the fact that we forced the obligors in the set $I_5 \setminus I_4$ to never default individually, but only in the event of the simultaneous default of all the surviving companies in the group $I_5 = \{1, \ldots, 125\}$. This Armageddon event is displayed as an isolated nonzero 'dot' at default nr 125 in the upper graph of Fig. 8.6. The gap over the set $\{80, \ldots, 124\}$ in the 2008 case is explained similarly due to our assumption regarding the companies in the group $I_{19} \setminus I_{18}$. Also note that the two 'dots' at default nr 125 in the top plot of Fig. 8.6 are manifested as spikes in the upper graph displayed in Fig. 8.5. The shape of the multimodal loss distributions presented in Fig. 8.5 and Fig. 8.6 are typical for models allowing simultaneous defaults (see for example Figure 2, page 59 in Brigo, Pallavicini, and Torresetti (2007) and Figure 2, page 710 in Elouerkhaoui (2007)).

8.4.4 Calibration Methodology and Results with Stochastic Intensities

We now consider the case where the shock intensities are stochastic CIR processes given as in example 8.2.12(ii). For simplicity, a and c are taken to be fixed a priori and we assume a piecewise-constant parametrization of every mean-reversion function $b_Y(t)$ (see expression (8.23)), so that for all k :

$$b_Y(t) = b_Y^{(k)}, \; t \in [T_{k-1}, T_k),$$

with $T_0 = 0$. The time grid (T_k) is the same one that we used in the previous section, i.e. $M = 2$, $T_1 = 3$, $T_2 = 5$. It corresponds to the set of standard CDS maturities that are lower or equal to the maturity of the fitted CDO tranches. In order to reduce the number of the parameters at hand, we consider that, for any $Y \in \mathcal{Y}$, the starting value of the corresponding intensity process is given by its first-pillar mean-reversion parameter, i.e. $X_0^Y = b_Y^{(1)}$. This specification guarantees that there is exactly the same number of parameters to fit as in Sect. 8.4.1 (piecewise-constant intensities and constant recovery). All other aspects of the model are the same as in Sect. 8.4.1, so we reproduce the same calibration methodology except that now individual mean-reversion parameters $\{b_i^{(k)} : i = 1, \ldots, n \text{ and } k = 1, 2\}$ play the role of former parameters $\{\lambda_i^{(k)} : i = 1, \ldots, n \text{ and } k = 1, 2\}$, and systemic parameters $\{b_{I_j}^{(k)} : j = 1, \ldots, m \text{ and } k = 1, 2\}$ play the role of the former parameters $\{\lambda_{I_j}^{(k)} : j = 1, \ldots, m \text{ and } k = 1, 2\}$.

We note that for such CIR processes X^Y, the survival shock probabilities

$$\mathbb{E}\left[\exp\left(-\int_0^t X_u^Y du\right)\right],$$

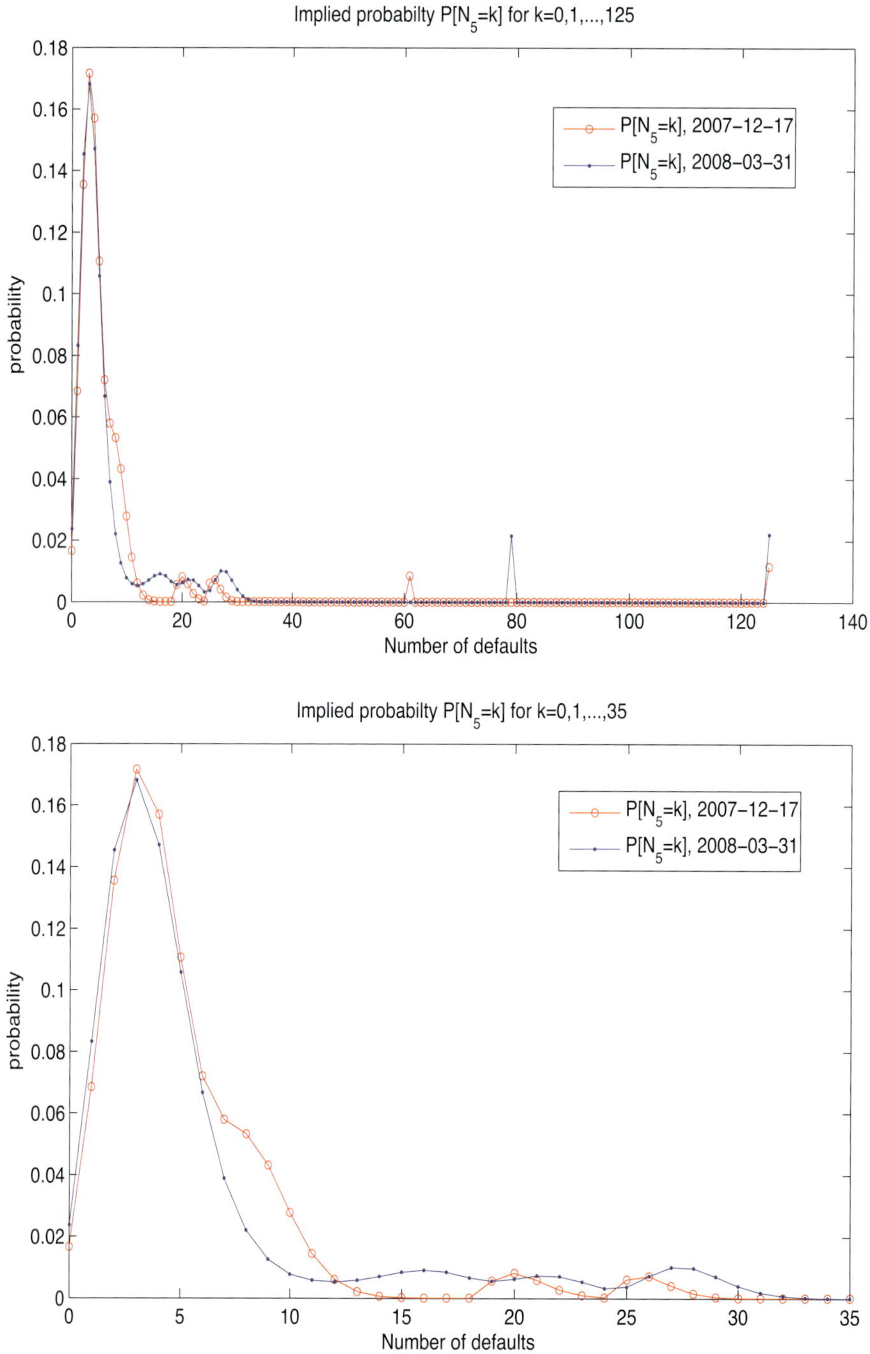

FIGURE 8.5: The implied distribution $\mathbb{Q}[N_5 = k]$ on $\{0, 1, \ldots, \nu\}$, for $\nu = 125$ *(top)* and zoomed for $\nu = 35$ *(bottom)*, where the model is calibrated against CDX.NA.IG Series 9, December 17, 2007 and iTraxx Europe Series 9, March 31, 2008.

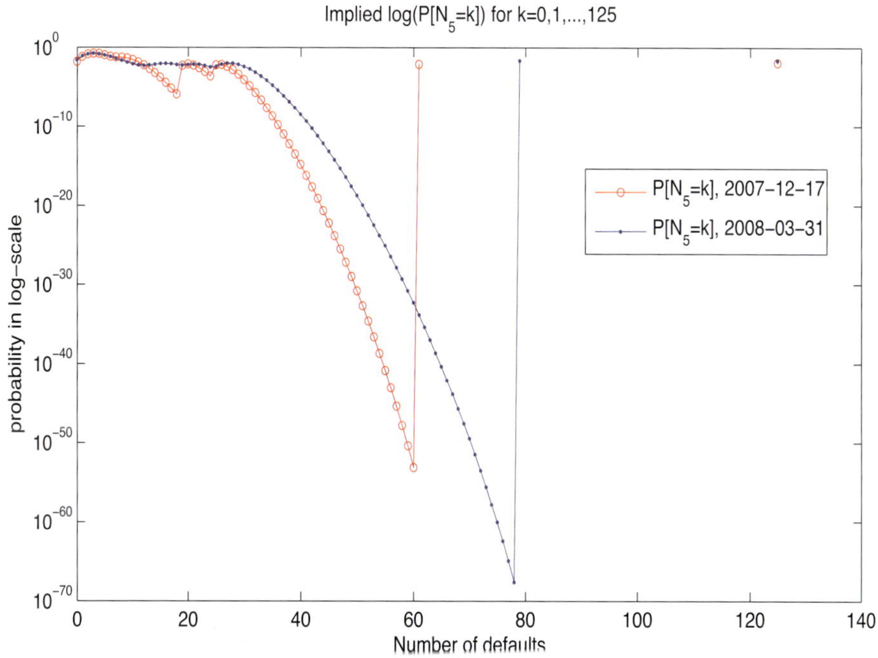

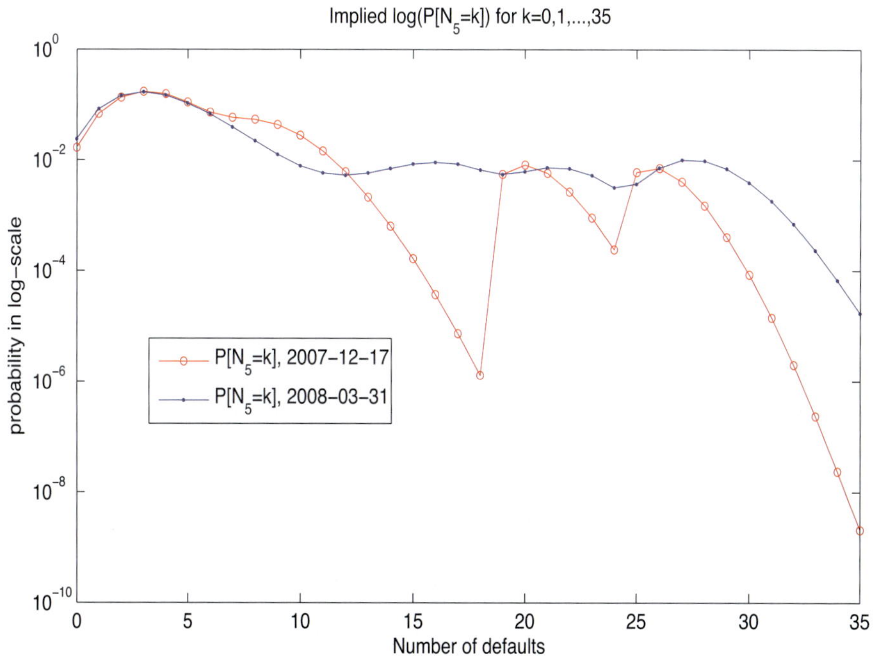

FIGURE 8.6: The implied log distribution $\ln(\mathbb{Q}\left[N_5 = k\right])$ on $\{0, 1, \ldots, \nu\}$, for $\nu = 125$ *(top)* and zoomed for $\nu = 35$ *(bottom)*, where the model is calibrated against CDX.NA.IG Series 9, December 17, 2007 and iTraxx Europe Series 9, March 31, 2008.

which are the "building blocks" of CDS and CDO prices, can be computed very efficiently (see Sect. 13.4.1).

We now study the calibration performance of this model specification. Playing with different values of parameters a (speed of mean-reversion) and c (volatility) may slightly affect the quality of the fit. We then consider the case $a = 3$ and $c = 0.5$, which rendered our best results.

As can be seen in Table 8.4, we obtain a correct fit for CDX 2007-12-17 even in the case where no name is removed from the calibration constraints. Here we use 5 groups $I_1, I_2, \ldots, I_5$, where $I_j = \{1, \ldots, i_j\}$, for $i_j = 8, 19, 27, 102, 125$. Note that, when the names from group I_5 are removed from the individual CDS calibration constraints, we are able to obtain a perfect fit, as already observed with the piecewise-constant intensity model with stochastic recovery.

TABLE 8.4: CDX.NA.IG Series 9, December 17, 2007. The market and model spreads and the corresponding absolute errors, both in bp and in percent of the market spread. The $[0, 3]$ spread is quoted in %. All maturities are for five years.

CDX 2007-12-17: Calibration with constant recovery

Tranche	$[0,3]$	$[3,7]$	$[7,10]$	$[10,15]$	$[15,30]$
Market spread	48.07	254.0	124.0	61.00	41.00
Model spread	50.37	258.01	124.68	61.32	41.91
Absolute error in bp	2.301	4.016	0.684	0.327	0.912
Relative error in %	4.787	1.581	0.552	0.536	2.225

For iTraxx Europe 2008-03-31, the calibration results were not improved with respect to the piecewise-constant intensity model with constant recovery.

8.4.5 Min-Variance Hedging Deltas

In this subsection we present some numerical results illustrating performance of the min-variance hedging strategies given in Proposition 8.3.4. We focus on hedging strategies for the data of CDX.NA.IG Series 9 on December 17, 2007, calibrated in the constant recovery model presented in Sect. 8.4.1. The aim of this subsection is to analyze the composition of the hedging portfolio at time $t = 0$ (the calibration date) when standardized CDO tranches are hedged with a group of d single-name rolling CDS contracts, for which the reference names are included in the underlying CDS index. Recall from the remark 8.3.3 that time $t = 0$ hedging ratios using a rolling CDS are the same as time $t = 0$ hedging ratios using the corresponding ordinary CDS contracts.

Since no spread factor $\mathbf{X}$ is used in the model, Proposition 8.3.4 implies that the min-variance hedging ratios at time $t = 0$ are given by

$$\zeta^{va}(0, \mathbf{H}_0) = \mathcal{C}(u, \mathbf{v})(\mathcal{C}(\mathbf{v}, \mathbf{v}))^{-1}(0, \mathbf{H}_0),$$

where

$$\mathcal{C}(u,\mathbf{v}) = \sum_{Y\in\mathcal{Y}} \lambda_Y(0)\Delta_Y u(\Delta_Y \mathbf{v})^{\mathsf{T}} \quad \text{and} \quad \mathcal{C}(\mathbf{v},\mathbf{v}) = \sum_{Y\in\mathcal{Y}} \lambda_Y(0)\Delta_Y \mathbf{v}(\Delta_Y \mathbf{v})^{\mathsf{T}}.$$

Hence, computing the min-variance hedging ratios involves a summation of the "jump differentials" $\lambda_Y(0)\Delta_Y u(\Delta_Y \mathbf{v})^{\mathsf{T}}$ and $\lambda_Y(0)\Delta_Y \mathbf{v}(\Delta_Y \mathbf{v})^{\mathsf{T}}$ over all possible shocks $Y \in \mathcal{Y}$, where $\mathcal{Y} = \{\{1\},\ldots,\{n\},I_1,\ldots,I_m\}$.

In the calibration of the CDX.NA.IG Series 9 we used $m = 5$ groups $I_1, I_2, \ldots, I_5$, where $I_j = \{1,\ldots,i_j\}$ for $i_j = 6, 19, 25, 61, 125$ and the obligors have been labeled by decreasing level of riskiness. At the calibration date $t = 0$ associated with December 17, 2007, no name has defaulted in CDX Series 9, so we set $\mathbf{H}_0 = \mathbf{0}$. In our empirical framework, the intensities $\lambda_Y(0)$, $Y \in \mathcal{Y}$ are computed from the constant default intensities $\lambda_i^\star$ that fit market spreads of 3-year maturity CDS contracts and from the 3-year horizon systemic intensities $\lambda_{I_j}^\star$ calibrated to CDO tranche quotes. The terms $\delta_Y u(0,\mathbf{H}_0)$ and $\delta_Y \mathbf{v}(0,\mathbf{H}_0)$ correspond to the change in value of the tranche and the single-name CDS contracts, at the arrival of the shock affecting all names in group Y. Recall from (8.29) that the cumulative change in value of the tranche is equal to

$$\Delta_Y u(0,\mathbf{H}_0) = u(0,\mathbf{H}_0^Y) - u(0,\mathbf{H}_0) + L_{a,b}(\mathbf{H}_0^Y) - L_{a,b}(\mathbf{H}_0),$$

where $\mathbf{H}_0^Y$ is the vector in $\{0,1\}^n$ such that only the components $i \in Y$ are equal to one.

Hence the tranche sensitivity $\Delta_Y u(0,\mathbf{H}_0)$ includes both the change in the ex-dividend price u of the tranche and the protection payment on the tranche associated with the default of group Y. The price sensitivity is obtained by computing the change in the present values of the default leg and the premium leg. The latter quantity involves the contractual spread that defines premium cash flows. As for CDX.NA.IG Series 9, the contractual spreads were respectively equal to 500 bps, 130 bps, 45 bps, 25 bps and 15 bps for the tranches [0-3%], [3-7%], [7-10%], [10-15%] and [15-30%]. We compute $u(0,\mathbf{H}_0^Y)$ and $u(0,\mathbf{H}_0)$ by the recursive convolution procedure of Sect. 8.3.3. Specifically, using the same notation as in Sect. 8.3.3, the CDO tranche price $u(0,\mathbf{H}_0^Y)$ (respectively $u(0,\mathbf{H}_0)$) is computed using the recursive procedure with $Z =$ the complement of Y (respectively $Z =$ all names). We let $i_1,\ldots i_d$ be the CDS contracts used in the min-variance hedging and assume that they all are initiated at time $t = 0$. Hence the market value at $t = 0$ for these CDS contracts are zero. As a result, when group Y defaults simultaneously, the change in value $\Delta_Y \mathbf{v}(0,\mathbf{H}_0)$ for buy-protection positions on these CDS is only due to protection payment associated with names in group Y. Hence, by (8.27), for one unit of nominal exposure on hedging CDS, the corresponding vector of sensitivities is equal to

$$\Delta_Y \mathbf{v}(0,\mathbf{H}_0) = ((1-R)\mathbb{1}_{i_1\in Y},\ldots,(1-R)\mathbb{1}_{i_d\in Y})^{\mathsf{T}},$$

where the recovery rate R is assumed to be constant and equal to 40%.

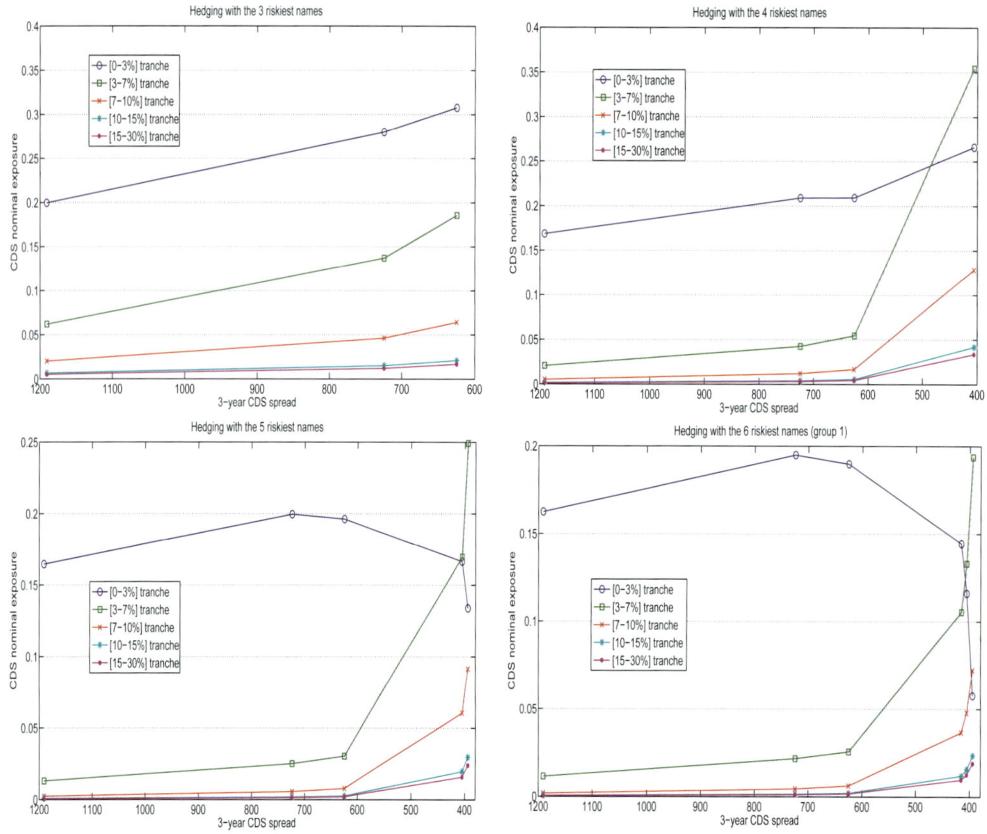

FIGURE 8.7: Min-variance hedging strategies associated with the d riskiest CDS contracts, $d = 3, 4, 5, 6$ for one unit of nominal exposure of different CDO tranches in a model calibrated to market spreads of CDX.NA.IG Series 9 on December 17, 2007.

TABLE 8.5: The names and CDS spreads (in bp) of the six riskiest obligors used in the hedging strategy displayed in Fig. 8.7.

Company (Ticker)	CCR-HomeLoans	RDN	LEN	SFI	PHM	CTX
3-year CDS spread	1190	723	624	414	404	393

Fig. 8.7 displays the nominal exposure for the d most riskiest CDS contracts when hedging one unit of nominal exposure in a CDO using the min-variance hedging strategy in Proposition 8.3.4. We use $d = 3, 4, 5$ and $d = 6$ in our computations. Furthermore, Table 8.5 displays the names and sizes of the 3-year CDS spreads used in the hedging strategy. Each plot in Fig. 8.7 should be interpreted as follows: in every pair (x, y) the x-component represents the size of the 3-year CDS spread at the hedging time $t = 0$, while the y-component is the corresponding nominal CDS-exposure computed via Proposition 8.3.4 using the d riskiest CDS contracts. The graphs are ordered from top to bottom, where the top panel corresponds to hedging with the $d = 3$ riskiest CDS and the bottom panel corresponds to hedging with the $d = 6$ riskiest names. Note that the x-axes are displayed from the riskiest obligor to the safest. Thus, hedge-sizes y for riskier CDS contracts are aligned to the left in each plot, while y-values for safer CDS contracts are consequently displayed more to the right. In doing this, going from the top to the bottom panel consists in observing the effect of including new safer names from the right part of the graphs. We have connected the pairs (x, y) with lines forming graphs that visualizes possible trends of the min-variance hedging strategies for the d riskiest CDS contracts.

For example, when the three riskiest names are used for hedging (top panel), we observe that the amount of nominal exposure in hedging instruments decreases with the degree of subordination, i.e. the [0-3%] equity tranche requires more nominal exposure in CDS contracts than the upper tranches. Note, moreover, that the min-variance hedging portfolio contains more CDS contracts referencing names with lower spreads. When lower-spread CDS contracts are added to the portfolio, the picture remains almost the same for the 3 riskiest names. For the remaining safer names however, the picture depends on the characteristics of the tranche. For the [0-3%] equity tranche, the quantity of the remaining CDS contracts required for hedging sharply decreases as additional safer names are added. One possible explanation is that adding too many names in the hedging strategy will be useless when hedging the equity tranche. This is intuitively clear since one expects that the riskiest obligors will default first and consequently reduce the equity tranche substantially, explaining the higher hedge-ratios for riskier names, while it is less likely that the safer names will default first and thus incur losses on the first tranche, which explains the lower hedge ratios for the safer names. We observe the opposite trend for the senior (safer) tranches: adding new (safer) names to the hedging portfolio seems to be useful for "non equity" tranches, since the nominal exposure required for these names increases when they are successively added.

Figs. 8.8 and 8.9 display min-variance hedging strategies when hedging a standard tranche with the 61 riskiest names, i.e. all names excepted names in group $I_5 \setminus I_4$. In contrast to Fig. 8.7, these graphs allow visualizing the effect of the "grouping structure" on the composition of the hedging portfolio. In this respect, we use different marker styles in order to distinguish names in the different disjoint groups I_1, $I_2 \setminus I_1$, $I_3 \setminus I_2$, $I_4 \setminus I_3$. We see that the min-variance hedging strategies are quite different among tranches. Moreover, whereas nominal exposures required for hedging are clearly monotone for names belonging to the same disjoint group, this tendency is broken when we consider names in different groups. This suggests that the grouping structure has a substantial impact on the distribution of names in the hedging portfolio. For the equity tranche, we observe in Fig. 8.7 that less safer-names are required for hedging. This feature is retained in Fig. 8.8 when we look at names in specific disjoint groups. Indeed, names in a given disjoint group are affected by the same common-shock, which in turn affect the equity tranche with the same severity. The only effect that may explain differences in nominal exposure among names in the same disjoint group is idiosyncratic defaults: names with wider spreads are more likely to default first, so we need them in greater quantity for hedging than names with tighter spreads.

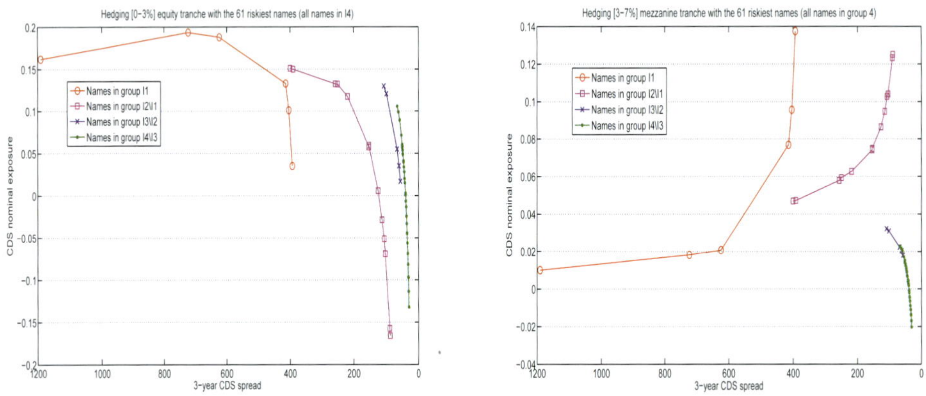

FIGURE 8.8: Min-variance hedging strategies when hedging one unit of nominal exposure in the [0-3%] equity tranche *(left)* and the [3-7%] mezzanine tranche *(right)* using the d riskiest CDS, $d = 61$ (all names excepted names in group $I_5 \setminus I_4$) for one unit of nominal exposure.

Note that nominal exposure in hedging CDS contracts even becomes negative for names within groups $I_2 \setminus I_1$ and $I_4 \setminus I_3$ when spreads are low. However, in Fig. 8.8 we observe that, for the equity tranche, some of the riskiest names in $I_4 \setminus I_3$ are more useful in the hedging than some of the safest names in group I_1, which may seem strange at first glance, given that the credit spread of the latter is much larger than the credit spread of the former. Recall that the equity tranche triggers protection payments corresponding to the few first defaults, if these occur before maturity. Even if names in group $I_4 \setminus I_3$ have a very low default probability, the fact that they can affect the tranche at the arrival of

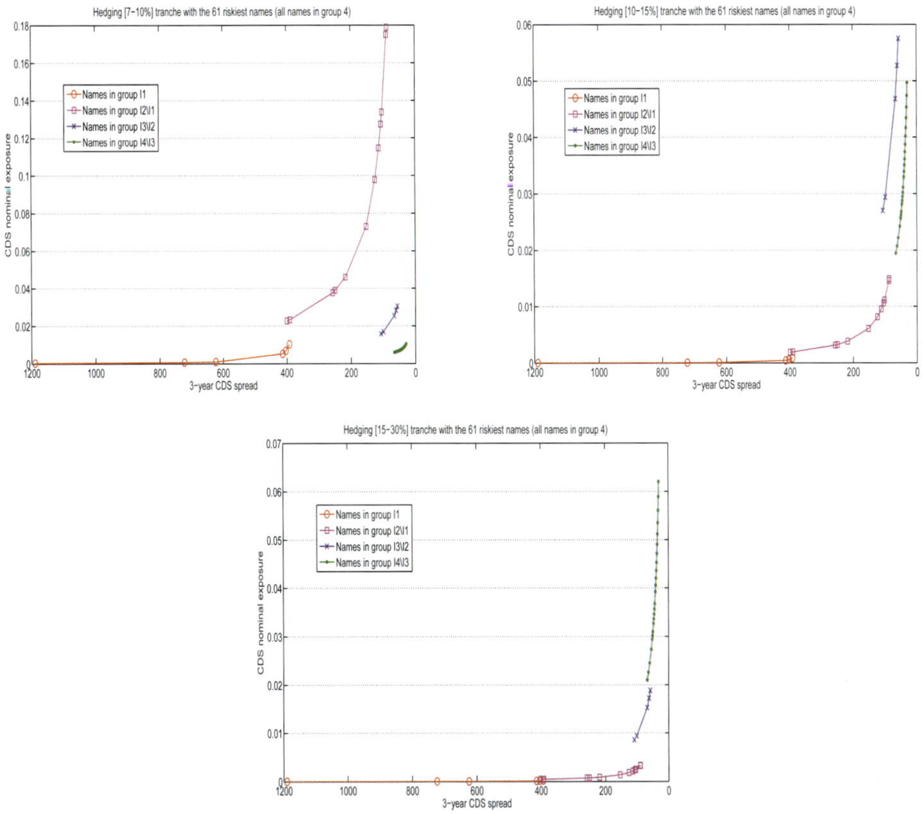

FIGURE 8.9: Min-variance hedging strategies when hedging one unit of nominal exposure in the [7-10%] tranche (top), the [10-15%] tranche (middle) and the [15-30%] tranche (bottom) with the d riskiest CDS contracts, $d = 61$ (all names excepted names in group $I_5 \setminus I_4$).

common-shock I_4 or I_5 makes these names appealing for hedging because they are less costly (they require less premium payments) than names in I_1.

Fig. 8.8 suggests that names with the lowest spreads should be ineffective for hedging the [0-3%] and the [3-7%] tranches. As can be seen in Fig. 8.9, the opposite holds for other tranches, i.e. the amount of low-spread names in the hedging portfolio increases as the tranche becomes less and less risky. For the [15-30%] super-senior tranche, we can see on the lowest graph of Fig. 8.9 that the safer a name is, the larger the quantity that is required for hedging. Furthermore, Fig. 8.9 also shows that in a consistent dynamic model of portfolio credit risk calibrated to a real data set, the [15-30%] super-senior tranche has significant (in fact, its greatest) sensitivity to very safe names with spreads less than a few dozens bps. For this tranche it is in fact likely that one could improve the hedge by inclusion of even safer names in the set of hedging instruments, provided these additional names could also be calibrated. Recall that on the data of CDX.NA.IG Series 9 on December 17, 2007, we calibrated our model to the 64 safest names in the portfolio.

8.5 CVA Pricing and Hedging

Our next aim is the study of (unilateral) CVA on credit derivatives (CDS contracts and/or CDO tranches) in a common-shock model for $(\tau_0, \ldots, \tau_n)$, where $\tau_0 = \tau_c$ is the default time of the counterparty. So we have:

$$\xi = (1 - R_0)\chi^+ \text{ with } \chi = P_\tau + \phi(\mathbf{H}_\tau) - \phi(\mathbf{H}_{\tau-}) - \Gamma_\tau,$$

where $P_\tau = u(\tau, \mathbf{X}_\tau, \mathbf{H}_\tau)$ for some clean pricing function u (by application of Proposition 8.3.1) and where ϕ is the default leg payoff function of the contract (in the notation set up in the introduction to this part of the book). Moreover, we assume a collateral process Γ such that, for any subset Z of names containing the counterparty (labeled 0), there exists a function $\Gamma_Z(t, \mathbf{x}, \mathbf{k})$ such that

$$\Gamma_\tau = \Gamma_Z(\tau, \mathbf{X}_{\tau-}, \mathbf{H}_{\tau-}) \tag{8.64}$$

on the event of a default of names in Z and only in Z (the event with indicator process H^Z). Recall that $\mathbf{H}_{t-}^Z$ below stands for the vector obtained from $\mathbf{H}_{t-}$ by replacing its components with indices in Z by ones, whereas M_t^Z represents the compensated martingale of H_t^Z (see (8.17)).

Proposition 8.5.1 *For $t \in [0, \bar{\tau}]$ we have:*

$$\Theta_t = \mathbb{E}_t\left[\mathbb{1}_{\tau < T}\xi\right] = \Theta(t, \mathbf{X}_t, \mathbf{H}_t) \tag{8.65}$$

for some suitable CVA function $\Theta(t, \mathbf{x}, \mathbf{k})$. The jump-to-counterparty-default martingale part of Θ_t is given by

$$d\Theta_t^{jd} = \sum_{0 \in Z \in \mathcal{Z}_t} \nu_t^Z dM_t^Z, \tag{8.66}$$

where for any $Z \in \mathcal{Z}$ containing 0 (the counterparty) [7]

$$\nu_t^Z = \delta_Z \Theta(t, \mathbf{X}_{t-}, \mathbf{H}_{t-}) = \xi_t^Z - \Theta_{t-}, \tag{8.67}$$

in which

$$\begin{aligned}\xi_t^Z &= (1 - R_0)\left(u(t, \mathbf{X}_{t-}, \mathbf{H}_{t-}^Z) + \phi(\mathbf{H}_{t-}^Z) - \phi(\mathbf{H}_{t-}) - \Gamma_Z(t, \mathbf{X}_{t-}, \mathbf{H}_{t-})\right)^+ \\ &=: \xi_Z(t, \mathbf{X}_{t-}, \mathbf{H}_{t-}).\end{aligned} \tag{8.68}$$

Proof. The second equality in (8.65) holds by the Markov property of the common-shock model. Moreover, we have that $\Delta\Theta_\tau = \xi - \Theta_{\tau-}$ on $\{\tau < T\}$, where $\xi = \xi_\tau^Z$ on the event of a default of names in Z and only in Z which, for any set Z of names containing the counterparty labelled "0", implies the representation (8.67) for the corresponding integrand $\delta_Z \Theta(t, \mathbf{X}_{t-}, \mathbf{H}_{t-})$.

□

We now discuss hedging jump-to-counterparty-default risk of the CVA by a clean rolling CDS referencing the counterparty. By Proposition 8.3.2, the compensated jump-to-counterparty-default martingale part $\widehat{Q}^{jd}$ of the gain process of a clean rolling CDS is such that, for $t \in [0, \bar{\tau}]$,

$$d\widehat{Q}_t^{jd} = (1 - R_0) \textstyle\sum_{Z \in \mathcal{Z}_t; 0 \in Z} dM_t^Z. \tag{8.69}$$

Let ζ be an $\mathbb{R}$-valued process representing the number of units of the rolling CDS contracts that are held, along with the constant savings asset, in a self-financing hedging strategy for the counterparty risk of the contract. Given (8.65) and (8.69), the jump-to-default component (e_t^{jd}) of the tracking error of the hedged portfolio satisfies $e_0 = 0$ and, for $t \in [0, \bar{\tau}]$ (see (3.28)):

$$\begin{aligned} de_t^{jd} &= d\Theta_t^{jd} - \zeta_t d\widehat{Q}_t^{jd} \\ &= \sum_{Z \in \mathcal{Z}_t; 0 \in Z} \left(\xi_t^Z - \Theta_{t-} - \zeta_t(1 - R_0)\right) dM_t^Z. \end{aligned}$$

Proposition 8.5.2 *The strategy ζ_t^{jd} that minimizes the risk-neutral variance of the jump-to-counterparty-default risk component of the hedging error is such that, for $t \le \bar{\tau}$:*

$$(1 - R_0)\zeta_t^{jd} = \sum_{Y \in \mathcal{Y}; 0 \in Y_t} w_t^Y \left(\xi_t^Y - \Theta_{t-}\right) = \mathcal{E}_t - \Theta_{t-}, \tag{8.70}$$

where $\mathcal{E}_t = \sum_{Y \in \mathcal{Y}; 0 \in Y_t} w_t^Y \xi_t^Y$ for weights w_t^Y given, for any $Y \in \mathcal{Y}$ with $0 \in Y_t$, by:

$$w_t^Y = \frac{\lambda_Y(t, \mathbf{X}_{t-})}{\sum_{Z \in \mathcal{Y}; 0 \in Z_t} \lambda_Z(t, \mathbf{X}_{t-})}. \tag{8.71}$$

---

[7] Other choices for Z are irrelevant here.

In particular, on $\{\tau < T\}$,

$$\zeta_\tau^{jd} = (1 - R_0)^{-1}\big(\varepsilon - \Theta_{\tau-}\big), \tag{8.72}$$

for the so called "modified expected positive exposure" $\varepsilon = \mathcal{E}_\tau = \mathbb{E}(\xi \,|\, \mathcal{G}_{\tau-})$.

Proof. By the sharp bracket regression formula (13.41), the strategy minimizing the risk-neutral variance of the counterparty jump-to-default risk component of the hedging error is given, for $t \le \tau$, by:

$$\zeta_t^{jd} = \frac{d\langle \Theta^{jd}, \widehat{Q}^{jd}\rangle_t}{d\langle \widehat{Q}^{jd}\rangle_t}.$$

So, in view of the dynamics for Θ^{jd} in (8.66) and $\widehat{Q}^{jd}$ in (8.69):

$$\zeta_t^{jd} = \frac{\sum_{Z\in\mathcal{Z}_t;\, 0\in Z} \ell_t^Z(\xi_t^Z - \Theta_{t-})(1 - R_0)}{\sum_{Z\in\mathcal{Z}_t;\, 0\in Z} \ell_t^Z(1 - R_0)^2},$$

from which (8.70) follows in view of the expression (8.18) for the intensities ℓ^Z of the M^Z, noting that $\xi_t^{Y_t} = \xi_t^Y$ holds for any $Y \in \mathcal{Y}$. To prove (8.72), it remains to show that

$$\mathbb{E}(\xi \,|\, \mathcal{G}_{\tau-}) = \sum_{Y\in\mathcal{Y};\, 0\in Y_\tau} w_\tau^Y \xi_\tau^Y. \tag{8.73}$$

In view of (8.5) and (8.71), conditional on $\mathcal{G}_{\tau-}$, we have that

$$\mathbf{H}_\tau \in \{\mathbf{H}_{\tau-}^Z;\, Z \in \mathcal{Z}_\tau, 0 \in Z\}$$

and it holds, for any $Z \in \mathcal{Z}_\tau$ containing 0, that

$$\mathbb{Q}\left(\mathbf{H}_\tau = \mathbf{H}_{\tau-}^Z \,|\, \mathcal{G}_{\tau-}\right) = \sum_{Y\in\mathcal{Y};\, Y_\tau = Z} w_\tau^Y.$$

So

$$\varepsilon = \mathbb{E}(\xi \,|\, \mathcal{G}_{\tau-}) = \sum_{Z\in\mathcal{Z}_\tau;\, 0\in Z} \left(\sum_{Y\in\mathcal{Y};\, Y_\tau = Z} w_\tau^Y\right) \xi_Z(\tau, \mathbf{X}_\tau, \mathbf{H}_{\tau-}),$$

and (8.73) follows from the fact that $\mathbf{H}_{\tau-}^Z = \mathbf{H}_{\tau-}^Y$, so that $\xi_Z(\tau, \mathbf{X}_{\tau-}, \mathbf{H}_{\tau-}) = \xi_Y(\tau, \mathbf{X}_{\tau-}, \mathbf{H}_{\tau-})$ for any $Y \in \mathcal{Y}$ such that $Y_\tau = Z$. □

A key ingredient of the hedging ratio (8.72) is the modified expected positive exposure $\varepsilon = \mathbb{E}(\xi \,|\, \mathcal{G}_{\tau-})$. From a dynamic point of view, this is the meaningful quantity, as opposed to the classical expected positive exposure $\mathbb{E}(\xi \,|\, \tau)$ of (3.26). Precisely, the proper hedging ratio is $(1 - R)^{-1}(\varepsilon - \Theta_{\tau-})$, where the second term in the difference accounts for the value of the CVA right before the counterparty default.

Observe that without hedging, so for $\zeta = 0$, the jump of the hedging error at the counterparty default τ (if $< T$) would be given by

$$\Delta e_\tau = \Delta e_\tau^0 = \xi - \Theta_{\tau-}.$$

With the ζ^{jd} strategy, we have instead

$$\Delta e_\tau = \Delta e_\tau^{jd} = \xi - \varepsilon = \xi - \mathbb{E}(\xi \,|\, \mathcal{G}_{\tau-}).$$

Note that in the common-shock model, the cumulative value process $\widehat{Q}$ of the rolling CDS only jumps at the default time τ of the counterparty, as opposed to jumps at other defaults in a "general" model of credit risk (model without any kind of immersion, such as the dynamic Gaussian copula model of the previous chapter). Moreover, the M^Z are pure jump processes that do not jump together, so that $\widehat{Q}$ is orthogonal to $\bar{\Theta}^{jd} = \int_0^\cdot \sum_{Z \in \mathcal{Z}_t; 0 \notin Z} \nu_t^Z dM_t^Z$. We thus also have that

$$\zeta_t^{jd} = \frac{d\langle \Theta^{jd} + \bar{\Theta}^{jd}, \widehat{Q} \rangle_t}{d\langle \widehat{Q} \rangle_t},$$

hence ζ^{jd} also minimizes the risk-neutral variance of the all-inclusive jump-to-default risk (not only the jump-to-counterparty-default) of the CVA hedging error. However, this strategy, which is "optimal" as far as the jump-to-default component of the counterparty risk is concerned, disregards the market risk component of the hedging error. In fact, this strategy typically creates some additional market risk. It needs to be stressed that in this section we only focus on hedging the jump-to-counterparty-default exposure of the CVA. Assuming a concrete specification of the intensities, it would also be possible to study hedging of the orthogonal component of the hedging error, corresponding to the market risk of the contract and the spread risk of the counterparty. For instance, one could derive the formula for the strategy that minimizes the risk-neutral variance of the hedging error altogether, using the above rolling CDS and other additional hedging instruments.

The computational issues regarding the Monte Carlo valuation of the common-shock CVA, based on the formula (8.66), are discussed in Sect. 10.2.

Chapter 9

CVA Computations for One CDS in the Common-Shock Model

9.1 Introduction

In this chapter we study counterparty risk embedded within in a CDS contract. This topic, which corresponds to the emblematic case of AIG selling CDS protection on the distressed Lehman in the 2008 credit crisis, has received a lot of attention in the literature. The following is a partial list of works studying it:

- Huge and Lando (1999) propose a rating-based approach,
- Hull and White (2001) use a static Gaussian copula model,
- Jarrow and Yu (2001) use an intensity contagion model, further considered in Leung and Kwok (2005),
- Brigo and Chourdakis (2008) work in a Gaussian copula model with CIR++ intensities, extended to bilateral counterparty credit risk in Brigo and Capponi (2008b),
- Blanchet-Scalliet and Patras (2011), Lipton and Sepp (2009b) and Lipton and Shelton (2012) develop

It can thus be considered as a benchmark topic in the area of counterparty risk.

Some preliminary results regarding counterparty risk relative to CDS contracts have already been discussed in Sect. 7.4.1. In this chapter we will, for the most part, work in the common-shock model of Chapter 8, in which the dependence between the counterparty in the contract and the reference name of the CDS is rendered by the possibility of their joint default. In the case of one CDS, the general computations of Chapter 8 can be pushed further and explicit formulas can be derived. We also provide comparative numerics regarding various specifications for the default intensities.

9.2 Generalities

In this section we briefly recall the basics of (unilateral) counterparty risk embedded in a CDS. Let there be given a CDS with maturity T and contractual spread S_1 referencing firm "1". We primarily consider a payer CDS, i.e. the CDS contract in which the bank is a buyer of credit protection against potential default of firm 1. In this instance, when we consider a receiver CDS, all the related quantities are denoted with "bar"; for example we'll write "$\overline{\text{CDS}}$" for a receiver CDS contract.

Following the convention of this part of the book, we use indices 0 (rather than c) and 1 to refer to quantities related to the counterparty and to the reference firm, respectively. In particular, τ_0 and τ_1 represent their respective default times and R_0 and R_1 stand for the corresponding (constant) recoveries. We assume no collateralization, i.e. $\Gamma = 0$ (see the next chapter for developments regarding collateralization). We assume a deterministic discount factor $\beta(t) = \exp(-rt)$.

By application of the definition 3.2.2(i), we obtain:

Lemma 9.2.1 *The clean price process P_t of the CDS is given by:*

$$\beta(t)tP_t = \mathbb{E}_t\left((1-R_1)\beta(t)\tau_1 \mathbb{1}_{t<\tau_1<T} - S_1 \int_t^{\tau_1 \wedge T} \beta(t)sds\right). \tag{9.1}$$

For $\overline{\text{CDS}}$, the clean price process is $\bar{P}_t = -P_t$.

We write

$$\begin{aligned}
&p_i(t) = \mathbb{Q}(\tau_i < t) \text{ and } q_i(t) = \mathbb{Q}(\tau_i > t) = 1 - p_i(t) \text{ , for } i = 0, 1,\\
&p_{(0)}(t) = \mathbb{Q}(\tau_0 \wedge \tau_1 < t), \;\; q_{(0)}(t) = \mathbb{Q}(\tau_0 \wedge \tau_1 > t),\\
&p_{(1)}(t) = \mathbb{Q}(\tau_0 \vee \tau_1 < t), \;\; q_{(1)}(t) = \mathbb{Q}(\tau_0 \vee \tau_1 > t),
\end{aligned}$$

so that

$$p_0(t) + p_1(t) + q_{(0)}(t) = 1 + p_{(1)}(t). \tag{9.2}$$

With calibration in view, note that target ("market") piecewise constant functions $p_i^\star(t)$ can be bootstrapped from the related CDS curves. Given a parameter $\rho \in [0, 1]$, target values $p_{(0)}^\star(t)$ can in turn be obtained by substituting ρ into a bivariate normal distribution function, i.e.

$$p_{(0)}^\star(t) := \Phi_\rho\left(\Phi^{-1}(p_0^\star(t)), \Phi^{-1}(p_1^\star(t))\right). \tag{9.3}$$

Remark 9.2.2 A value of ρ can be retrieved from ("calibrated to") the Basel II correlations per asset class (see pages 63 to 66 in Basel Committee on Banking Supervision (2006)), so that ρ is called "asset correlation" henceforth.

In view of (9.1), the following model-free result is a direct consequence of (3.23), (3.24).

Proposition 9.2.3 *For $t \in [0, \bar{\tau}]$, we have:*

$$\beta(t)t\Theta_t = \mathbb{E}_t\left[\beta(t)\tau_0 \mathbb{1}_{\tau_0<T}\xi\right], \quad \beta(t)t\bar{\Theta}_t = \mathbb{E}_t\left[\beta(t)\tau_0 \mathbb{1}_{\tau_0<T}\bar{\xi}\right], \tag{9.4}$$

for the $\mathcal{G}_{\tau_0}$-measurable exposures at default defined by

$$\begin{aligned}
\xi &= (1-R_0)\left(\mathbb{1}_{\tau_0<\tau_1}P_{\tau_0}^+ + \mathbb{1}_{\tau_0=\tau_1}(1-R_1)\right),\\
\bar{\xi} &= (1-R_0)\mathbb{1}_{\tau_0<\tau_1}P_{\tau_0}^-.
\end{aligned} \tag{9.5}$$

Remark 9.2.4 For a payer CDS in the common-shock model used later in this chapter, the wrong-way risk of a CDS can be accounted for by the two terms in ξ: the "large" term $(1 - R_1)$, as well as by possibly large term $P^+_{\tau_0}$. Roughly speaking, the wrong-way risk manifests itself in two regimes:

Regime 1:

$\mathbb{Q}(\tau_0$ is small) is large and $\mathbb{Q}(\tau_0 < \tau_1)$ is large. Then it would be the (large) term $P^+_{\tau_0}$ that accounts for the wrong-way risk.

Regime 2:

$\mathbb{Q}(\tau_0$ is small) is large and $\mathbb{Q}(\tau_0 = \tau_1)$ is large. Then it would be the (large) term $(1 - R_1)$ that accounts for the wrong-way risk. For the receiver CDS there is no wrong-way risk, at least not of this type.

9.2.1 Specification-Free Results in a Common-Shock Setup

We now state general results related to clean valuation and to CVA hedging with regard to one CDS in the common-shock model of Chapter 8. Recall, however, that in the present chapter we work with constant (but nonzero) interest rates r. Various intensity specifications will be considered in the next sections. Since in this chapter we are only concerned with the univariate counterparty risk, we consider here a common-shock model $(\mathbf{X}_t, \mathbf{H}_t)$ with $n = 2$ names, i.e. the counterparty and the CDS reference firm, and with the set of shocks $\mathcal{Y} = \{\{0\}, \{1\}, \{0, 1\}\}$.

9.2.1.1 Clean Valuation

Under the affine intensity structure postulated in the assumption 8.2.10(i), the clean price process P_t in (9.5) can be represented in the form

$$P_t = J^1_t v(t, \lambda^1_t),$$

where v is a pre-default pricing function[1] in the variables (t, λ_1), given as

$$v(t, \lambda^1_t) = \mathbb{E}\left(\int_t^T e^{-\int_t^s (r+\lambda^1_\zeta)d\zeta}\left((1 - R_1)\lambda^1_s - S_1\right) ds \,\middle|\, \lambda^1_t\right).$$

Therefore

$$\beta(t)v(t, \lambda_1) = \int_t^T \beta(s)\big((1 - R_1)g(t, s, \lambda^1_t) - S_1\big)\Gamma(t, s, \lambda^1_t)ds, \tag{9.6}$$

where

$$\Gamma(t, s, \lambda^1_t) = \mathbb{E}\left(e^{-\int_t^s \lambda^1_\zeta d\zeta} \,\middle|\, \lambda^1_t\right) \text{ and } \Gamma(t, s, \lambda^1_t)g(t, s, \lambda^1_t) = \mathbb{E}\left(e^{-\int_t^s \lambda^1_\zeta d\zeta}\lambda^1_s \,\middle|\, \lambda^1_t\right)$$

are the pre-default values[2] of $\mathbb{Q}(\tau_1 > s|\mathcal{G}_t)$ and $\frac{\mathbb{Q}(\tau_1 \in ds|\mathcal{G}_t)}{ds}$, respectively.

---

[1]Pricing function used for $\{t < \tau\}$.
[2]Cf. Lemma 13.7.2.

9.2.1.2 Min-Variance Hedging of the CVA Jump-To-Counterparty-Default Exposure

Writing the payoff function of the protection leg of the CDS in the form $\phi(\mathbf{k}) = (1-R_1)k_1$ and denoting by v_1 the pricing function in the $(t, \mathbf{x}, \mathbf{k})$-variables as in Sect. 8.5, we obtain:

$$\begin{aligned}
&v_1(t, \mathbf{X}_t, \mathbf{H}_{t-}^{\{0\}}) + \phi(t, \mathbf{H}_{t-}^{\{0\}}) - \phi(t, \mathbf{H}_{t-}) = \mathbb{1}_{t\leq\tau_1} v(t, \lambda_t^1), \\
&v_1(t, \mathbf{X}_t, \mathbf{H}_{t-}^{\{0,1\}}) + \phi(t, \mathbf{H}_{t-}^{\{0,1\}}) - \phi(t, \mathbf{H}_{t-}) = 0 + (1 - R_1) - (1 - R_1)H_{t-}^1 \\
&= \mathbb{1}_{t\leq\tau_1}(1 - R_1).
\end{aligned}$$

Consequently, we obtain in the notation of (8.68) (with $\Gamma = 0$ here):

$$\begin{aligned}
\xi_t^{\{0\}} &= (1 - R_0)\mathbb{1}_{t\leq\tau_1} v^+(t, \lambda_t^1), \\
\xi_t^{\{0,1\}} &= (1 - R_0)\mathbb{1}_{t\leq\tau_1}(1 - R_1).
\end{aligned}$$

Therefore the formula (8.70) in Proposition 8.5.2 yields the following expression for the hedging strategy that minimizes the variance of the jump-to-default exposure of the bank, using a clean rolling CDS on the counterparty as hedging instrument:

$$\zeta_t^{jd} = (1 - R_0)^{-1}(\mathcal{E}_t - \Theta_{t-}), \ t \leq \tau_1 \wedge T$$

(and $\zeta^{jd} = 0$ on $(\tau_1 \wedge T, \bar{\tau}]$), where

$$\begin{aligned}
\mathcal{E}_t &= \sum_{Y\in\mathcal{Y};\, 0\in Y_t} w_t^Y \xi_t^Y \\
&= \mathbb{1}_{t\leq\tau_1}\Big(w_t^{\{0\}} v^+(t, \lambda_t^1) + w_t^{\{0,1\}}(1 - R_1)\Big) \qquad (9.7)
\end{aligned}$$

with

$$w_t^{\{0\}} = \frac{\lambda_{\{0\}}(t, \mathbf{X}_t)}{\lambda_{\{0\}}(t, \mathbf{X}_t) + \lambda_{\{0,1\}}(t, \mathbf{X}_t)}, \quad w_t^{\{0,1\}} = \frac{\lambda_{\{0,1\}}(t, \mathbf{X}_t)}{\lambda_{\{0\}}(t, \mathbf{X}_t) + \lambda_{\{0,1\}}(t, \mathbf{X}_t)}.$$

9.3 Common-Shock Model with Deterministic Intensities

In this section we consider the case of deterministic intensities. This is the fully deterministic subcase of the example 8.2.12(i) (so there is no factor $\mathbf{X}_t$ involved here), which we refer to by (0F) henceforth. We write

$$\lambda_i(t) = \lambda_{\{i\}}(t) + \lambda_{\{0,1\}}(t), \ i = 0, 1, \ \lambda(t) = \lambda_{\{1\}}(t) + \lambda_{\{0\}}(t) + \lambda_{\{0,1\}}(t). \qquad (9.8)$$

Proposition 9.3.1 **(i)** *For every* $Y \in \mathcal{Y}$, *the* $\mathbb{H}$*-intensity of* H^Y *is of the form* $\lambda_Y(t, \mathbf{H}_t)$ *for a suitable function* $\lambda_Y(t, \mathbf{k})$, *namely*

$$\lambda_{\{0\}}(t, \mathbf{k}) = \mathbb{1}_{k_0=0}\left(\mathbb{1}_{k_1=0}\lambda_{\{0\}}(t) + \mathbb{1}_{k_1=1}\lambda_0(t)\right),$$
$$\lambda_{\{1\}}(t, \mathbf{k}) = \mathbb{1}_{k_1=0}\left(\mathbb{1}_{k_0=0}\lambda_{\{1\}}(t) + \mathbb{1}_{k_0=1}\lambda_1(t)\right),$$
$$\lambda_{\{0,1\}}(t, \mathbf{k}) = \mathbb{1}_{\mathbf{k}=(0,0)}\lambda_{\{0,1\}}(t).$$

Consequently, the process M^Y *defined by*

$$M_t^Y = H_t^Y - \int_0^t \lambda_Y(s, H_s)ds \tag{9.9}$$

is an $\mathbb{H}$*-martingale.*
(ii) *For* $i = 0, 1$, *the* $\mathbb{H}$*-intensity process of* H^i *is given by* $J_t^i\lambda_i(t)$. *In other words, the process* M^i *defined by*

$$M_t^i = H_t^i - \int_0^t J_s^i\lambda_i(s)ds$$

is an $\mathbb{H}$*-martingale.*
(iii) *For every* s, t, *we have:*

$$\mathbb{Q}(\tau_0 > s, \tau_1 > t) = e^{-\int_0^s \lambda_{\{1\}}(\zeta)d\zeta - \int_0^t \lambda_{\{0\}}(\zeta)d\zeta - \int_0^{s\vee t} \lambda_{\{0,1\}}(\zeta)d\zeta} \tag{9.10}$$

and therefore

$$\begin{aligned} q_0(t) &= e^{-\int_0^t \lambda_0(\zeta)d\zeta}, \quad q_1(t) = e^{-\int_0^t \lambda_1(\zeta)d\zeta}, \\ q_{(0)}(t) &= e^{-\int_0^t \lambda(\zeta)d\zeta} = q_0(t)q_1(t)e^{\int_0^t \lambda_{\{0,1\}}(s)ds}. \end{aligned} \tag{9.11}$$

(iv) *The correlation of* H_t^1 *and* H_t^0 *(default correlation at the time horizon t) is*

$$\varrho(t) = \frac{\mathbb{C}\mathrm{ov}(H_t^0, H_t^1)}{\sqrt{\mathbb{V}\mathrm{ar}(H_t^0)\mathbb{V}\mathrm{ar}(H_t^1)}} = \frac{e^{\int_0^t \lambda_{\{0,1\}}(s)ds} - 1}{\sqrt{\left(e^{\int_0^t \lambda_0(s)ds} - 1\right)\left(e^{\int_0^t \lambda_1(s)ds} - 1\right)}}. \tag{9.12}$$

Proof. **(i)** follows by an application of the formula (8.18).
(ii) follows by an application of the formula (8.6).
(iii) All the formulas in (9.11) follow from (9.10), which itself derives from (8.14).
(iv) Since H_t^i is a Bernoulli random variable with parameter $p_i(t)$,

$$\mathbb{V}\mathrm{ar}(H_t^i) = p_i(t)(1 - p_i(t))$$

and

$$\begin{aligned} \mathbb{C}\mathrm{ov}(H_t^0, H_t^1) &= \mathbb{C}\mathrm{ov}(J_t^0, J_t^1) = \mathbb{E}\left(J_t^0 J_t^1\right) - \mathbb{E}J_t^0\mathbb{E}J_t^1 = q_{(0)}(t) - q_0(t)q_1(t) \\ &= e^{-\int_0^t \lambda(s)ds} - e^{-\int_0^t \lambda_0(s)ds}e^{-\int_0^t \lambda_1(s)ds}. \end{aligned}$$

Thus, after some algebraic simplifications, the desired formula (9.12) follows for $\varrho(t)$. □

Remark 9.3.2 By application of the formula (8.21) in Proposition 8.2.8 , the generator of $\mathbf{H}_t = (H_t^0, H_t^1)$ is given as follows in matrix form, where the first to fourth rows (or columns) correspond to the four possible states $(0,0)$, $(1,0)$, $(0,1)$ and $(1,1)$ of $\mathbf{H}_t$:

$$A(t) = \begin{bmatrix} -\lambda(t) & \lambda_{\{0\}}(t) & \lambda_{\{1\}}(t) & \lambda_{\{0,1\}}(t) \\ 0 & -\lambda_0(t) & 0 & \lambda_0(t) \\ 0 & 0 & -\lambda_1(t) & \lambda_1(t) \\ 0 & 0 & 0 & 0 \end{bmatrix}. \tag{9.13}$$

In accordance with the Markov copula theory recalled in Chapter 14, for each $i = 0, 1$, the process H^i is $\mathbb{H}$-Markov with generator given in matrix form by:

$$A_i(t) = \begin{bmatrix} -\lambda_i(t) & \lambda_i(t) \\ 0 & 0 \end{bmatrix}. \tag{9.14}$$

In the next proposition we compute the pre-default clean pricing function v of (9.1) and, via the EPE defined by (3.26), the pre-default CVA, here deterministic and denoted accordingly by $\Theta(t)$ (for $0 \le t \le T$).

Proposition 9.3.3 *Under (0F):*
(i) *The pre-default clean price of the CDS is given by*

$$\beta(t)tv(t) = \int_t^T \beta(t)se^{-\int_t^s \lambda_1(\zeta)d\zeta} \left((1 - R_1)\lambda_1(s) - S_1\right) ds. \tag{9.15}$$

(ii) *For* $t \in [0, T]$, *we have:*

$$\text{EPE}(t) = (1 - R_0)\Big((1 - R_1)\frac{\lambda_{\{0,1\}}(t)}{\lambda_0(t)} + v^+(t)\frac{\lambda_{\{0\}}(t)}{\lambda_0(t)}\Big)e^{-\int_0^t \lambda_{\{1\}}(\zeta)d\zeta} \tag{9.16}$$

and

$$\Theta(t) = (1 - R_0)\int_t^T \beta(t)s\Big((1 - R_1)\lambda_{\{0,1\}}(s) + v^+(s)\lambda_{\{0\}}(s)\Big)e^{-\int_t^s \lambda(\zeta)d\zeta}ds \tag{9.17}$$

Proof. In the case (0F), the formula (9.15) directly follows from (9.1).
(ii) Set

$$\Phi(\tau_0) = \mathbb{E}(\mathbb{1}_{\tau_1=\tau_0<T}|\tau_0), \;\; \Psi(\tau_0) = \mathbb{E}(\mathbb{1}_{\tau_0<\tau_1\wedge T}|\tau_0),$$

which are characterized by:

$$\begin{aligned} \mathbb{E}\big(\Phi(\tau_0)\varphi(\tau_0)\big) &= \mathbb{E}\big(\varphi(\tau_0)\mathbb{1}_{\tau_1=\tau_0\le T}\big),, \\ \mathbb{E}(\Psi(\tau_0)\varphi(\tau_0)) &= \mathbb{E}(\varphi(\tau_0)\mathbb{1}_{\tau_0<\tau_1\wedge T}), \end{aligned} \tag{9.18}$$

for any function φ. In particular, for t fixed in $(0,T]$ we take $\varphi(s) = \mathbb{1}_{s\le t}$. Using the law of τ_0 (cf. first line in (9.12)), the left-hand sides in (9.18) are given by:

$$\mathbb{E}\big(\Phi(\tau_0)\mathbb{1}_{\tau_0\le t}\big) = \int_0^t \Phi(s)\lambda_0(s)e^{-\int_0^s \lambda_0(\zeta)d\zeta}ds,$$
$$\mathbb{E}\big(\Psi(\tau_0)\mathbb{1}_{\tau_0\le t}\big) = \int_0^t \Psi(s)\lambda_0(s)e^{-\int_0^s \lambda_0(\zeta)d\zeta}ds.$$

Regarding the right-hand-sides in (9.18), we have by Proposition 9.3.1(i)(iii):

$$\mathbb{E}\big(\mathbb{1}_{\tau_0\le t}\mathbb{1}_{\tau_1=\tau_0\le T}\big) = \mathbb{E}\big(\int_0^t dH_s^{\{0,1\}}\big)$$
$$= \int_0^t \mathbb{E}(J_s^0 J_s^1)\lambda_{\{0,1\}}(s)ds = \int_0^t e^{-\int_0^s \lambda(\zeta)d\zeta}\lambda_{\{0,1\}}(s)ds$$

and

$$\mathbb{E}\big(\mathbb{1}_{\tau_0\le t}\mathbb{1}_{\tau_0<\tau_1,\,\tau_0\le T}\big) = \mathbb{E}\Big(\int_0^t \mathbb{1}_{s\le\tau_1}dH_s^{\{0\}}\Big) = \mathbb{E}\Big(\int_0^t \mathbb{1}_{s\le\tau_1}\lambda_{\{0\}}(s,\mathbf{H}_s)ds\Big)$$
$$= \mathbb{E}\Big(\int_0^t J_s^0 J_s^1\lambda_{\{0\}}(s)ds\Big) = \int_0^t e^{-\int_0^s \lambda(\zeta)d\zeta}\lambda_{\{0\}}(s)ds,$$

where the second identity in the first line uses that $H^{\{0\}}$ does not jump at τ_1. Therefore for $\varphi(s) = \mathbb{1}_{s\le t}$ the identities in (9.18) can be rewritten as

$$\int_0^t \Phi(s)\lambda_0(s)e^{-\int_0^s \lambda_0(\zeta)d\zeta}ds = \int_0^t \lambda_{\{0,1\}}(s)e^{-\int_0^s \lambda(\zeta)d\zeta}ds,$$
$$\int_0^t \Psi(s)\lambda_0(s)e^{-\int_0^s \lambda_0(\zeta)d\zeta}ds = \int_0^t \lambda_{\{0\}}(s)e^{-\int_0^s \lambda(\zeta)d\zeta}ds.$$

Differentiating these equations with respect to t yields that for almost all t (with respect to the Lebesgue measure on the real line) we have:

$$\Phi(t) = \frac{\lambda_{\{0,1\}}(t)e^{-\int_0^t \lambda(\zeta)d\zeta}}{\lambda_0(t)}e^{\int_0^t \lambda_0(\zeta)d\zeta}\ ,\quad \Psi(t) = \frac{\lambda_{\{0\}}(t)e^{-\int_0^t \lambda(\zeta)d\zeta}}{\lambda_0(t)}e^{\int_0^t \lambda_0(\zeta)d\zeta}.$$

Hence (9.16) follows (after adjusting the EPE function on a null Lebesgue measure set, if needed). Finally, (3.25) yields that, for $t \in [0,T]$:

$$\beta(t)t\Theta(t) = \int_t^T \beta(t)s\mathrm{EPE}(s)e^{-\int_0^s \lambda_0(\zeta)d\zeta}\lambda_0(s)ds,$$

so that (9.17) follows from (9.16). □

For comparison with the EPE in (9.16), note that the "modified EPE" $\mathcal{E}_t$ of (9.7) (which is deterministic here as there is no factor process $\mathbf{X}$ involved, and therefore written as $\mathcal{E}(t)$ below) is given by

$$\mathcal{E}(t) = (1-R_0)\Big((1-R_1)\frac{\lambda_{\{0,1\}}(t)}{\lambda_0(t)} + \frac{\lambda_{\{0\}}(t)}{\lambda_0(t)}v^+(t)\Big) = e^{\int_0^t \lambda_{\{1\}}(\zeta)d\zeta}\mathrm{EPE}(t).$$

9.3.1 Implementation

9.3.1.1 Linear Intensities

With calibration in mind, it is more suitable to restate the model primitives in terms of the individual intensity functions λ_0, λ_1 and the systemic intensity function $\lambda_{\{0,1\}}$. The idiosyncratic intensities $\lambda_{\{0\}}$ and $\lambda_{\{1\}}$ are then retrieved from these by (9.8). We parameterize

$$\lambda_i(t) = a_i + b_i t, \lambda_{\{0,1\}}(t) = a_{\{0,1\}} + b_{\{0,1\}} t, \tag{9.19}$$

with

$$a_{\{0,1\}} = \alpha(a_0 \wedge a_1), \ \ b_{\{0,1\}} = \alpha(b_0 \wedge b_1),$$

for some parameter $\alpha \in [0,1]$ (so that the idiosyncratic intensities retrieved from (9.8) are nonnegative). In view of (9.1), the time-0 fair spread S_0^i of a clean CDS on name i is then given, for $i = 0, 1$, by:

$$S_0^i = (1 - R_i)\frac{ds \int_0^T \beta(t)t(a_i + b_i t)\exp(-a_i t - \frac{b_i}{2}t^2)dt}{ds \int_0^T \beta(t)t\exp(-a_i t - \frac{b_i}{2}t^2)dt}. \tag{9.20}$$

Proposition 9.3.1(v) yields that

$$\varrho = \varrho(T) = \frac{e^{a_{\{0,1\}}T + b_{\{0,1\}}T^2/2} - 1}{\sqrt{\left(e^{a_1 T + b_1 T^2/2} - 1\right)\left(e^{a_0 T + b_0 T^2/2} - 1\right)}}. \tag{9.21}$$

9.3.1.2 Calibration Issues

The a_i and b_i can be calibrated, based on (9.20), to the market CDS curves of the reference and the counterparty. Recall that market CDS curves can be considered as "clean CDS curves", as they typically refer to collateralized transactions.

As for the model dependence parameter α, in case the market price of an instrument sensitive to the dependence structure of default times (basket credit instrument on the reference and the counterparty) is available, one can use it to calibrate α. An alternative procedure consists of "calibrating" α to a target value $p^\star_{(0)}(T)$ deduced from an asset correlation ρ, as explained in the remark 9.2.2.

9.3.1.3 Constant Intensities

In the special case where $b_0 = b_1 = b_{\{0,1\}} = 0$, we have that

$$\lambda_0(t) = a_0 \ , \ \lambda_1(t) = a_1 \ , \ \lambda_{\{0,1\}}(t) = a_{\{0,1\}}.$$

The expression (9.21) for the correlation coefficient ϱ reduces to

$$\varrho = \frac{e^{a_{\{0,1\}}T} - 1}{\sqrt{(e^{a_0 T} - 1)(e^{a_1 T} - 1)}},$$

from which $a_{\{0,1\}}$ can be calculated as

$$a_{\{0,1\}} = \frac{1}{T} \ln \left(1 + \varrho \sqrt{\left(e^{a_0 T} - 1\right)\left(e^{a_1 T} - 1\right)} \right).$$

As is well known, in case of a constant intensity, the clean price of a CDS set up at par is null on $[0, T]$, i.e. $v(t) = 0$ when $b_1 = 0$. The EPE formula (9.16) reduces to

$$\text{EPE}(t) = (1 - R_0)(1 - R_1)\frac{a_{\{0,1\}}}{a_0} e^{-(a_1 - a_{\{0,1\}})t}.$$

Finally, from (9.4), we get for low values of the coefficients:

$$\begin{aligned} \text{CVA}(0) &\simeq (1 - R_0)(1 - R_1) a_{\{0,1\}} T && (9.22) \\ &= (1 - R_0)(1 - R_1) \ln \left[1 + \varrho \sqrt{\left(e^{a_0 T} - 1\right)\left(e^{a_1 T} - 1\right)} \right] \\ &\simeq (1 - R_0)(1 - R_1) \sqrt{a_0 a_1} T \varrho. && (9.23) \end{aligned}$$

9.4 Numerical Results with Deterministic Intensities

Our next aim is to assess the numerical impact of the asset correlation ρ and of the counterparty's fair spread $S_0^\star$ on the bank's CVA. We fix the general data of Table 9.1 (linear intensities), respectively Table 9.3 (constant intensities with all b equal to 0), and we consider twelve alternative sets of values for a_0, b_0 and ρ given as in columns one, two and four of Table 9.2 (linear intensities), respectively for a_0 and ρ given as in columns one and three of Table 9.4 (constant intensities).

r	R_0	R_1	T	a_1	b_1	$S_1^\star$
5%	40%	40%	10 years	.0095	.0010	84 bp

TABLE 9.1: Linear intensities: fixed data.

In the case of linear intensities, the corresponding spreads $S_0^\star$, default correlation ϱ, model dependence parameter α and joint default probabilities $p_{(0)}(T)$ are respectively displayed in the third, fifth, sixth and seventh columns of Table 9.2, the last column of which (to be discussed later) gives the corresponding time-0 CVA. The analogous results in the case of constant intensities are displayed in Table 9.4.

Figs. 9.1, 9.2 and 9.3 show the related EPE as a function of time, the CVA as a function of time and the time-0 CVA as a function of ρ. The four curves on each graph correspond to $S_0^\star = 50$, 75, 100 and 150bps.

Observe that the EPE decreases with $S_0^\star$ on each graph of Fig. 9.1. This is in line with the stylized features expected by practitioners. [The EPE is the expectation of the bank's exposure conditional on a default of the counterparty at time t. A default of a

a_0	b_0	$S_0^\star$	ρ	ϱ	α	$p_{(0)}(T)$	CVA(0)
.0056	.0006	50 bp	10%	.0378	.0520	.0147	.0013
.0085	.0009	75 bp	10%	.0418	.0472	.0211	.0018
.0122	.0010	100 bp	10%	.0444	.0522	.0269	.0021
.0189	.0014	150 bp	10%	.0476	.0702	.0376	.0028
.0056	.0006	50 bp	40%	.1859	.2531	.0286	.0056
.0085	.0009	75 bp	40%	.1998	.2230	.0388	.0074
.0122	.0010	100 bp	40%	.2074	.2406	.0472	.0087
.0189	.0014	150 bp	40%	.2145	.3107	.0616	.0110
.0056	.0006	50 bp	70%	.4020	.5406	.0489	.0119
.0085	.0009	75 bp	70%	.4256	.4673	.0640	.0153
.0122	.0010	100 bp	70%	.4336	.4937	.0754	.0178
.0189	.0014	150 bp	70%	.4306	.6100	.0925	.0214

TABLE 9.2: Linear intensities: variable data.

r	R_0	R_1	T	a_1	$S_1^\star$
5%	40%	40%	10 years	.0140	84 bp

TABLE 9.3: Constant intensities: fixed data.

a_0	$S_0^\star$	ρ	ϱ	α	$p_{(0)}$	CVA(0)
.0083	50 bp	10%	.0372	.0510	.0138	.0011
.0125	75 bp	10%	.0411	.0464	.0198	.0015
.0167	100 bp	10%	.0438	.0515	.0254	.0018
.0250	150 bp	10%	.0470	.0690	.0355	.0023
.0083	50 bp	40%	.1839	.2501	.0272	.0054
.0125	75 bp	40%	.1977	.2207	.0368	.0070
.0167	100 bp	40%	.2056	.2387	.0451	.0084
.0250	150 bp	40%	.2128	.3073	.0587	.0104
.0083	50 bp	70%	.3998	.5372	.0469	.0117
.0125	75 bp	70%	.4231	.4650	.0613	.0150
.0167	100 bp	70%	.4315	.4921	.0726	.0175
.0250	150 bp	70%	.4288	.6063	.0889	.0210

TABLE 9.4: Constant intensities: variable data.

counterparty with a lower spread is "worse news" than a default of a counterparty with a higher spread. Therefore the related EPE should be larger.]

In contrast, the CVA increases with $S_0^\star$ on each graph of Fig. 9.1, in line again with the stylized features expected by practitioners. Also note that the CVA decreases with time (the less time to maturity, the less counterparty risk on the CDS).

Finally, observe in Fig. 9.3 that the time-0 CVA is increasing in the asset correlation ρ, consistent, in the case of constant coefficients, with the explicit formula (9.23) in terms of the default correlation ϱ.

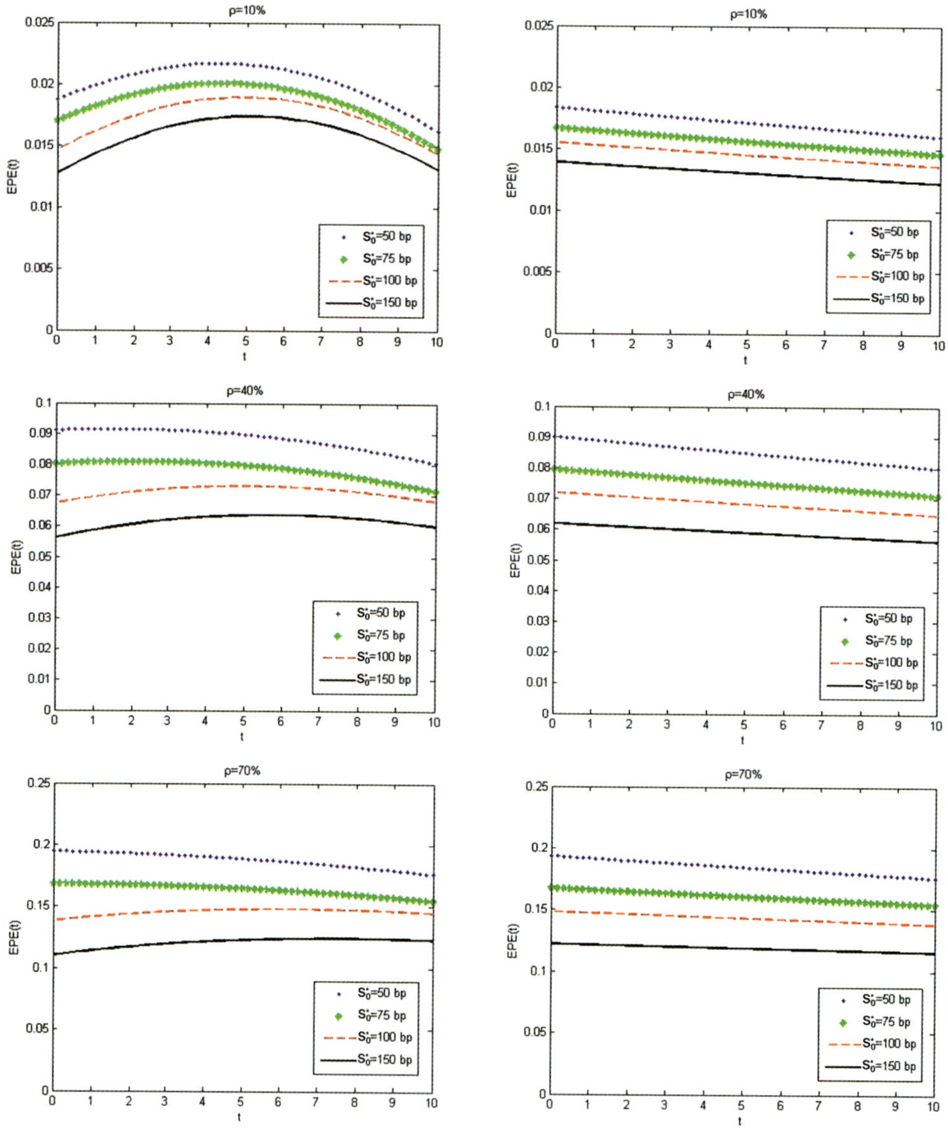

FIGURE 9.1: EPE(t). From the top to down $\rho = 10\%$, $\rho = 40\%$ and $\rho = 70\%$. *Left*: linear intensities. *Right*: constant intensities.

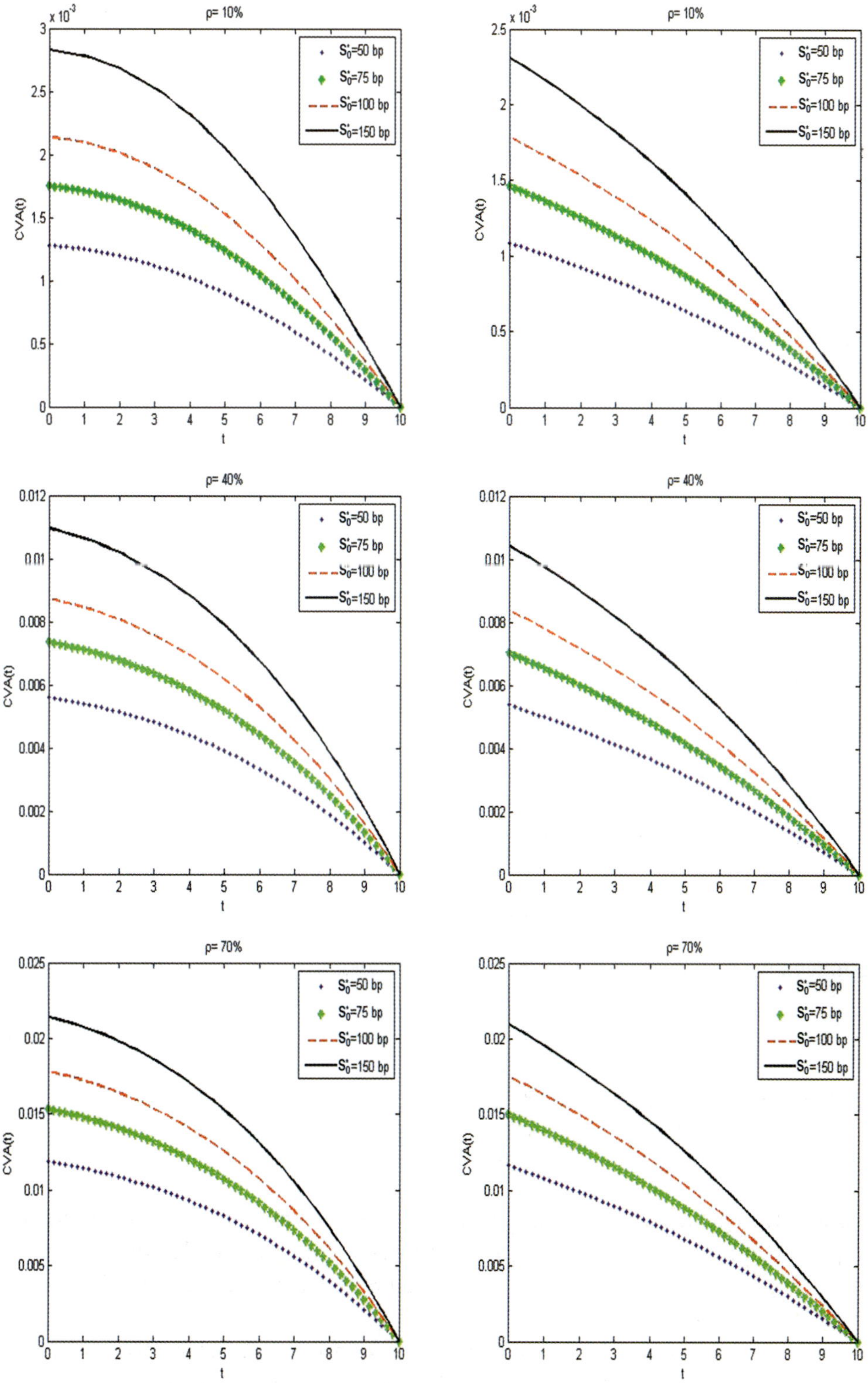

FIGURE 9.2: CVA(t). From the top to down $\rho = 10\%$, $\rho = 40\%$ and $\rho = 70\%$. *Left*: linear intensities. *Right*: constant intensities.

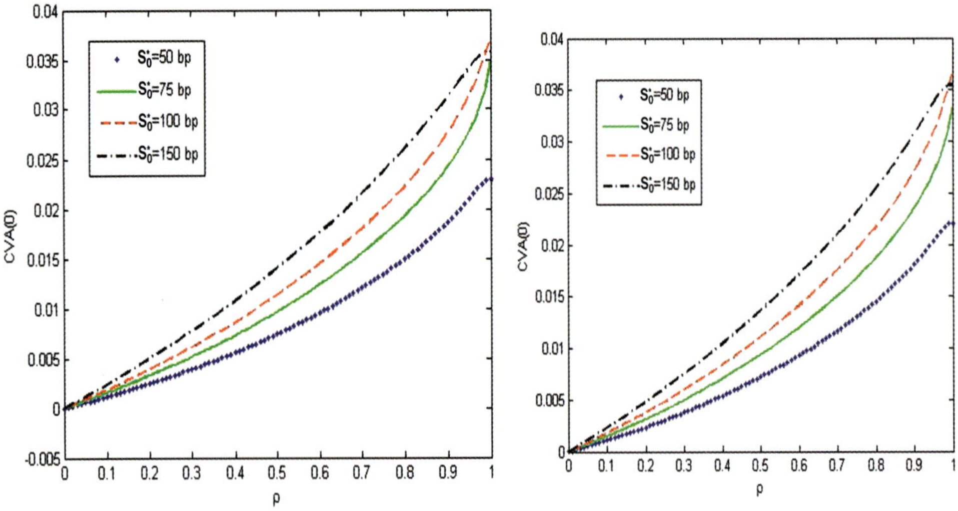

FIGURE 9.3: CVA(0) as a function of ρ for $S_0^\star = 50$ bp, 75 bp, 100 bp and 150 bp. *Left*: linear intensities. *Right*: constant intensities.

9.5 Common-Shock Model with Stochastic Intensities

9.5.1 CIR++ Intensities

In this section we use a deterministic group intensity specification, as in the example 8.2.12(i), for CIR idiosyncratic intensities of the form

$$dX_t^{\{i\}} = a(b_{\{i\}} - X_t^{\{i\}})dt + c\sqrt{X_t^{\{i\}}}dW_t^{\{i\}},$$

for nonnegative coefficients a, b_i and c and independent Brownian motions $W^{\{i\}}$. We set

$$\lambda_t^i = \eta_i(t) + X_t^{\{i\}}, \tag{9.24}$$

for functions $\eta_i(t) \geq \lambda_{\{0,1\}}(t)$, so that

$$\lambda_t^{\{i\}} = \eta_i(t) + X_t^{\{i\}} - \lambda_{\{0,1\}}(t) \geq 0.$$

This intensity specification is referred to as (2F) henceforth. Note that, since we don't want to limit the range of the volatility in the model (cf. Sect. 9.6.4 and Brigo and Capponi (2008b), Brigo and Chourdakis (2008)), we do not restrict ourselves to the inaccessible origin case $2ab_{\{i\}} > c^2$.

In view of Proposition 13.4.1 (special case of a constant b there), we have the following result, which specializes to (2F) the CDS clean value formula (9.6).

Lemma 9.5.1 *For $s \geq t$, writing $x_{\{1\}} = \lambda_1 - \eta_1(t)$ in (2F), we have:*

$$\begin{aligned} \Gamma(t, \lambda_1, s) &= e^{-\left(\int_t^s \eta_1(\zeta)d\zeta + \Phi_0(s-t)x_{\{1\}} + \Psi_0(s-t)b_{\{1\}}\right)}, \\ g(t, \lambda_1, s) &= \eta_1(s) + \Phi_0(s-t)ab_{\{1\}} + \dot{\Phi}_0(s-t)x_{\{1\}}, \end{aligned} \tag{9.25}$$

in which the functions Φ_0 and Ψ_0 are those of Proposition 13.4.1 (for $y = 0$ there).

Also note that under both parameterizations (0F) and (2F) we have, by Proposition 8.2.4:

$$q_{(0)}(t) = q_0(t)q_1(t)e^{\int_0^t \lambda_{\{0,1\}}(s)ds}. \tag{9.26}$$

9.5.1.1 Calibration Methodology

We calibrate the model to individual CDS curves of the counterparty and the reference firm, and to an asset correlation ρ (see the remark 9.2.2). The speed of mean-reversion a and the volatility parameter c are assumed to be given (as follows from the findings of Sect. 9.6.5, these can be calibrated to CDS spread options data if available). We denote by $(T_0 = 0, T_1, \ldots, T_m)$ the tenor term structure of the CDS contracts and we set $\delta_j = T_j - T_{j-1}$. We assume that η_0, η_1 and $\lambda_{\{0,1\}}$ are constant on each time interval $[T_{j-1}, T_j)$, with corresponding values denoted by a superscript ^{j}. We proceed in four steps as follows:

- We bootstrap the CDS curve for each name $i = 0, 1$ into a c.d.f. $p_i^\star(\cdot)$ such that $p_i^\star(t) = p_i^\star(T_j)$ on each interval $[T_{j-1}, T_j)$.
- Given $p_0^\star(\cdot)$, $p_1^\star(\cdot)$ and the target correlation ρ, we compute target values $p_{(0)}^\star(\cdot)$ as explained in Sect. 9.3.1.2, so that target values $q_{(0)}^\star(t)$ follow from (9.2).
- The relation (9.26) yields the following system of m linear equations in the m unknowns $\lambda^1_{\{0,1\}}, \ldots, \lambda^m_{\{0,1\}}$:

$$\begin{cases} \delta_1 \lambda^1_{\{0,1\}} + \cdots + \delta_j \lambda^j_{\{0,1\}} = \ln \dfrac{q^\star_{(0)}(T_j)}{q_0^\star(T_j)q_1^\star(T_j)}, \\ \text{subject to } \lambda^j_{\{0,1\}} \geq 0\,,\ j = 1, \ldots, m. \end{cases} \tag{9.27}$$

- Finally, the first line in (9.25) (which, for $t = 0$, reduces to $q_1(s)$) for the reference firm and the analogous expression regarding the counterparty result in the following two systems of m linear equations in the $(m+2)$ unknowns $X_0^{\{i\}}, b_{\{i\}}, \eta_i^1, \ldots, \eta_i^m$: for $i = 0, 1$,

$$\begin{cases} \delta_1 \eta_i^1 + \cdots + \delta_j \eta_i^j + \Phi_0(T_j)X_0^{\{i\}} + \Psi_0(T_j)b_{\{i\}} = -\ln\, q_i^\star(T_j), \\ \text{subject to } X_0^{\{i\}} \geq 0\,,\ b_{\{i\}} \geq 0\,,\ \eta_i^j \geq \lambda^j_{\{0,1\}},\ j = 1, \ldots, m. \end{cases} \tag{9.28}$$

In practice, the equations (9.27) and (for $i = 0, 1$) (9.28) are solved in the sense of mean-square minimization under constraints. Note that this is also the approach we use for calibrating (0F) in Sect. 9.6.1, additionally fixing the $X_0^{\{i\}}$ and the $b_{\{i\}}$ equal to zero in the above description.

9.5.2 Extended CIR Intensities

In this section we use the extended CIR intensities specification of the example 8.2.12(ii). By comparison with (2F), we also let the joint default intensity be stochastic, via a third factor $X^{\{0,1\}}$. We model the factors X^Y as affine processes of the form

$$dX_t^Y = a(b_Y(t) - X_t^Y)dt + c\sqrt{X_t^Y}dW_t^Y,$$

for independent Brownian motions W^Y, so that $\lambda^i = X^{\{i\}} + X^{\{0,1\}}$, for $i = 0, 1$, is again an extended CIR process, with parameters a, $b_i(t) = b_{\{i\}}(t) + b_{\{0,1\}}(t)$ and c. The intensity matrix of $\mathbf{H}$ is now given in functional form as

$$A(t,x) = \begin{bmatrix} -\lambda & x_{\{0\}} & x_{\{1\}} & x_{\{0,1\}} \\ 0 & -\lambda_0 & 0 & \lambda_0 \\ 0 & 0 & -\lambda_1 & \lambda_1 \\ 0 & 0 & 0 & 0 \end{bmatrix},$$

where

$$\lambda_i = x_{\{i\}} + x_{\{0,1\}},\ i = 0, 1,$$
$$\lambda = x_{\{0\}} + x_{\{1\}} + x_{\{0,1\}}.$$

9.5.2.1 Implementation

We assume that $b_{\{i\}}(t) = b_{\{i\}}^{(j)}$ on each time interval $[T_{j-1}, T_j)$. The individual intensity processes $\lambda_t^i = \lambda_t^i$ are extended CIR processes with $b_i(t) = b_{\{i\}}^{(j)} + b_{\{0,1\}}^{(j)}$ on each time interval $[T_{j-1}, T_j)$. We refer to this model parametrization as (3F). Proposition 13.4.1 yields explicit formulas for g and Γ in the CDS clean value formula (9.6). Likewise, Proposition 8.2.4 yields:

$$q_i(T_j) = \mathbb{E}e^{-\int_0^{T_j}\lambda_s^i ds} = e^{-\left(\Phi_0(T_j)\lambda_0^i + a\sum_{k=1}^{j}(\Psi_0(T_j-T_{k-1})-\Psi_0(T_j-T_k))b_i^{(k)}\right)},$$

$$\begin{aligned} q_{(0)}(T_j) &= \mathbb{E}e^{-\int_0^{T_j}\left(X_s^{\{0\}}+X_s^{\{1\}}+X_s^{\{0,1\}}\right)ds} \\ &= q_0(T_j)q_1(T_j)e^{\Phi_0(T_j)X_0^{\{0,1\}}+a\sum_{k=1}^{j}(\Psi_0(T_j-T_{k-1})-\Psi_0(T_j-T_k))b_{\{0,1\}}^{(k)}}. \end{aligned}$$

Having fixed some values for a and c, one can follow the same calibration lines as in Sect. 9.5.1.1, obtaining the following three systems (for $Y = \{0\}, \{1\}, \{0,1\}$) of m linear equations in $(m+1)$ unknowns, i.e. X_0^Y and the $b_Y^{(k)}$, under the related nonnegativity constraints:

$$\begin{cases} \Phi_0(T_j)X_0^{\{0,1\}} + a\sum_{k=1}^{j}(\Psi_0(T_j - T_{k-1}) - \Psi_0(T_j - T_{(k)}))b_{\{0,1\}}^{(k)} \\ \qquad = \ln \dfrac{q_{(0)}^\star(T_j)}{q_0^\star(T_j)q_1^\star(T_j)},\ j = 1, \ldots, m, \\ \text{subject to } X_0^{\{0,1\}} \geq 0,\ b_{\{0,1\}}^{(k)} \geq 0,\ k = 1, \ldots, m, \end{cases}$$

and then, for $i = 0, 1$,

$$\begin{cases} \Phi_0(T_j)(X_0^{\{i\}} + X_0^{\{0,1\}}) + a \sum_{k=1}^{j} (\Psi_0(T_j - T_{k-1}) - \Psi_0(T_j - T_{(k)}))(b_{\{i\}}^{(k)} + b_{\{0,1\}}^{(k)}) \\ \qquad = -\ln\, q_i^{\star}(T_j)\,,\ j = 1, \ldots, m, \\ \text{subject to } X_0^{\{i\}} \geq 0\,,\ b_{\{i\}}^{(k)} \geq 0\,,\ k = 1, \ldots, m. \end{cases}$$

These equations are solved in the sense of mean-square minimization under constraints.

9.6 Numerics

The numerical tests below have been done using the following model parameterizations:

(0F) deterministic intensities constant on each interval $[T_{j-1}, T_j)$, corresponding to omitting $X^{\{i\}}$ or setting it to 0 in the (2F) specification (9.24),

(2F) two independent CIR++ factors as in Sect. 9.5.1,

(3F) three independent extended CIR factors as in Sect. 9.5.2.

The mean-reversion parameter a (whenever relevant) is fixed to 10%. The recovery rates are set to 40% and the risk-free rate r is set equal to 5%. All the CVA numbers are computed by Monte Carlo simulations based on (3.25).

9.6.1 Calibration Results

We consider March 30, 2008 CDS contracts on UBS AG issued by four different counterparties: Gaz de France, Carrefour, AXA and Telecom Italia SpA, so ordered by increasing risk as assessed on the basis of their time-0 5y CDS spreads, henceforth referred to as CP1, CP2, CP3 and CP4 for brevity. The corresponding piecewise-constant c.d.f., bootstrapped from the related CDS curves, are represented in Table 9.6. Note the slight inversion at the long-end of the Ref and AXA curves, quite typical of the subprime crisis period (in fact, much more spectacular inversions could sometimes be seen, especially in the US at the peak of the crisis). The quality of the fits is shown in Tables 9.7 through 9.9.

9.6.2 CVA Stylized Features

Fig. 9.4 shows the time-0 CVA of a risky CDS with maturity $T = 10$yrs on UBS AG, as a function of the volatility parameter c of the CIR factors $X^{\{i\}}$. The clean value of the default leg is equal to $\mathrm{DL}_0 = 0.1031$. The four curves on each graph correspond to the

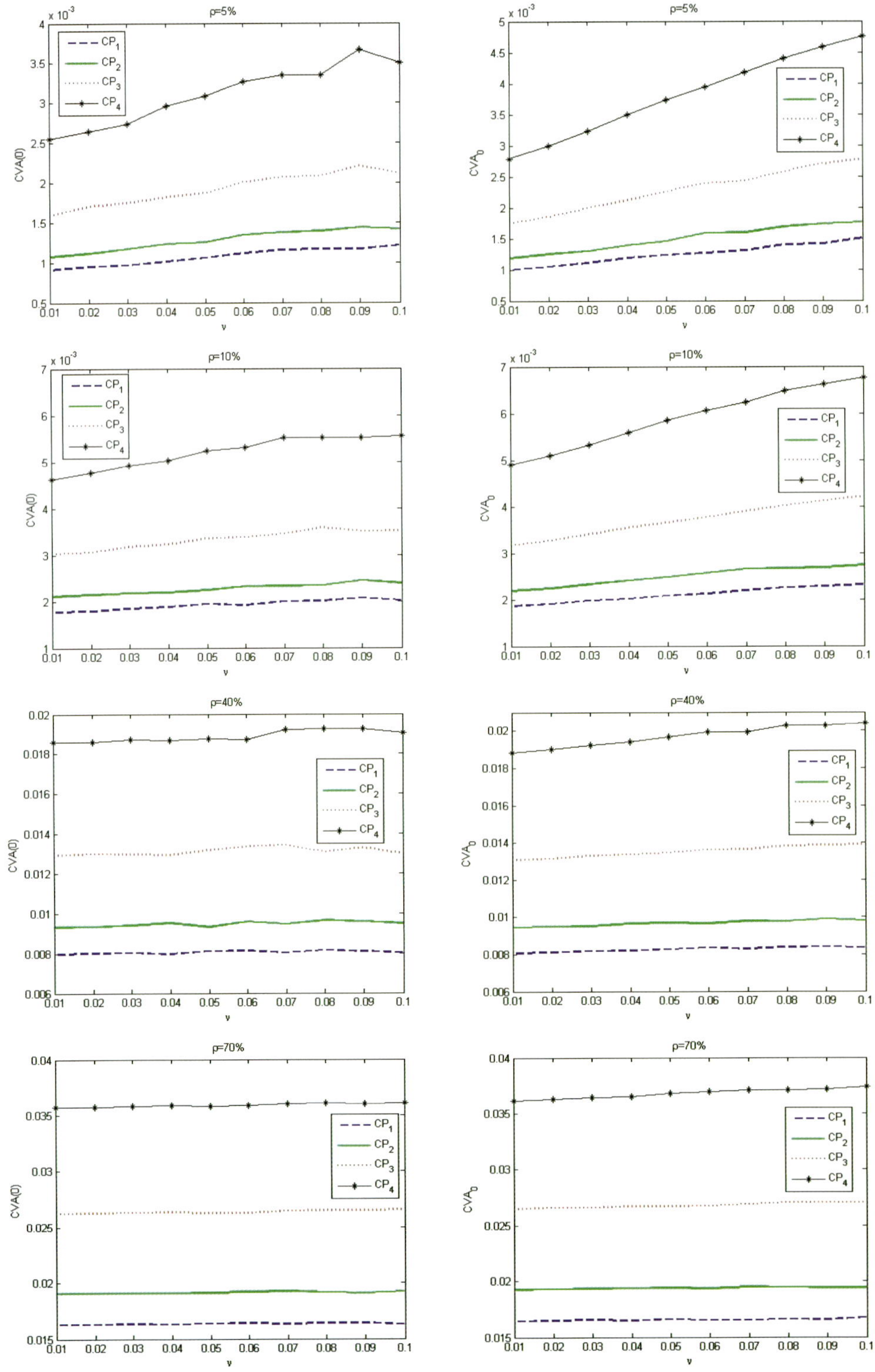

FIGURE 9.4: Θ_0 versus c for the payer CDS on Ref. The graphs of the left column correspond to the case (2F) and those of the right column correspond to (3F). In each graph, ρ is fixed with, from the top to down, $\rho = 5\%$, $\rho = 10\%$, $\rho = 40\%$ and $\rho = 70\%$.

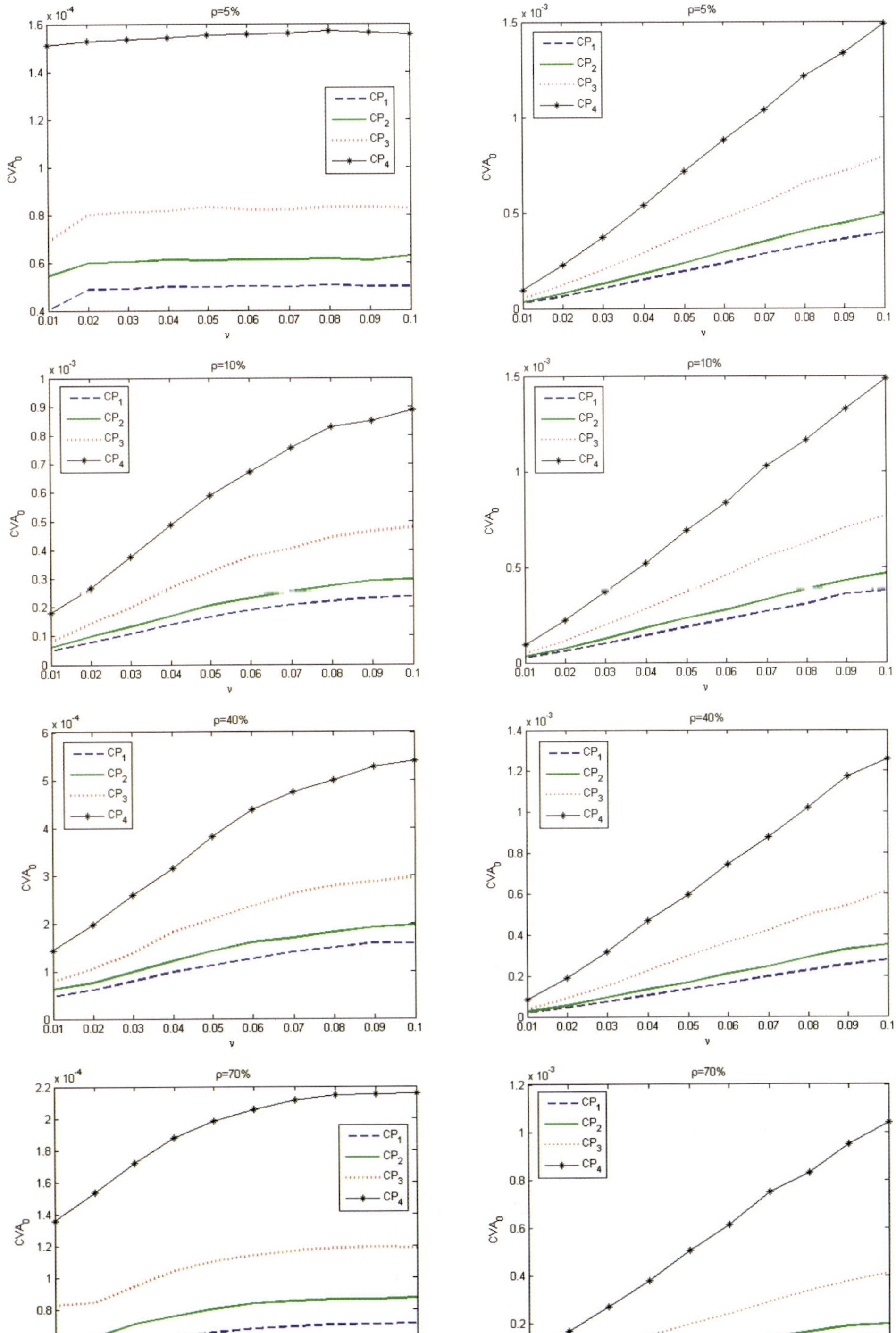

FIGURE 9.5: $\bar{\Theta}_0$ versus c for the receiver CDS on Ref. The graphs of the left column correspond to the case (2F) and those of the right column correspond to (3F). In each graph ρ is fixed with, from the top to down, $\rho = 5\%$, $\rho = 10\%$, $\rho = 40\%$ and $\rho = 70\%$.

		1 year	2 years	3 years	5 years	7 years	10 years
Ref	UBS AG	90	109	129	147	148	146
CP1	Gaz de France	27	35	42	53	57	61
CP2	Carrefour	34	42	53	67	71	76
CP3	AXA	72	83	105	128	129	128
CP4	Telecom Italia SpA	99	157	210	243	255	262

TABLE 9.5: Market CDS spreads in bps for different time horizons on March 30, 2008.

	1 year	2 years	3 years	5 years	7 years	10 years
Ref	146	355	631	1185	1612	2193
CP1	44	116	212	445	664	1005
CP2	56	138	264	558	822	1246
CP3	118	269	517	1042	1434	1964
CP4	155	504	1026	1903	2662	3670

TABLE 9.6: Cumulative default probabilities in bps for different time horizons.

four counterparties. The graphs of the left panels show the results using the parametrization (2F), while the graphs on the right panels correspond to (3F). On each graph the asset correlation ρ is fixed with, from the top to down, $\rho = 5\%$, 10%, 40% and 70%. Observe that Θ_0 is increasing in the default risk of the counterparty, increasing in the asset correlation ρ and slowly increasing in the volatility c of the factors. Table 9.10 shows the values of Θ_0 calculated with (0F).

In the receiver case, $\bar{\Theta}_0$ is shown in Fig. 9.5 as a function of the volatility parameter, c, for both parameterizations (2F) and (3F). Note that $\bar{\Theta}_0$ is much smaller and more dependent on c than is Θ_0. This is due to the absence of the common jump term in $\bar{\Theta}_0$ (see the remark 9.2.4). Also, $\bar{\Theta}_0$ is decreasing in the asset correlation ρ.

Regarding the execution times of the three model parameterizations:

- a calibration takes about 0.01, 0.30 and 0.35 seconds for (0F), (2F) and (3F), respec-

	1 year	2 years	3 years	5 years	7 years	10 years	mean	max
Ref	0.2920	0.1622	0.3957	0.3296	0.2537	0.1964	0.2716	0.3957
CP1	0.1285	0.0693	0.1556	0.1080	0.0829	0.0639	0.1014	0.1556
CP2	0.1067	0.0576	0.1897	0.1370	0.1054	0.0819	0.1130	0.1897
CP3	0.0096	0.0052	0.3108	0.2665	0.2052	0.1586	0.1593	0.3108
CP4	0.0098	0.0060	0.4711	0.5125	0.4097	0.3310	0.2900	0.5125

TABLE 9.7: Differences in bps between market spreads and calibrated spreads in the case (0F), with $\rho = 40\%$.

	1 year	2 years	3 years	5 years	7 years	10 years	mean	max
Ref	0.3343	0.2167	0.4315	0.3980	0.3120	0.3489	0.3402	0.4315
CP1	0.0719	0.0781	0.0131	0.0150	0.0305	0.1040	0.0521	0.1040
CP2	0.0345	0.0028	0.1352	0.0852	0.0598	0.0833	0.0668	0.1352
CP3	0.0203	0.0088	0.2876	0.2426	0.1461	0.1855	0.1485	0.2876
CP4	0.0698	0.0537	0.5219	0.5614	0.4584	0.3976	0.3438	0.5614

TABLE 9.8: Differences in bps between market spreads and calibrated spreads in the case (2F) with $\rho = 40\%$.

	1 year	2 years	3 years	5 years	7 years	10 years	mean	max
Ref	1.8110	1.6440	0.6820	0.8820	0.4950	0.4790	0.9988	1.8110
CP1	0.7730	0.6560	0.4300	0.2370	0.2130	0.0750	0.3973	0.7730
CP2	0.7400	0.8190	0.4300	0.2030	0.2140	0.1250	0.4218	0.8190
CP3	1.1690	0.9320	1.4230	0.5160	0.6940	0.4710	0.8675	1.4230
CP4	5.6840	3.5300	1.7190	0.6740	0.5720	0.4910	2.1117	5.6840

TABLE 9.9: Differences in bps between market spreads and calibrated spreads in the case of (3F) with $\rho = 40\%$.

	$\rho = 5\%$	$\rho = 10\%$	$\rho = 40\%$	$\rho = 70\%$
CP1	.0009	.0018	.0080	.0163
CP2	.0011	.0021	.0093	.0190
CP3	.0016	.0030	.0129	.0262
CP4	.0025	.0047	.0186	.0358

TABLE 9.10: Θ_0 for CDS contracts written on Ref in the case (0F).

tively;
• a Monte Carlo CVA computation takes about 0.015, 5 and 12 seconds for (0F), (2F) and (3F), respectively.

9.6.3 Case of a Low-Risk Reference Entity

In the previous example, except for the lowest value of ρ, the dependence of Θ_0 on c was quite limited (see Fig. 9.4). For a low-risk reference entity, however, c is expected to have more impact on Θ_0, including when ρ is larger. To show this numerically we consider a low-risk obligor, referred to as Ref′, whose piecewise-constant c.d.f. is given in Table 9.11. The clean value of the default leg is equal to $\mathrm{DL}'_0 = 0.0240$. On each graph

1 year	2 years	3 years	5 years	7 years	10 years
100	150	200	300	400	500

TABLE 9.11: Cumulative default probabilities of Ref′ (in bps).

of Fig. 9.6, the asset correlation is fixed to $\rho = 5\%, 10\%, 40\%$ or 70%. Table 9.12 shows the values of Θ_0 calculated within the parametrization (0F).

	$\rho = 5\%$	$\rho = 10\%$	$\rho = 40\%$	$\rho = 70\%$
CP1	.0002	.0006	.0031	.0073
CP2	.0003	.0007	.0035	.0080
CP3	.0004	.0009	.0046	.0096
CP4	.0007	.0014	.0061	.0108

TABLE 9.12: Θ_0 for CDS contracts on Ref′ in the case (0F).

9.6.4 CDS Options-Implied Volatilities

The goal of this subsection is to assess the level of CDS spreads volatility implied, in the sense of the Black-implied volatility of CDS options, by various intensity specifications. A payer (respectively receiver) CDS option gives the right to enter at time T_a a payer (respectively receiver) CDS with maturity T_b and contractual spread (strike) K on the reference firm. The time-0 price of the option is given by

$$\mathbb{E}\big[J^1_{T_a}\beta(T_a)v(T_a,\lambda^1_{T_a})^+\big] \text{ (respectively } \mathbb{E}\big[J^1_{T_a}\beta(T_a)v(T_a,\lambda^1_{T_a})^-\big] \text{)}, \quad (9.29)$$

where $v(t, \lambda_1)$ is the CDS pricing function of (9.6) (depending on the specification of the intensities). The Black formula for the price of CDS options can be found as equation (28) in Brigo (2005).

For the numerical tests we consider CDS options on CP1, Ref and CP4 (see Table 9.5), with $T_a = 3$, $T_b = 10$ and respective strikes of $K = 65$, 150 and 250 bps. Fig. 9.7 shows

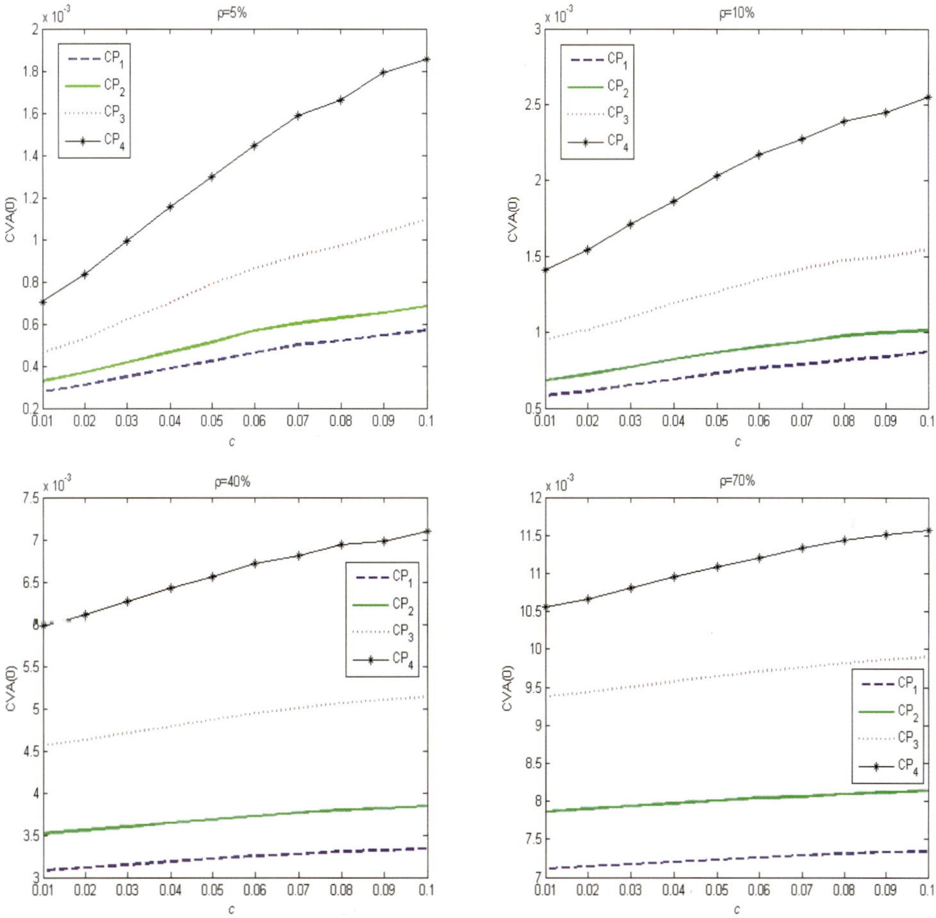

FIGURE 9.6: Θ_0 versus c for a CDS on Ref′ in the case (2F). In each graph ρ is fixed.

the implied volatility, as a function of the model volatility parameter c, of payer CDS options on each of the three names (from top to bottom) and for both parameterizations (2F) (left panels) and (3F) (right panels). The case of (0F) would essentially correspond to what happens at low c in (2F). Fig. 9.8 displays the analogous graphs in the case of receiver CDS options.

Observe that, for a given level of c, the implied volatility is typically much higher with (3F) than with (2F). This was expected, since the joint defaults intensity $\lambda_{\{0,1\}}$ is deterministic in (2F), whereas intensities are "fully stochastic" in (3F). Also, for a fixed level of c, the implied volatility is decreasing in the riskiness of the underlying name. With (3F), the implied volatility curves are nondecreasing in c, in both the payer and the receiver cases. This is not so with (2F).

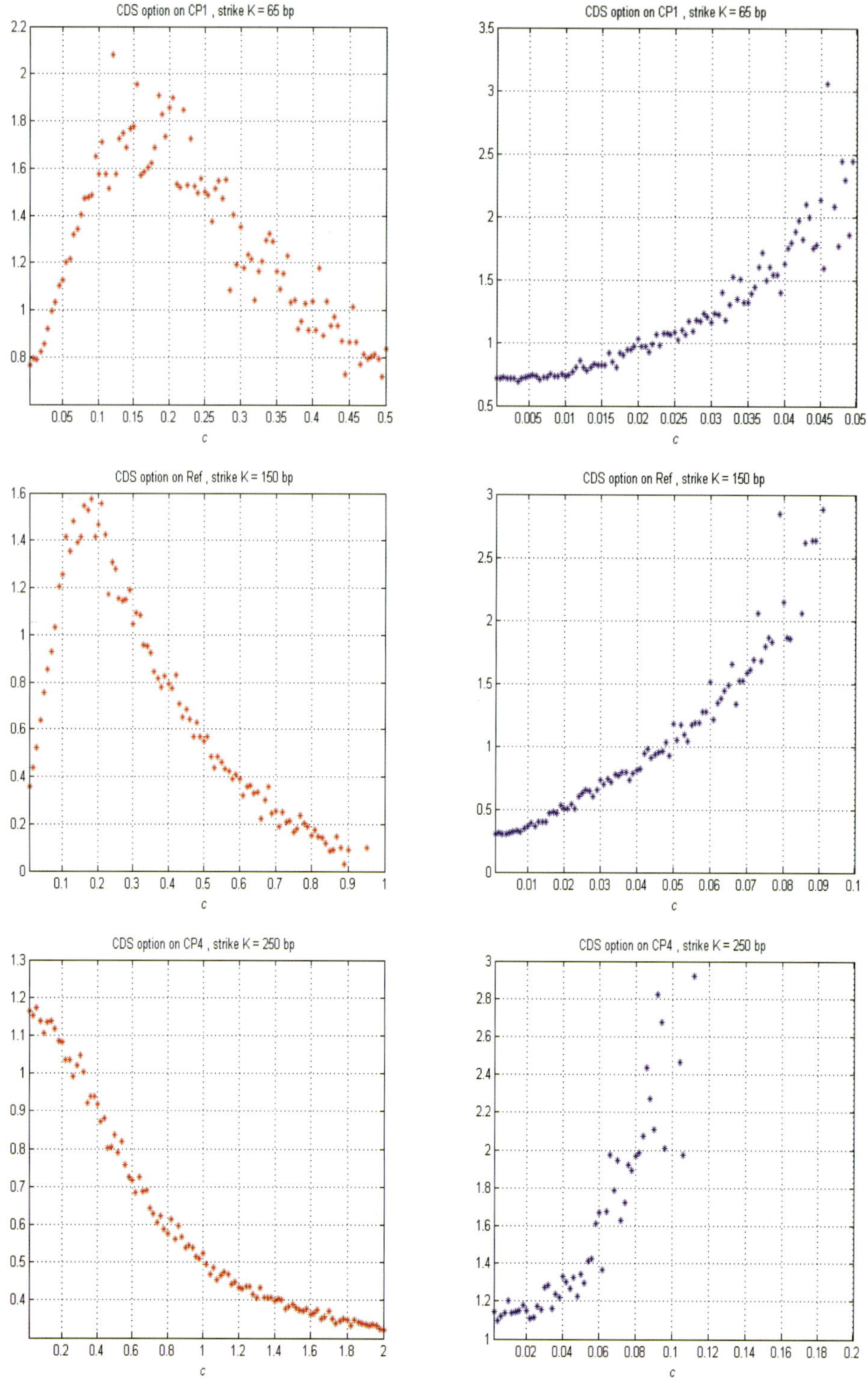

FIGURE 9.7: Implied volatility as a function of c for payer CDS options on CP1 *(top)*, Ref *(middle)* and CP4 *(bottom)*, using the model specifications (2F) *(left)* and (3F) *(right)*.

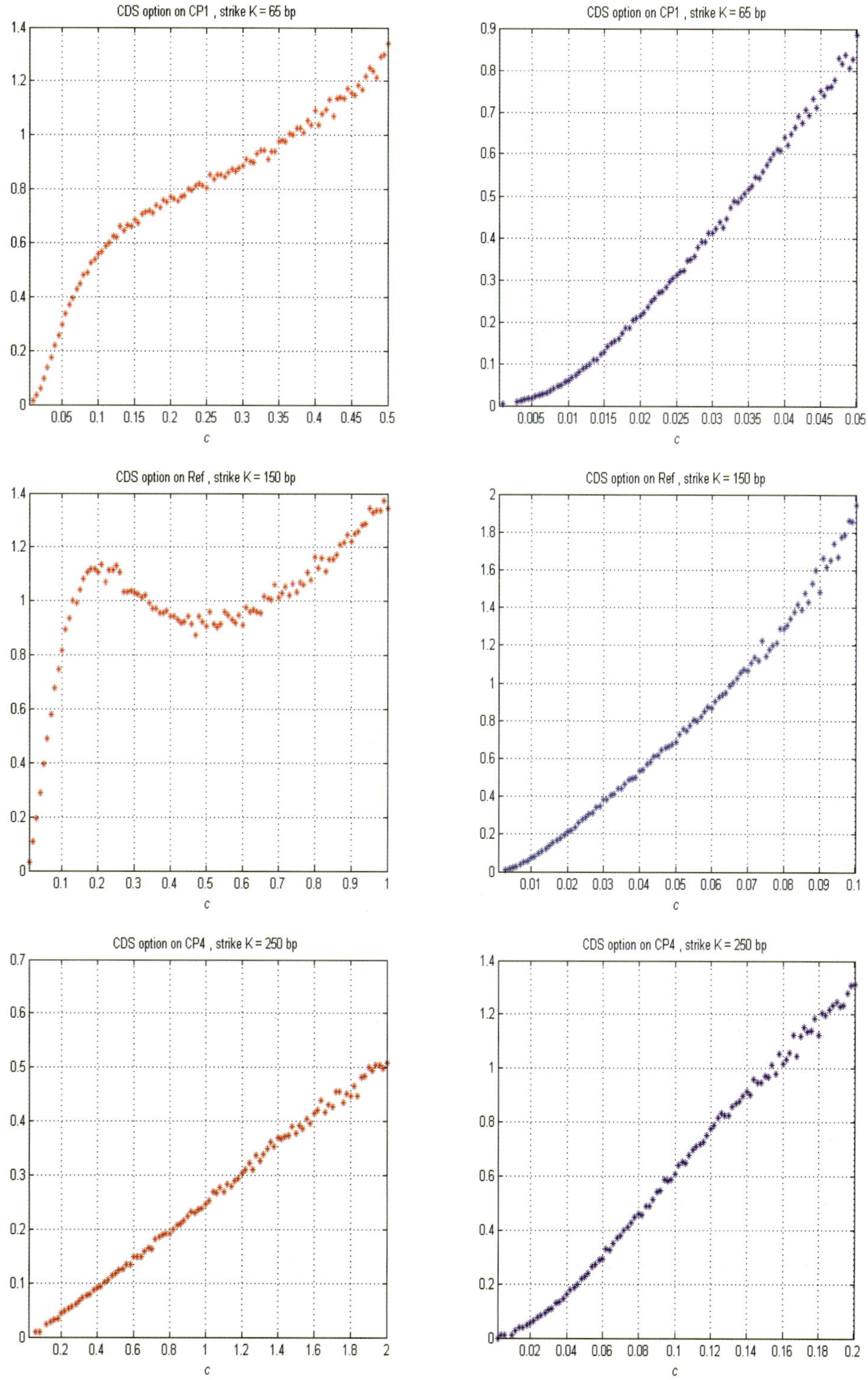

FIGURE 9.8: Implied volatility as a function of c for receiver CDS options on CP1 *(top)*, Ref *(middle)* and CP4 *(bottom)*, using the model specifications (2F) *(left)* and (3F) *(right)*.

9.6.5 Contribution of the Joint Default

In this subsection, we assess the contribution of the joint default to the CVA (parameter $\mu^\flat$ below) and to the default scenarios of the counterparty (parameter $\mu^\sharp$ below).

Recall that in the common-shock model Θ_0 has two components: joint default and mark-to-market. We introduce the parameter $\mu^\flat$ which measures the contribution of the joint default to Θ_0:

$$\mu^\flat = \frac{\Theta_0^\flat}{\Theta_0}, \text{ where } \Theta_0^\flat = \mathbb{E}\big[\mathbb{1}_{\{\tau_0<T\}}\beta(t)\tau_0\xi^\flat\big] \text{ with } \xi^\flat = (1-R_0)(1-R_1)\,\mathbb{1}_{\tau_0=\tau_1}.$$

In relation with the remark 9.2.4, it is interesting to investigate the behavior of the parameter $\mu^\flat$ with respect to the asset correlation ρ in the case where both $\mathbb{Q}(\tau_0$ is small) and $\mathbb{Q}(\tau_0 < \tau_1)$ are large. Fig. 9.9 represents $\mu^\flat$, as a function of ρ, for the most risky counterparty CP4 and for the reference entity Ref (left panel) and Ref′ (right panel). The three curves on each graph correspond to the specifications (0F), (2F) and (3F). Observe that $\mu^\flat$ always increases in ρ. Moreover, with (0F) and (2F) (and also, though to a lesser extent, with (3F)), we have that $\mu^\flat \to 1$ as $\rho \to 1$. This means that, in our model, it is the joint default that plays the essential role in determining the value of Θ_0.

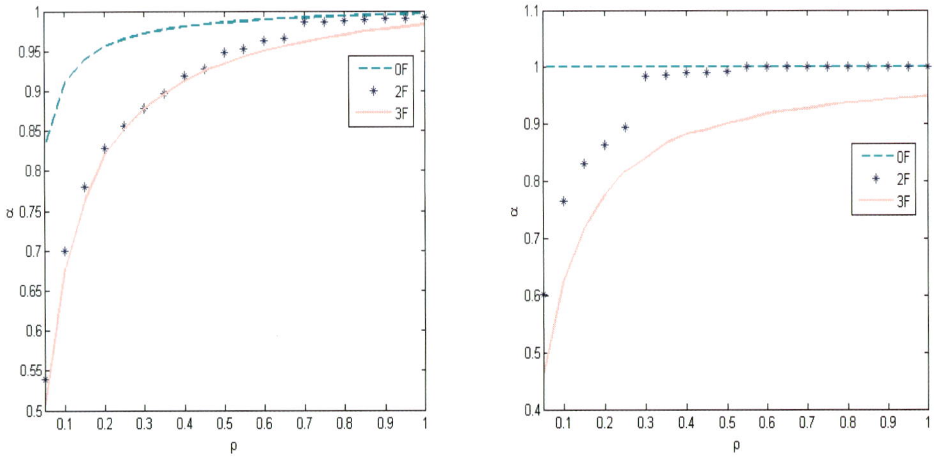

FIGURE 9.9: $\mu^\flat$ as a function of ρ for the counterparty CP4, in the case of deterministic intensities (0F) and stochastic intensities (2F) and (3F), both with $c = 0.1$. The left graphs correspond to the reference name Ref and the right graphs correspond to Ref′.

We define another parameter, $\mu^\sharp$, which represents the proportion of default scenarios of the counterparty due to its joint default with the reference entity:

$$\mu^\sharp = \frac{\sharp\{\tau_0 = \tau_1 < T\}}{\sharp\{\tau_0 = \tau_1 < T\} + \sharp\{\tau_0 < \tau_1 \wedge T\}} \approx \frac{\mathbb{Q}(\tau_0 = \tau_1 < T)}{\mathbb{Q}(\tau_0 = \tau_1 < T) + \mathbb{Q}(\tau_0 < \tau_1 \wedge T)}.$$

As already observed in Figs. 9.4 and 9.6, for a payer CDS, the volatility parameter has a greater impact on Θ_0 in the case of a safer reference name and of a lower correlation between the counterparty and the reference name. Table 9.13 shows $\mu^\sharp$ in the case of deterministic intensities (0F), for the reference name Ref′, different counterparties and different levels of the correlation ρ. Note that $\mu^\sharp$ always increases in ρ.

	$\rho = 5\%$	$\rho = 10\%$	$\rho = 40\%$	$\rho = 70\%$
CP1	.0105	.0220	.1160	.2636
CP2	.0099	.0208	.1062	.2333
CP3	.0087	.0180	.0857	.1725
CP4	.0070	.0141	.0596	.1023

TABLE 9.13: $\mu^\sharp$, as a function of ρ for Ref′ and the parametrization (0F).

The curves in Fig. 9.10 represent $\mu^\sharp$ as a function of the asset correlation ρ, for the less risky reference entity, Ref′, and the four counterparties. The graph on the left corresponds to (2F) and the graph on the right corresponds to (3F), with the model volatility parameter c set equal to 0.1 in both cases. Observe that $\mu^\sharp$ increases in ρ. This is in line with our previous observation: for a safer reference entity and a lower level of the correlation ρ, there are more scenarios where the counterparty defaults prior to the reference entity (and hence the impact on the CVA of the volatility parameter, c, is greater).

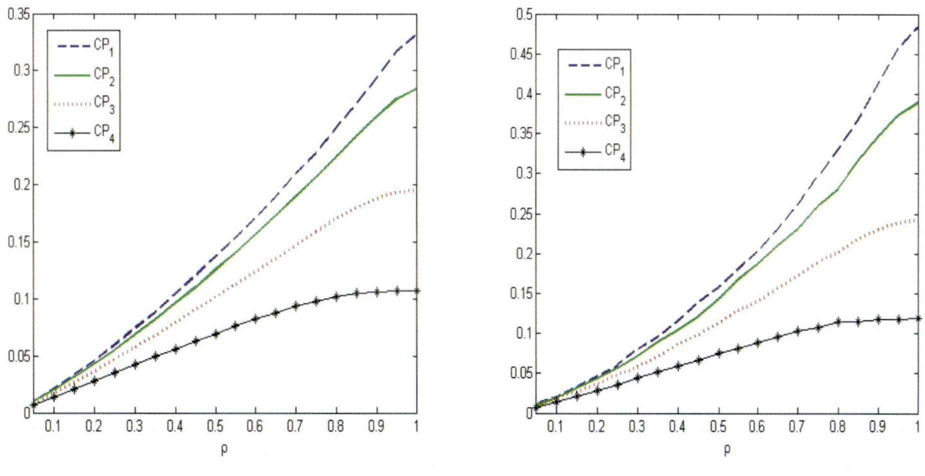

FIGURE 9.10: $\mu^\sharp$ as a function of ρ for Ref′. The left graph corresponds to (2F) and the right one corresponds to (3F). In each graph $c = 0.1$.

Conclusions

In this chapter we applied the common-shock model to the valuation of counterparty risk embedded in one CDS. The issue of "wrong-way risk", which is particularly important in the case of a payer CDS, is represented by the possibility of a simultaneous default of the counterparty and of the reference firm of the CDS.

Three specifications of the shock intensities were considered. The numerical results show that, in the case of a payer CDS on a "sufficiently risky" reference entity and for a sufficient level of correlation between the counterparty and the reference entity, a time-deterministic specification of intensities does a quick and good job.

For a receiver CD or for a payer CDS with a safe reference entity or a low correlation between the counterparty and the reference entity, a time-deterministic specification of intensities "misses" a nonnegligible component of the CVA, due to CDS spreads volatility. In this case, stochastic specifications of the intensities are preferred: CIR++ intensities obtained as sums of affine processes and deterministic functions of time or, better, fully stochastic extended CIR intensities.

Chapter 10

CVA Computations for Credit Portfolios in the Common-Shock Model

In this chapter we provide a numerical illustration of CVA computations relative to credit portfolio derivatives (portfolio of CDS contracts or CDO tranches) in the common-shock model. The notation $N = \{0, \ldots, n\}$ will be used throughout the chapter.

10.1 Portfolio of CDS

In this section we consider the situation where, with regard to each reference name $i = 1, \ldots, n$, the following products are contracted between the bank and the counterparty: either a payer CDS (default protection bought by the bank from the counterparty) or a receiver CDS (default protection sold by the bank to the counterparty) with maturity $T_i \geq T$, contractual spread S_i and recovery R_i. We write P^i for the clean price of a payer CDS on the i-th firm, so that the clean portfolio value is

$$P = \Big(\sum_{i \text{ pay}} - \sum_{i \text{ rec}}\Big) P^i.$$

Proposition 10.1.1 *For a (netted) portfolio of CDS contracts with unilateral counterparty risk, we have:*

$$\beta_t \Theta_t = \mathbb{E}_t\big[\beta_\tau \mathbb{1}_{\tau < T} \xi\big], \ t \in [0, \bar{\tau}], \tag{10.1}$$

where

$$\xi = (1 - R_0)\Big(P_\tau + \Big(\sum_{i\ pay} - \sum_{i\ rec}\Big)\mathbb{1}_{T_i = \tau < T_i}(1 - R_i) - \Gamma_\tau\Big)^+, \tag{10.2}$$

with $\Gamma = 0$ in the no collateralization case and with $\Gamma_\tau = P^+_{\tau-}$ in the unilateral continuous collateralization case. In the non-netted and non-collateralized case, we have:

$$\xi = (1 - R_0)\left(\sum_{i\ pay}\Big(P^i_\tau + \mathbb{1}_{\tau = \tau_i < T_i}(1 - R_i)\Big)^+ + \sum_{i\ rec} P^{i,-}_\tau\right). \tag{10.3}$$

Proof. An application of Proposition 3.2.8 yields (10.1), whereas, in the non-netted and non-collateralized case, (10.3) is obtained by summation of the individual exposures regarding the different CDS contracts composing the portfolio. □

10.1.1 Common-Shock Model Specification

Since we are not interested in CDO tranches in this section (these will be dealt with in the next one), we can use the following variant of the intensity structure of Sect. 8.2.3. The individual default intensities λ_t^i are assumed to be of the affine form

$$\lambda_t^i = \eta_i + X_t^{\{i\}}, \tag{10.4}$$

where η_i is a constant and the $X^{\{i\}}$ are independent homogenous CIR processes, specified as

$$dX_t^{\{i\}} = a_{\{i\}}(b_{\{i\}} - X_t^{\{i\}})\,dt + c_{\{i\}}\sqrt{X_t^{\{i\}}}\,dW^{\{i\}}. \tag{10.5}$$

Writing $\mathbf{X}_t = (X_t^{\{i\}})_{i\in N}$, the systemic shock intensities are defined, for $I \in \mathcal{I}$, as

$$\lambda_I(t, \mathbf{X}_t) = \alpha_I \inf_{i\in I} \lambda_t^i, \tag{10.6}$$

for nonnegative constant multipliers α_I such that

$$\sum_{I\in\mathcal{I}} \alpha_I \leq 1. \tag{10.7}$$

Therefore, the idiosyncratic intensities are given by:

$$\lambda_{\{i\}}(t, \mathbf{X}_t) = \lambda_t^i - \sum_{i\in I} \lambda_I(t, \mathbf{X}_t) \geq 0,$$

where the nonnegativity is ensured by the condition (10.7).

So, individual names have affine intensities λ_t^i (for CDS tractability), but, in this section, we do not have a global affine structure in the sense of the assumption 8.2.10(i) (since it is not needed here, as opposed to cases involving CDO tranches).

10.1.2 Numerical Results

We conduct numerical experiments for the three cases considered in Proposition 10.1.1. All the computations are carried out with $m = 2 \times 10^6$ Monte Carlo simulations. The recovery rates are fixed at 40%. We consider an underlying portfolio of one hundred 5yr CDS contracts, with the following market spreads (in bps):

$$
\begin{aligned}
(S_1^\star, \dots, S_{100}^\star) = \\
&(405.936, 225.937, 620.786, 195.083, 37.97, 32.17,\\
&1743.673, 348.411, 399.788, 297.902, 3013.286,\\
&359.909, 327.962, 2085.618, 145.519, 234.948, 212.135,\\
&120.000, 124.000, 304.845, 225.904, 39.78,\\
&28.320, 229.577, 291.100, 349.071, 132.982, 889.620,\\
&28.110, 25.110, 311.131, 210.919, 368.858,\\
&480.993, 359.483, 200.581, 164.500, 127.000, 456.170,\\
&130.027, 229.912, 343.500, 361.515, 300.346,\\
&583.736, 342.688, 133.451, 141.984, 2440.458, 579.000,\\
&306.745, 324.709, 647.019, 433.597, 201.960,\\
&192.860, 243.031, 296.210, 333.747, 295.873, 374.750,\\
&270.432, 436.182, 430.537, 127.000, 145.043,\\
&52.270, 29.070, 123.000, 45.880, 29.930, 31.560,\\
&190.000, 361.990, 36.980, 31.440, 33.880,\\
&16.570, 39.000, 73.560, 70.470, 24.840, 21.270, 47.880,\\
&53.450, 31.190, 499.517, 1092.000, 26.770, 49.810,\\
&60.400, 180.000, 27.960, 130.876, 123.000, 47.110,\\
&190.000, 38.170, 42.730, 200.000).
\end{aligned}
\tag{10.8}
$$

We divide the portfolio into two categories, respectively composed of 70 payer CDS contracts and 30 receiver CDS contracts. The choice of the names for the payer CDS and receiver CDS is made by taking the first 70 names from (10.8) and declaring them to represent payer contracts and declaring the remaining 30 names to represent receiver contracts. Next, setting $n = 100$ and indexing the counterparty by 0, we define I_1, I_2 and I_3 as the sets of the 20, 70 and 101 riskiest obligors, respectively, as measured by the spread of the corresponding five year market CDS quotes (based on the list (10.8) sorted in decreasing order). Thus, we consider a nested grouping of the reference names and of the counterparty. The names in I_1, $I_2 \setminus I_1$ and $I_3 \setminus I_2$ are referred to as high, middle and low credit risk names, respectively. Moreover, in the default intensities (10.4)-(10.5), we assume a common factor process $X^{\{i\}} = X^j$ with parameters a_j, b_j, c_j for all names in $I_j \setminus I_{j-1}$, for $j = 1, 2, 3$ (where $I_0 = \emptyset$). This setup corresponds to taking a common factor process for each group of obligors. We assume that all obligors with high credit risk have the same credit risk parameters, i.e. I_1 is a homogenous group with the CIR parameters given in the upper row of Table 10.1. Similarly, all the obligors with middle credit risk, i.e. belonging to $I_2 \setminus I_1$, have the same CIR parameters shown in the middle row of Table 10.1; all the obligors with low credit risk, i.e. belonging to $I_3 \setminus I_2$, have the same CIR parameters shown in the bottom row of Table 10.1.

Credit Risk Level	a_j	b_j	c_j	X_0^j
high ($j = 1$)	0.50	0.05	0.2	0.05
middle ($j = 2$)	0.80	0.02	0.1	0.02
low ($j = 3$)	0.9	0.001	0.01	0.001

TABLE 10.1: Parameters for CIR processes for the portfolio of CDS contracts

Given the factors processes X^j specified as in Table (10.1), the constants η_i of (10.4) for the reference names are fitted to the spreads listed in (10.8), yielding:

$$(\eta_1^\star, \ldots, \eta_{100}^\star) =$$

$$\begin{aligned}(&0.0194, 0.0178, 0.0552, 0.0126, 0.0053, 0.0044, 0.2424, 0.0382, 0.0184, 0.0297, 0.4540,\\ &0.0401, 0.0348, 0.2993, 0.0044, 0.0193, 0.0155, 0.0001, 0.0008, 0.0309, 0.0177, 0.0056,\\ &0.0037, 0.0184, 0.0286, 0.0383, 0.0023, 0.1000, 0.0037, 0.0032, 0.0320, 0.0153, 0.0132,\\ &0.0319, 0.0400, 0.0135, 0.0075, 0.0013, 0.0278, 0.0018, 0.0184, 0.0373, 0.0404, 0.0302,\\ &0.0490, 0.0372, 0.0023, 0.0038, 0.3585, 0.0482, 0.0312, 0.0342, 0.0596, 0.0240, 0.0138,\\ &0.0122, 0.0206, 0.0295, 0.0357, 0.0294, 0.0142, 0.0252, 0.0244, 0.0235, 0.0013, 0.0043,\\ &0.0077, 0.0038, 0.0006, 0.0066, 0.0040, 0.0043, 0.0118, 0.0404, 0.0052, 0.0042, 0.0046,\\ &0.0018, 0.0055, 0.0113, 0.0107, 0.0031, 0.0025, 0.0070, 0.0079, 0.0042, 0.0350, 0.1337,\\ &0.0035, 0.0073, 0.0091, 0.0101, 0.0037, 0.0019, 0.0006, 0.0069, 0.0118, 0.0054, 0.0061,\\ &0.0134).\end{aligned} \tag{10.9}$$

We consider a counterparty with a stochastic default intensity and CDS spread $S_0^\star = 10, 20, 60, 100, 120, 300, 400$ or 500 bps. Table 10.2 gives the behavior of the CVA (with the standard deviation of each Monte Carlo CVA estimate in parentheses) with respect to the default intensity of the counterparty, in the three cases of no netting and no margining, netting and no margining and both netting and margining. The heading "Counterparty risk type" for the first column refers to the classification of the counterparty into one of the homogenous groups $I_1, I_2 \setminus I_1$ or $I_3 \setminus I_2$, whose respective default intensities are modeled using CIR parameters corresponding to high, middle and low credit risk regimes, respectively. As expected, the CVA increases with the riskiness of the counterparty. More precisely, since increasing the riskiness of the counterparty does not increase $\lambda_{I_3}(t, \mathbf{x})$ as long as the counterparty stays in the "low" risk group, the joint default intensity of the counterparty is constant for all rows in this scenario and the CVA only increases rather slowly. However, in subsequent rows of the table, with spreads $S_0^\star$ of 120 or 300 bps, the counterparty belongs to the "middle" risk group, its joint default intensity is higher and the net effect is a big increase of the CVA, even in the case of continuous collateralization. In Fig. 10.1.2 we give the EPE curves (functions of time (3.26)), obtained by fitting a quadratic polynomial through the values of ξ for $\tau < T$, for the cases of no netting with no margining, netting with no margining and netting with margining. In each case the EPE curves are given for $S_0^\star = 20$ bps, $S_0^\star = 60$ bps and $S_0^\star = 100$ bps. To explain the pattern of the EPE curve, we need to consider the ratios $\mathbb{E}[\lambda_{I_j}(t, \mathbf{X_t})/\lambda_t^0]$ (see Table 10.3), which give the probability of a joint default of the counterparty, conditional on its default. When the CDS spread of the counterparty $S_0^\star$ increases, as long as the counterparty risk profile (low, middle or high) does not change, the numerator in this formula remains unchanged, whereas the denominator increases. Therefore the ratio decreases, meaning that the conditional probability of a joint default decreases. Since joint defaults are responsible for most of the exposure, as $S_0^\star$ increases the EPE curves become lower, as we can see in the figures. However, if the riskiness of the counterparty increases such that it becomes part of the group I_2 and/or I_3, then the joint default probability will increase. Hence, the EPE curve becomes higher, as we can see in the figures.

Table 10.2 also shows that, if netting does have a significant mitigation impact on the CVA, this does not seem to be the case with margining. Indeed, due to joint defaults, counterparty risk on CDS contracts simply can't be mitigated by the kind of margining that we use here.

Counterparty risk type	$S_0^\star$	CVA No Nett. No Marg.	CVA Nett. No Marg.	CVA Nett. with Marg.
low	10	488.4 (7.0)	262.1 (3.8)	256.7 (3.8)
low	20	808.4 (8.9)	433.9 (4.9)	423.0 (4.8)
low	60	834.2 (8.9)	448.5 (4.8)	415.9 (4.8)
low	100	860.4 (8.8)	463.2 (4.8)	409.8 (4.8)
middle	120	5440.2 (21.4)	4338.1 (17.2)	4256.6 (17.1)
middle	300	5364.1 (21.0)	4243.1 (16.9)	4076.8 (16.8)
high	400	8749.9 (22.1)	7211.3 (18.1)	6943.0 (18.0)
high	500	8543.5 (21.8)	7017.9 (17.8)	6713.8 (17.7)

TABLE 10.2: CVA for the portfolio of CDS contracts ($(\alpha_{I_1}, \alpha_{I_2}, \alpha_{I_3}) = (0.3, 0.3, 0.3)$).

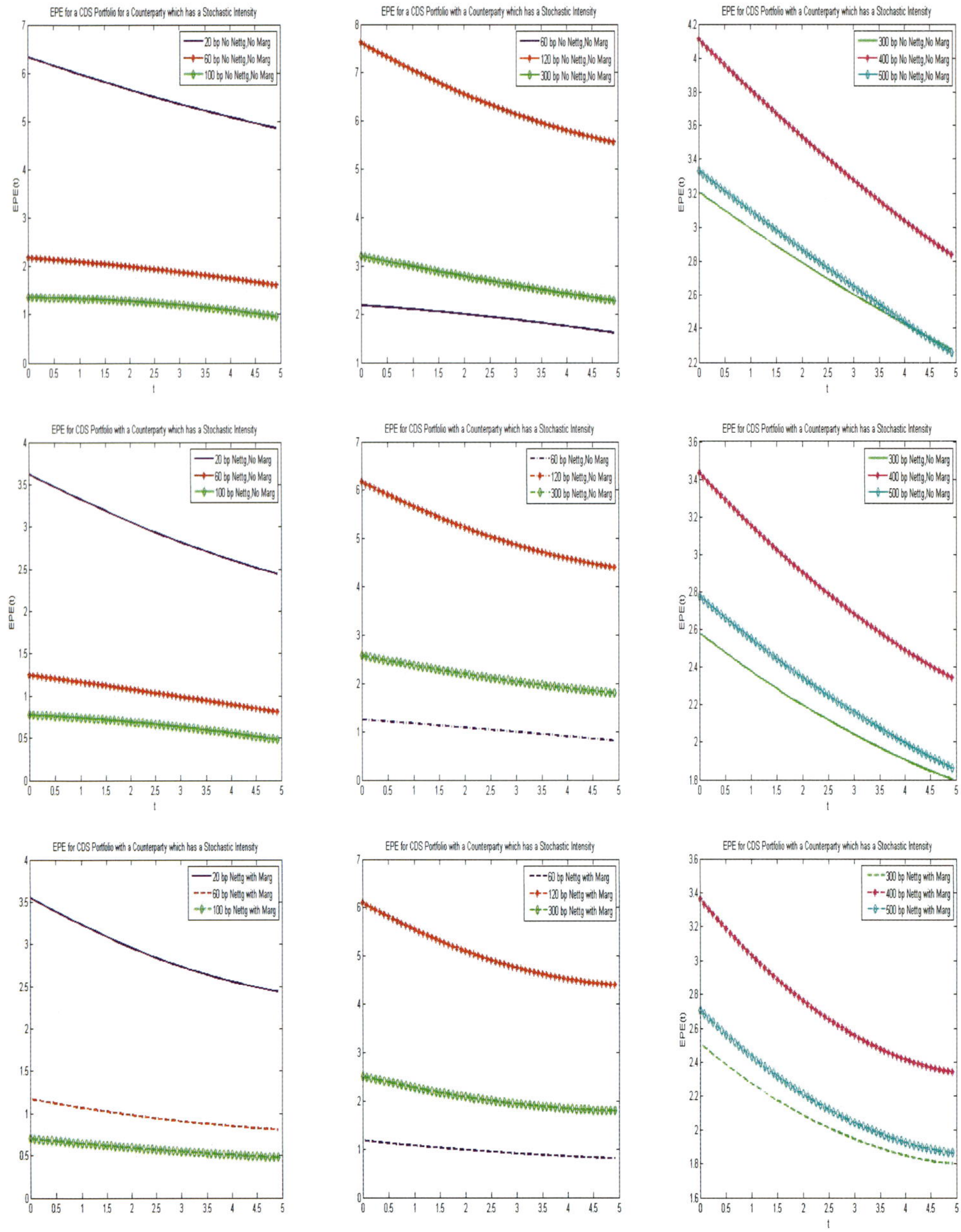

FIGURE 10.1: EPE for no netting and no margining *(top)*, netting and no margining *(middle)* and both netting and margining *(bottom)*.

Counterparty risk type	$S_0^\star$	$\mathbb{1}_{0\in I_1}$ × $\mathbb{E}\left[\lambda_{I_1}(t,\mathbf{X}_t)/\lambda_t^0\right]$	$\mathbb{1}_{0\in I_2}$ × $\mathbb{E}\left[\lambda_{I_2}(t,\mathbf{X}_t)/\lambda_t^0\right]$	$\mathbb{1}_{0\in I_3}$ × $\mathbb{E}\left[\lambda_{I_3}(t,\mathbf{X}_t)/\lambda_t^0\right]$
low	20	0	0	0.2485
low	60	0	0	0.0828
low	100	0	0	0.0497
middle	120	0	0.3000	0.0429
middle	300	0	0.1222	0.0166
high	400	0.2809	0.0937	0.0126
high	500	0.2247	0.0744	0.0100

TABLE 10.3: Dependence of EPE on $\mathbb{E}\left[\lambda_{I_j}(t,\mathbf{X}_t)/\lambda_t^0\right]$ with $t=\frac{1}{12}$, for the portfolio of CDS contracts.

10.2 CDO Tranches

Let N_t represent the cumulative default process, i.e. the number of firms defaulted by time t within a reference pool of n credit names. For simplicity we consider protection legs of CDO tranches with cumulative payoff process given, for $t \in [0,T]$, by:

$$\phi(N_t) = ((1-R)\frac{N_t}{n} - a)^+ \wedge (b-a), \tag{10.10}$$

for a homogenous recovery R of the reference names and where the attachment and detachment points a and b are such that $0 \le a \le b \le 1$. Assuming $r=0$, the related clean price process P_t is written, for $t \in [0,T]$, as:

$$P_t = \mathbb{E}_t(\phi(N_T) - \phi(N_t)). \tag{10.11}$$

By application of Proposition 3.2.8, the CVA of the bank can be represented by the process Θ_t given, for $t \in [0,\bar{\tau}]$, by:

$$\Theta_t = \mathbb{E}_t(\mathbb{1}_{\tau<T}\xi), \tag{10.12}$$

for an exposure ξ given as

$$\xi = (1-R_0)\left(P_\tau + \Delta_\tau - \Gamma_\tau\right)^+, \tag{10.13}$$

where $\Delta_t = \phi(N_t) - \phi(N_{t-})$ is the jump process of the payoff process and where Γ_t is the collateral. So, under the so-called naked (no collateralization) scheme $\Gamma = 0$ and the (fully collateralized) scheme $\Gamma = P_-$, we have, respectively:

$$\begin{aligned} \Theta_t &= \Theta_t^0 = \mathbb{E}_t(\mathbb{1}_{\tau<T}\xi^0) \text{ with } \xi^0 = (1-R_0)\left(P_\tau + \Delta_\tau\right)^+, \\ \Theta_t &= \Theta_t^1 = \mathbb{E}_t(\mathbb{1}_{\tau<T}\xi^1) \text{ with } \xi^1 = (1-R_0)\left(P_\tau - P_{\tau-} + \Delta_\tau\right)^+. \end{aligned} \tag{10.14}$$

In the common-shock model of Chapter 8 (here with no factor process $\mathbf{X}$), the simulation of the counterparty's default time $\tau = \tau_0$ and the computation of P_τ, Δ_τ and $P_{\tau-}$ are fast and exact. This results in the following Monte Carlo estimates $\widehat{\Theta}_0^0$ and $\widehat{\Theta}_0^1$ for Θ_0^0 and Θ_0^1, based on m_0 and m_1 simulated trajectories:

$$
\begin{aligned}
\widehat{\Theta}_0^0 &= (1 - R_0)\frac{1}{m_0}\sum_{j=1}^{m_0} \mathbb{1}_{\tau^j < T}\left(P_{\tau^j} + \Delta_{\tau^j}\right)^+ , \\
\widehat{\Theta}_0^1 &= (1 - R_0)\frac{1}{m_1}\sum_{j=1}^{m_1} \mathbb{1}_{\tau^j < T}\left(P_{\tau^j} - P_{\tau^j -} + \Delta_{\tau^j}\right)^+ .
\end{aligned}
\tag{10.15}
$$

Note that these Monte Carlo estimates involve no time-discretization nor approximation error regarding the valuation of the CDO tranche at time τ. The only error is the $O(m^{-\frac{1}{2}})$-statistical error.

10.2.1 Numerical Results

We consider a stylized credit portfolio of 100 obligors, for CDO tranches with maturity $T = 2$ years and nominal Nom $= 100$, for recovery rates $R_0 = R = 40\%$. The individual default intensities are taken as $\lambda_{\{i\}} = 10^{-4} \times (200 - i)$ (decreasing from ~ 200 bps to 100 bps as i increases from 0 to 99). We use four nested groups of joint defaults, i.e. $\nu = 3$, respectively consisting of the riskiest 3, 9, 21 and 100 (i.e. all) names, with corresponding joint default intensities $\lambda_I = 10^{-3} \times \frac{2}{1+l}$ (decreasing from about ~ 20 bps to 5 bps as ν increases from 0 to 3). The counterparty is taken as the fiftieth riskiest name (name with median risk) in the portfolio. The results are presented in Table 10.4, where:

- the CVA numbers correspond to a notional of one hundred per obligor; so, for instance, the maximal loss on the equity tranche 0-5 is $Nom \times n \times 5\% = 500$,
- σ is the standard error associated with each Monte Carlo CVA estimate,
- $\%\sigma$ is the percentage error in the sense of $10^2 \times \frac{\sigma}{\mathrm{CVA}}$.

Observe that the impact of the collateralization is only significant for the equity tranche. So, for the equity tranche 0-5, the naked CVA is $\widehat{\Theta}_0^0 = 4.78$, whereas the collateralized CVA is $\widehat{\Theta}_0^1 = 3.41$. For the 35+ tranche these respective numbers become 2.44 and 2.26. That's because the senior tranche CVA is due less to the clean price term P_τ in (10.13) and more to the joint defaults term Δ_τ, where the latter is hardly collateralizable (see also Sect. 10.1.2 in this regard).

A challenging question is how to extend the CVA computations of this chapter to the presence of funding costs, in a bilateral counterparty risk setup. Then one has to deal with nonlinear and very high-dimensional TVA equations. See Chapter 12 and Crépey and Song (2014).

	Naked			Collateralized		
Tranche	0-5	5-35	35+	0-5	5-35	35+
CVA	4.78	2.96	2.44	3.41	2.73	2.26
σ	0.08	0.24	0.20	0.05	0.16	0.14
$\%\sigma$	1.6	8.1	8.2	1.4	6.0	6.0

TABLE 10.4: *Naked versus collateralized CVA on CDO tranches ($m_0 = 1.5 \times 10^5, m_1 = 3 \times 10^5$).*

Part V

Further Developments

Chapter 11

Rating Triggers and Credit Migrations

11.1 Introduction

Modeling, managing and mitigating counterparty risk is a crucial task for all financial institutions. One of the most popular mitigation techniques used by market participants is the inclusion of additional termination events (ATE) in OTC transactions. As defined in Section 5(b)(vi) of International Swaps and Derivatives Association (2002), ATEs allow institutions to terminate and close out the derivatives transactions with a counterparty if a termination event occurs. Here, we will consider a particular and in fact the most common termination event: rating triggers.

A rating trigger is defined as a threshold credit rating level that agreed to upon the of the contract. If the credit rating of the counterparty or the bank decreases below the trigger level, the contract is terminated and closed out without MtM loss. Therefore, from the perspective of the bank, rating triggers provide additional protection from a potential default of a counterparty with a deteriorating credit rating by allowing the bank to terminate the contract prior to a default event, and analogously from the perspective of the counterparty.

Even though mitigating of the counterparty risk via rating triggers is recognized as a very important risk controlling tool, there is but little literature on counterparty risk modeling with rating triggers. In Yi (2011), CVA valuation with rating triggers is studied for optional and mandatory termination events, and a compound Poisson model is introduced for modeling rating transitions and default probabilities. Zhou (2013) considers practical problems regarding CVA valuation with additional termination events under a simple model and from a practitioner's point of view. Mercurio, Caccia, and Cutuli (2012) studied a similar problem and introduced a valuation model by proposing several assumptions to simplify the CVA computations (in the case of unilateral counterparty risk). However, a comprehensive approach which involves the joint modeling of rating transitions in a risk-neutral setting and the dynamic, ratings-dependent collateralization has not been studied in the literature.

In this chapter, we consider the problem of collateralized bilateral TVA valuation with rating triggers and credit migrations. We first find the TVA representation in the presence of rating triggers. We show that the value of the contract also needs to be adjusted for the rating triggers. This new adjustment term is called the rating valuation adjustment (RVA). We show that RVA represents the expected loss in case of a default event that is preceded by a trigger event. In the bilateral case, we see that RVA is decomposed into two components: CRVA and DRVA, representing the rating valuation adjustments for the counterparty's and the bank's rating triggers. Furthermore, we consider dynamic collateralization using the rating transitions. In this framework, the collateral thresholds are defined as the functionals of the current credit ratings of the counterparty and the bank. In practice such rating-dependent margin agreements are standard and they are described in the Credit Support Annex (CSA).

One of the key issues in valuation of TVA in the presence of rating triggers is the issue of modeling of ratings migrations of the two parties and, in particular, the issue of modeling of dependence between their ratings migrations. We employ the Markov copula approach (cf. Chapter 14) for modeling the joint rating migrations and the default probabilities of the counterparty and the bank.

We illustrate our results with numerical examples, analyzing the impact of early termination clauses and dynamic collateralization on the TVA in case of an IRS and a CDS contract.

This chapter is organized as follows. In Sect. 11.2 we present a general framework for the valuation of collateralized total valuation adjustment in the presence of rating triggers. In Sect. 11.3 we employ the Markov copula approach for modeling the joint rating transitions of the counterparty and the bank. In Sect. 11.4 we present numerical results in case of a CDS contract and an IRS contract.

11.2 Credit Value Adjustment and Collateralization under Rating Triggers

In the model we propose in this chapter, the counterparty risk associated with the contract between the bank and the counterparty will be sensitive to the current creditworthiness of the two parties. We postulate that the creditworthiness of each party is represented by the same $\mathcal{K} = \{1, 2, \ldots, K\}$ rating categories. We will use the convention that the ratings are ordered from the best, i.e. 1, to the worst, i.e. K, and that the level K corresponds to a default.

To model the evolution of the credit worthiness of the bank and of the counterparty we introduce two $\mathcal{K}$-valued processes X^c and X^b on $(\Omega, \mathbb{G}, \mathbb{Q})$. The processes X^c and X^b represent the evolution of the credit ratings of the counterparty and the bank. The two parties are default prone and the respective default times are given as

$$\tau_i = \inf\{t > 0 : X^i_t = K\}, \quad i = c, b.$$

The case $K = 2$, allowing only the default and the pre-default states prevail, corresponds to the models presented and discussed in the rest of the book. We consider CSA contracts with $Q = P$. The mechanics and the modeling of the collateral process Γ in the present set-up are discussed in Sect. 11.2.2.

11.2.1 Pricing Bilateral Counterparty Risk with Rating Triggers

We now proceed with introducing the rating trigger times and the closeout cash flows in the TVA valuation. We also show how the clean price of our OTC contract must be adjusted for the counterparty risk and the rating triggers.

We consider the following rating trigger clause. We assume that if the bank's or the counterparty's credit rating deteriorates to or below the trigger level (except the default level), the contract is terminated and closed out. In particular, there are no MtM losses associated with the trigger events. The trigger levels are set as K_c for the counterparty and K_b for the bank, where $1 < K_c, K_b \leq K$. Let τ_i' represent the first time that the i-th party's credit rating crosses its rating trigger, i.e.

$$\tau_i' = \inf\{t > 0 : X_t^i \geq K_i\}, \quad i = c, b,$$

assuming that $X_0^i < K_i$ for $i = c, b$. We set

$$\tau' = \tau_c' \wedge \tau_b', \ \bar{\tau}' = \tau' \wedge T, \ H_t' = \mathbb{1}_{\{\tau' \leq t\}}.$$

The closeout of the cumulative dividend process of the counterparty risky contract needs to account for the MtM exchange without incurring any losses at a trigger time other than default. On the other hand, if a trigger event occurs simultaneously with a default event, the deal will be settled according to the default event. Consequently, defining a $\mathcal{G}_{\tau'}$-measurable random variable χ' as (compare with (3.1))

$$\chi' = P_{\tau'} + \Delta_{\tau'} - \Gamma_{\tau'},$$

we propose the following definition of the cumulative dividend process of the counterparty risky contract in the presence of rating triggers (compare with (3.2)):

Definition 11.2.1 *The CSA closeout cash flow of an OTC contract subject to rating triggers is defined as*

$$\begin{aligned} \mathfrak{R}' = \Gamma_{\tau'} + \mathbb{1}_{\tau'=\tau_b}\big(R_b\chi'^{,+} - \chi'^{,-}\big) - \mathbb{1}_{\tau'=\tau_c}\big(R_c\chi'^{,-} - \chi'^{,+}\big) \\ - \mathbb{1}_{\tau'=\tau_b=\tau_c}\chi' + \mathbb{1}_{\tau'<\tau}\chi'. \end{aligned} \tag{11.1}$$

Accordingly, the price process associated with a counterparty risky contract with rating triggers is defined as follows,

Definition 11.2.2 *The price process Π_t' and the cumulative value process $\widehat{\Pi}_t'$ of a counterparty risky contract with rating triggers are defined as*

$$\beta_t\Pi_t' = \mathbb{E}_t\Big[\int_t^{\bar{\tau}'} \beta_s dC_s + \beta_{\bar{\tau}'}\mathbb{1}_{\tau'<T}\mathfrak{R}'\Big], \ \beta_t\widehat{\Pi}_t' = \beta_t\Pi_t' + \Big[\int_0^t \beta_u dC_u\Big],$$

for all $t \in [0, \bar{\tau}']$.

We now introduce the total valuation adjustment term when the underlying contract is subject to rating triggers.

Definition 11.2.3 *The bilateral total valuation adjustment with rating triggers is defined, for* $t \in [0, \bar{\tau}']$, *by:*

$$\Theta'_t = \widehat{P}_t - \widehat{\Pi}'_t. \tag{11.2}$$

Note that

$$\Theta'_t = P_t - \Pi'_t,$$

for $t \in [0, \bar{\tau}')$.

Let the modified exposure at default ξ' be defined as (compare (3.13), recalling we use $Q = P$ in this chapter):

$$\begin{aligned}\xi' = P_{\tau'} + \Delta_{\tau'} - \mathfrak{R}' &= \chi' - \mathbb{1}_{\tau'=\tau_c}\left(R_c\chi'^{,+} - \chi'^{,-}\right) + \mathbb{1}_{\tau'=\tau_b}\left(R_b\chi'^{,-} - \chi'^{,+}\right) \\ &\quad + \mathbb{1}_{\tau'=\tau_b=\tau_c}\chi' - \mathbb{1}_{\tau'<\tau}\chi' \\ &= \mathbb{1}_{\tau'=\tau_c}(1 - R_c)\chi^{+} - \mathbb{1}_{\tau'=\tau_b}(1 - R_b)\chi^{-}.\end{aligned} \tag{11.3}$$

The following representation generalizes Proposition 3.2.8 (for $Q = P$ here).

Proposition 11.2.4 *The bilateral total valuation adjustment defined in* (11.2) *can be represented, for* $t \in [0, \bar{\tau}']$, *as:*

$$\beta_t\Theta'_t = \mathbb{E}_t\left[\beta_\tau \mathbb{1}_{\tau'<T}\xi'\right]. \tag{11.4}$$

Proof. The proof is the same as that of Proposition 3.2.8, replacing τ there by τ'. □

Note that since there are no losses associated with the trigger events other than defaults, and since the TVA only reflects the expected losses, these cases do not appear in (11.4).

Now, similar to (3.15), we can decompose Θ' as

$$\Theta' = \mathrm{CVA}' + \mathrm{DVA}',$$

where

$$\begin{aligned}\mathrm{CVA}'_t &= \beta_t^{-1}\mathbb{E}_t[\mathbb{1}_{\{\tau'=\tau_c\le T\}}\beta_{\tau'}(1 - R_c)(P_{\tau'} + \Delta_{\tau'} - \Gamma_{\tau'})^{+}], \\ \mathrm{DVA}'_t &= -\beta_t^{-1}\mathbb{E}_t[\mathbb{1}_{\{\tau'=\tau_b\le T\}}\beta_{\tau'}(1 - R_b)(P_{\tau'} + \Delta_{\tau'} - \Gamma_{\tau'})^{-}],\end{aligned}$$

for $t \in [0, \bar{\tau}']$.

It is important to observe the difference between TVA and TVA′ processes, which indicates the change in the TVA due to rating triggers. This leads us to introduce the following concept,

Definition 11.2.5 *The Rating Valuation Adjustment (RVA) process is defined as*

$$\mathrm{RVA}_t = \Theta_t - \Theta'_t, \tag{11.5}$$

for $t \in [0, \bar{\tau}']$.

The rating valuation adjustment term defined above has the following representation.

Proposition 11.2.6 *The* RVA *process can be represented as*

$$\begin{aligned}\mathrm{RVA}_t &= \beta_t^{-1}\mathbb{E}_t[\mathbb{1}_{\{\tau'<\tau=\tau_c<T\}}\beta_\tau(1-R_c)(P_\tau+\Delta_\tau-\Gamma_\tau)^+] \\ &\quad - \beta_t^{-1}\mathbb{E}_t[\mathbb{1}_{\{\tau'<\tau=\tau_b<T\}}\beta_\tau(1-R_b)(P_\tau+\Delta_\tau-\Gamma_\tau)^-],\end{aligned}$$

for all $t \in [0, \bar{\tau}')$.

Proof. From (3.15) and (11.4) we obtain

$$\begin{aligned}\beta_t(\Theta_t - \Theta'_t) &= \mathbb{E}_t\Big[\mathbb{1}_{\{\tau=\tau_c<T\}}\beta_{\tau_c}(1-R_c)(P_{\tau_c}+\Delta_{\tau_c}-\Gamma_{\tau_c})^+\Big] \\ &\quad - \mathbb{E}_t\Big[\mathbb{1}_{\{\tau=\tau_b<T\}}\beta_{\tau_b}(1-R_b)(P_{\tau_b}+\Delta_{\tau_b}-\Gamma_{\tau_b})^-\Big] \\ &\quad - \mathbb{E}_t\Big[\mathbb{1}_{\{\tau'=\tau_c<T\}}\beta_{\tau_c}(1-R_c)(P_{\tau_c}+\Delta_{\tau_c}-\Gamma_{\tau_c})^+\Big] \\ &\quad + \mathbb{E}_t\Big[\mathbb{1}_{\{\tau'=\tau_b<T\}}\beta_{\tau_b}(1-R_b)(P_{\tau_b}+\Delta_{\tau_b}-\Gamma_{\tau_b})^-\Big],\end{aligned}$$

which reduces to

$$\begin{aligned}\beta_t(\Theta_t - \Theta'_t) &= \mathbb{E}_t\Big[(\mathbb{1}_{\{\tau=\tau_c<T\}} - \mathbb{1}_{\{\tau'=\tau_c<T\}})\beta_{\tau_c}(1-R_c)(P_{\tau_c}+\Delta_{\tau_c}-\Gamma_{\tau_c})^+\Big] \\ &\quad - \mathbb{E}_t\Big[(\mathbb{1}_{\{\tau=\tau_b<T\}} - \mathbb{1}_{\{\tau'=\tau_b<T\}})\beta_{\tau_b}(1-R_b)(P_{\tau_b}+\Delta_{\tau_b}-\Gamma_{\tau_b})^-\Big],\end{aligned}$$

which in turn is equivalent to

$$\begin{aligned}\beta_t(\Theta_t - \Theta'_t) &= \mathbb{E}_t\Big[\mathbb{1}_{\{\tau'<\tau=\tau_c<T\}}\beta_{\tau_c}(1-R_c)(P_{\tau_c}+\Delta_{\tau_c}-\Gamma_{\tau_c})^+\Big] \\ &\quad - \mathbb{E}_t\Big[\mathbb{1}_{\{\tau'<\tau=\tau_b<T\}}\beta_{\tau_b}(1-R_b)(P_{\tau_b}+\Delta_{\tau_b}-\Gamma_{\tau_b})^-\Big].\end{aligned}$$

□

Remark 11.2.7 Note that RVA can be positive or negative. If RVA is positive (in particular, in the case of unilateral counterparty risk with $\tau_b = \infty$ a.s.), then there is a decrease in the total adjustment, i.e. a benefit to the bank. If RVA is negative, then this indicates an increase in the total adjustment, i.e. a cost to the bank, due to adding rating triggers.

Note that, for $t \in [0, \bar{\tau}']$, we can define:

$$\begin{aligned}\mathrm{CRVA}_t &= \beta_t^{-1}\mathbb{E}_t[\mathbb{1}_{\{\tau'<\tau=\tau_b<T\}}\beta_{\tau'}(1-R_b)(P_\tau+\Delta_\tau-\Gamma_\tau)^-], \\ \mathrm{DRVA}_t &= -\beta_t^{-1}\mathbb{E}_t[\mathbb{1}_{\{\tau'<\tau=\tau_c<T\}}\beta_{\tau'}(1-R_c)(P_\tau+\Delta_\tau-\Gamma_\tau)^+].\end{aligned}$$

Then RVA has the following decomposition:

$$\mathrm{RVA}_t = -(\mathrm{CRVA}_t + \mathrm{DRVA}_t), \quad t \in [0, \bar{\tau}'].$$

Here CRVA is the expected loss in case the bank defaults first after a rating trigger. Similarly, (-DRVA) represents the expected benefit if the counterparty defaults first, preceded by a rating trigger. Therefore, including a rating triggers provision in a CSA provides protection to the bank (respectively the counterparty) from losses due to default events of the counterparty (respectively the bank) that happen after a credit downgrade. Accordingly, the value of the contract is adjusted for this protection, as shown in the following result.

Corollary 11.2.8 *We have the following decomposition for the counterparty risky price process*

$$\begin{aligned}\Pi'_t &= P_t - \Theta'_t \\ &= P_t - \Theta_t + \mathrm{RVA}_t \\ &= P_t - CVA_t - \mathrm{DVA}_t - \mathrm{CRVA}_t - \mathrm{DRVA}_t\,,\end{aligned}$$

for $t \in [0, \bar{\tau}')$.

11.2.2 Dynamic Collateralization

In bilateral margin agreements, counterparties are required to post collateral as soon as the clean price of the contract exceeds thresholds, η_t^c and η_t^b, which are defined in CSA (see Sect. 3.3.2). In particular, these thresholds are defined in terms of the credit ratings of the counterparties. Specifically, the collateral threshold of a counterparty decreases as a result of a credit rating downgrade and increases as a result of a credit rating upgrade. Consequently, a counterparty with a higher credit rating will have higher threshold than a counterparty with a lower credit rating.

It is important to note that there is an adverse relation between the margin requirements and the credit ratings. A credit downgrade along with higher borrowing rates and exposures forces companies to post increasing amounts of collateral to their counterparties, which can be fatal. For example, the ratings-linked collateral thresholds, coupled with the rehypothecation discussed in the next subsection, have been considered to be one of the key drivers of AIG's collapse in 2008. Before 2007, as a "AAA" rated company, AIG had not been required to post any collateral for most of its derivatives transactions. However, after the Lehman turmoil and several downgrades, AIG had posted more than $40 billion in collateral as of November 2008 (see International Swaps and Derivatives Association (2009) for details).

Thus one of the key issues in modeling of the collateral process is the issue of modeling of the thresholds. In what follows, we will model the collateral threshold for the counterparty at time t as $\eta_t^c = \eta_c(t, X_t^c, P_t)$, where $\eta_c : [0,T] \times \mathcal{K} \times \mathbb{R} \to \mathbb{R}_+$ is a measurable function. Similarly, we will model the collateral threshold for the bank at time t as $\eta_t^b = \eta_b(t, X_t^b, P_t)$, where $\eta_b : [0,T] \times \mathcal{K} \times \mathbb{R} \to \mathbb{R}_-$ is a measurable function. In the sequel we use the first variant of the ISDA collateralization scheme of Sect. 3.3.2, in conjunction with the following two possible specifications of collateral rates:

- The linear case:

$$\rho_l^i(t,x) = \frac{K-x}{K-1},$$

for all $i = c, b$. In particular, $\rho^i(t,1) = 1$ and $\rho^i(t,K) = 0$.

- The exponential case:

$$\rho_e^i(t,x) = \begin{cases} e^{1-x} & \text{if } x < K, \\ 0 & \text{if } x = K. \end{cases}$$

for all $i = c, b$.

Note that, in practice, the threshold levels are set in CSA documents for different rating levels. Here we propose two alternative methods for determining the collateral threshold levels. In the linear case, collateral thresholds increase or decrease linearly with the credit qualities of the counterparties. Similarly, in the exponential case, the collateral thresholds exponentially increase or decrease with the credit ratings. Therefore, the collateral rates in the exponential case are always less than the ones in the linear case, which leads to lower collateral thresholds and, as a result, more collateral posted in the margin account.

11.3 Markov Copula Approach for Rating-Based Pricing

In this section, we employ Markov copulas for modeling joint evolution of credit ratings of the two parties, the counterparty and the bank, in our framework. Consider two Markov chains Y^c and Y^b on $(\Omega, \mathbb{G}, \mathbb{Q})$ with the infinitesimal generators $A_c = [a^c_{ij}]$ and $A_b = [a^b_{hk}]$, respectively. Next, consider the system of equations, in unknowns $a_{ih,jk}$,

$$\sum_{k\in\mathcal{K}} a_{ih,jk} = a^c_{ij}, \quad \forall i,j,h \in \mathcal{K},\ i \neq j, \tag{11.6}$$

$$\sum_{j\in\mathcal{K}} a_{ih,jk} = a^b_{hk}, \quad \forall i,h,k \in \mathcal{K},\ h \neq k. \tag{11.7}$$

We know from Sect. 14.4.2.1 that the above system admits at least one solution such that matrix $A = [a_{ih,jk}]_{i,h,j,k\in\mathcal{K}}$, where diagonal elements are defined appropriately (cf. (14.12)), satisfies the conditions for a generator matrix of a bivariate time-homogeneous Markov chain, say $X = (X^c, X^b)$, whose components are Markov chains with the same laws as Y^c and Y^b. The system (11.6)–(11.7) serves as a Markov copula between the Markov margins Y^c, Y^b and the bivariate Markov chain X. Note that the system (11.6)–(11.7), containing more unknowns than the number of equations, is underdetermined. Therefore, as proposed by Bielecki, Vidozzi, and Vidozzi (2008a), we impose additional constraints on the variables in the system (11.6)–(11.7). We postulate that

$$a_{ih,jk} = \begin{cases} 0\,, & \text{if } i \neq j, h \neq k, j \neq k, \\ \alpha \min(a^c_{ij}, a^b_{hk})\,, & \text{if } i \neq j, h \neq k, j = k, \end{cases} \tag{11.8}$$

where $\alpha \in [0,1]$. We interpret the constraint (11.8) as follows. X^c and X^b migrate according to their marginal laws. Nevertheless, they can have the same values. The intensity of migrating to the same rating category is measured by the parameter α. If $\alpha = 0$, then the components X^c and X^b of X migrate independently. However, if $\alpha = 1$, the tendency of X^c and X^b migrating to the same categories is at maximized. Using the constraints (11.8), the system (11.6)–(11.7) becomes fully decoupled and we can obtain the generator of the joint process.

Remark 11.3.1 Note that in practice the rating transition matrices typically indicate the historical default probabilities, so that, in practice, we need switch to the risk-neutral probabilities. The change of measure needs to be done in such a way that the resulting risk-neutral probabilities are consistent with the default probabilities inferred from the quoted CDS spreads. Toward this aim, we need to apply changes of measure, while preserving the Markov structure of the model X, which is a bivariate time-homogeneous Markov chain. Therefore, the process X, that is Markov under the statistical measure, will remain Markov under the risk-neutral measure as well. The Markov consistency and Markov copulae are only needed under the statistical measure if one starts from modeling under this measure. The Markov change of measure is calibrated to market data, resulting in a risk-neutral generator for the bivariate process X. We refer, for the discussion, to (Bielecki, Vidozzi, and Vidozzi 2008a). In the next section, in order to avoid discussing the Markov change of measure, calibration, etc., we will postulate that we are given marginal generators for both parties under the risk neutral measure and that the generator for the bivariate ratings process is obtained via the conditions (11.6)–(11.8).

11.4 Applications

In this section, we illustrate our results in the context of a CDS and an interest rate swap (IRS) contract. We postulate that our CDS and IRS contracts are subject to rating triggers, so that they are terminated in case a trigger event occurs. We compute, for different rating trigger levels, the time-0 RVA, RVA_0, and triggers-inclusive TVA, Θ'_0, and we report the related TVA mitigation in the sense of RVA_0/Θ_0.

For the sake of simplicity, we carry out our analysis with $K = 4$ rating categories: A, B, C and D. The level A represents the highest rating level, whereas D corresponds to the default state. We assume that the counterparty initially has rating A. In what follows, we suppose that the 1-year rating transition matrix, e^{A_c}, is given in Table 11.1. Likewise,

	A	B	C	D
A	0.9	0.08	0.017	0.003
B	0.05	0.85	0.09	0.01
C	0.01	0.09	0.8	0.1
D	0	0	0	1

TABLE 11.1: Counterparty's rating transition matrix

we assume that the current rating of the bank is A. The bank's 1-year rating transition matrix, e^{A_b}, is given as in Table 11.2.

	A	B	C	D
A	0.8	0.1	0.05	0.05
B	0.04	0.9	0.03	0.03
C	0.015	0.1	0.7	0.185
D	0	0	0	1

TABLE 11.2: Bank's rating transition matrix

In oreder to simplify the presentation, we assume that the rating transition matrices given above are already risk-neutral, so that no Markov change of measure is needed. We also assume deterministic recovery rates; $R_c = R_b = 0.4$.

11.4.1 Interest Rate Swap with Rating Triggers

In this section, we consider a fixed-for-float payer 10-year contract with \$1 notional, in the presence of rating triggers as break clauses. We assume that the payments are done every quarter and the fixed leg pays the swap rate, while the floating leg pays the Libor rate. We also assume that the swap is initiated at $T_0 = 0$ and we denote by $T_1 < T_2 < \cdots < T_n$ the collection of payment dates and by S the fixed rate. The cumulative dividend process of the IRS contract at time T_i is given by

$$D_{T_i} = \sum_{k=1}^{i} (L(T_k) - S)\delta_k \,,$$

where $L(T_i)$ is time-T_i Libor rate and $\delta_k = T_k - T_{k-1}$ for $k = 1, 2, \ldots, n$. We also suppose that the instantaneous interest rate r follows the Vasicek dynamics

$$dr_t = (\theta - \nu\, r_t)dt + \sigma dW_t.$$

where we set $r_0 = 0.05$, $\theta = 0.1$, $\nu = 0.05$ and $\sigma = 0.01$. We find the corresponding swap rate as $S = 0.0496$. We carry out our analysis for uncollateralized, linearly collateralized and exponentially collateralized cases for α in (11.8) equal to 0 or 1. The TVA mitigation results are displayed in Fig. 11.1–11.6 and Tables 11.3–11.8. Depending on the trigger level of the counterparty in particular, the TVA mitigation is quite substantial.

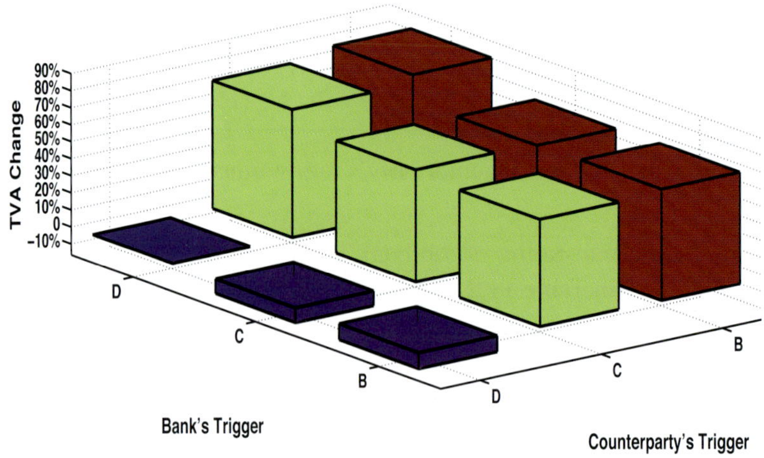

FIGURE 11.1: Mitigation in the TVA of an IRS (in %), $\alpha = 0$, no collateralization.

(B,B)	(B,C)	(C,B)	(C,C)	(B,D)	(D,B)	(C,D)	(D,C)
65.42%	62.98 %	64.80%	65.79%	-9.40%	80.25%	-8.61%	75.12%

TABLE 11.3: Numerical values corresponding to Fig. 11.1.

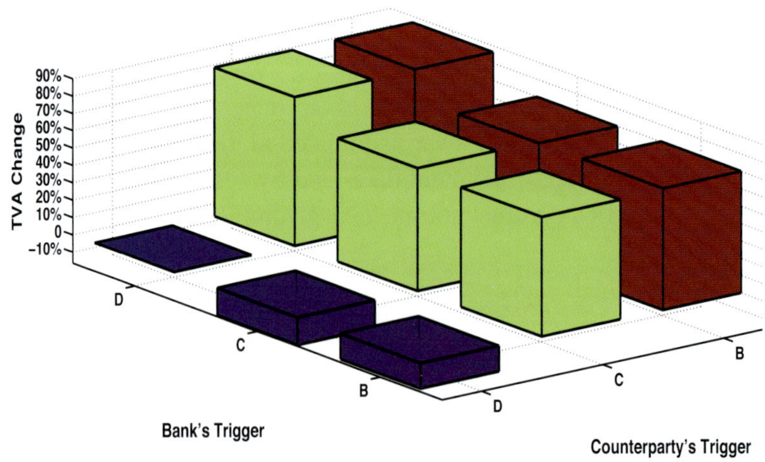

FIGURE 11.2: Mitigation in the TVA of an IRS (in %), $\alpha = 1$, no collateralization.

(B,B)	(B,C)	(C,B)	(C,C)	(B,D)	(D,B)	(C,D)	(D,C)
70.12%	68.98%	69.83%	70.90%	-14.67%	86.38%	-15.81%	85.76%

TABLE 11.4: Numerical values corresponding to Fig. 11.2.

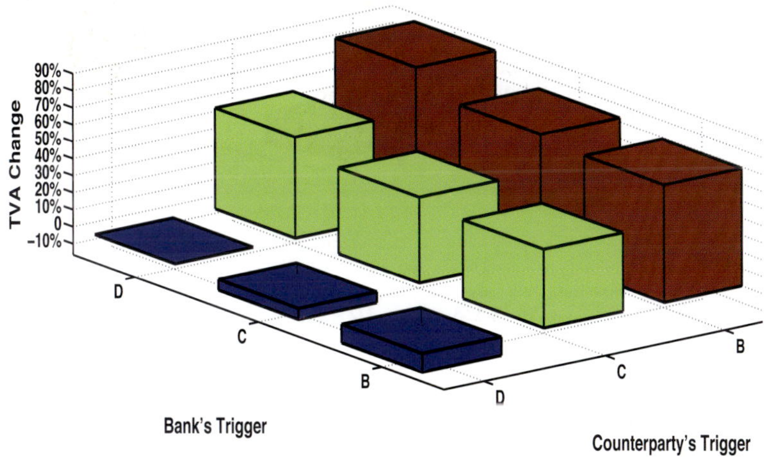

FIGURE 11.3: Mitigation in the TVA of an IRS (in %), $\alpha = 0$, linear collateral rate ρ_l^i.

(B,B)	(B,C)	(C,B)	(C,C)	(B,D)	(D,B)	(C,D)	(D,C)
69.05%	46.36%	72.34%	50.37%	-10.97%	85.25%	-6.43%	59.33%

TABLE 11.5: Numerical values corresponding to Fig. 11.3.

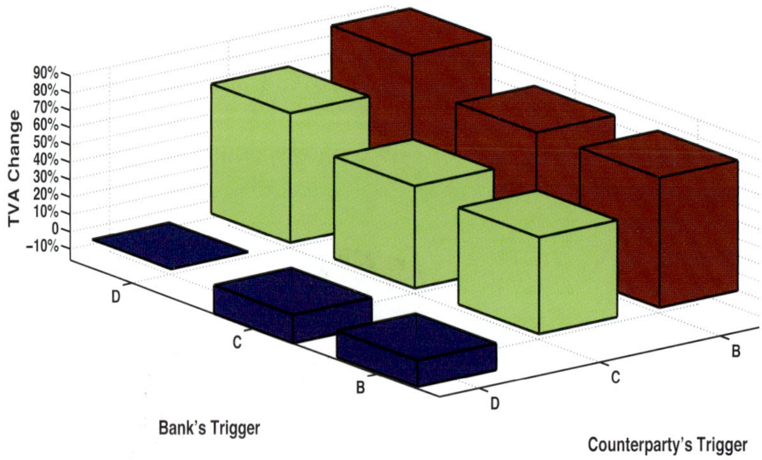

FIGURE 11.4: Mitigation in the TVA of an IRS (in %), $\alpha = 1$, linear collateral rate ρ_l^i.

(B,B)	(B,C)	(C,B)	(C,C)	(B,D)	(D,B)	(C,D)	(D,C)
75.03%	55.84%	74.68%	59.01%	-15.38%	92.95%	-16.39%	74.75%

TABLE 11.6: Numerical values corresponding to Fig. 11.4.

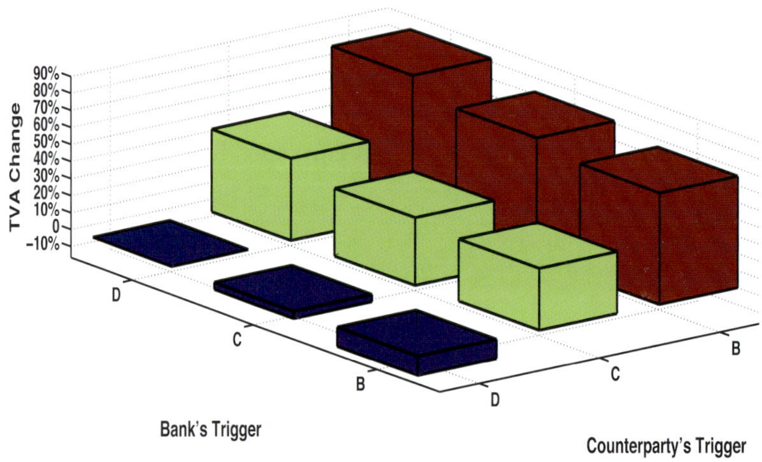

FIGURE 11.5: Mitigation in the TVA of an IRS (in %), $\alpha = 0$, exponential collateral rate ρ_e^i.

(B,B)	(B,C)	(C,B)	(C,C)	(B,D)	(D,B)	(C,D)	(D,C)
65.47%	36.25%	71.38%	39.88%	-11.40%	81.88%	-4.27%	48.33%

TABLE 11.7: Numerical values corresponding to Fig. 11.5.

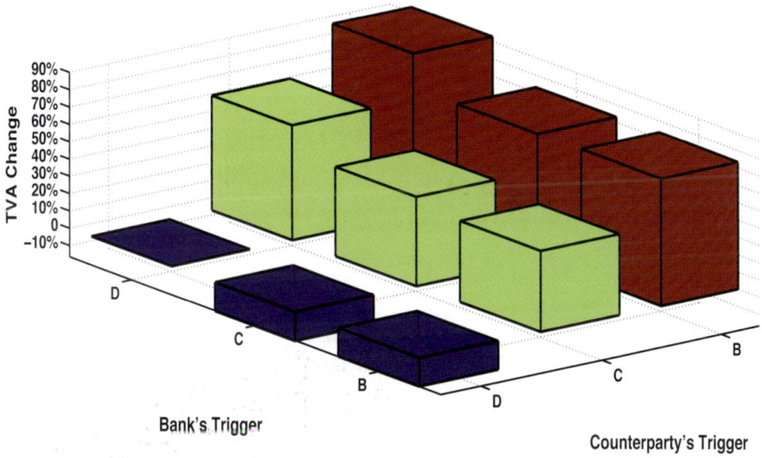

FIGURE 11.6: Mitigation in the TVA of an IRS (in %), $\alpha = 1$, exponential collateral rate ρ_e^i.

(B,B)	(B,C)	(C,B)	(C,C)	(B,D)	(D,B)	(C,D)	(D,C)
73.44%	46.90%	72.96%	51.48%	-16.50%	92.78%	-16.81%	66.35%

TABLE 11.8: Numerical values corresponding to Fig. 11.6.

11.4.2 CDS with Rating Triggers

In this section, we consider a CDS contract in the presence of rating triggers as break clauses. We assume that the reference entity is free of any trigger events. We denote by τ_1 the default time of the reference entity and by R_1 the recovery rate of the reference entity. We assume that the CDS contract has spread S and expires at T. Consequently, the risk-free cumulative dividend process of the CDS contract is given, for all $t \in [0, T]$, by:

$$D_t = (1 - R_1)\mathbb{1}_{\{\tau_1 \le t\}} - S_1(t \wedge \bar{\tau}).$$

We also assume that the underlying entity's 1-year rating transition matrix is given as in Table 11.9.

Similar to the IRS example, we carry out our analysis for uncollateralized, linearly

	A	B	C	D
A	0.95	0.03	0.019	0.001
B	0.04	0.85	0.107	0.003
C	0.01	0.19	0.791	0.009
D	0	0	0	1

TABLE 11.9: Underlying entity's rating transition matrix

collateralized and exponentially collateralized CDS contracts where $\alpha = 0$ and $\alpha = 1$. We display our results in Figs. 11.7 through 11.12 and Tables 11.10 through 11.15. Again, depending on the trigger level of the counterparty in particular, the TVA mitigation can be quite substantial.

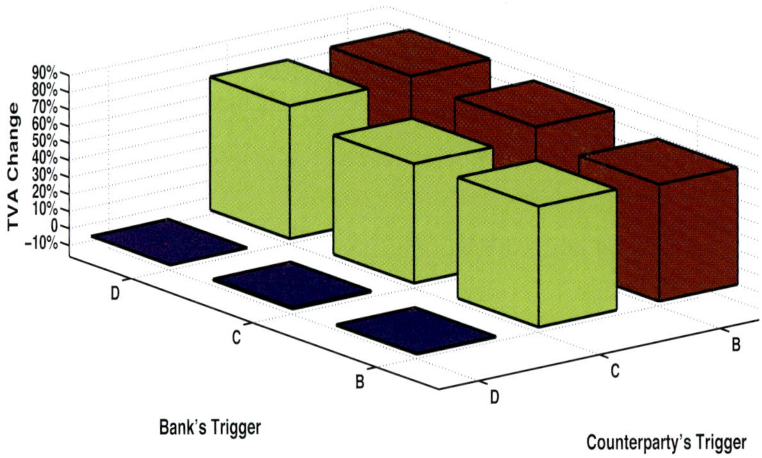

FIGURE 11.7: Mitigation in the TVA of a CDS (in %), $\alpha = 0$, no collateralization.

(B,B)	(B,C)	(C,B)	(C,C)	(B,D)	(D,B)	(C,D)	(D,C)
68.58%	71.15%	76.52%	70.01%	-0.41%	80.21%	0.87%	78.22%

TABLE 11.10: Numerical values corresponding to Fig. 11.7.

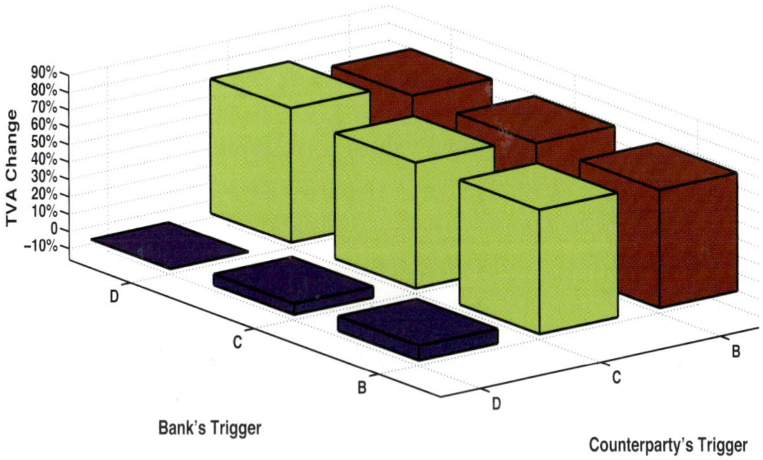

FIGURE 11.8: Mitigation in the TVA of a CDS (in %), $\alpha = 1$, no collateralization.

(B,B)	(B,C)	(C,B)	(C,C)	(B,D)	(D,B)	(C,D)	(D,C)
68.02%	71.73%	68.65%	72.50%	8.39%	69.65%	6.25%	77.56%

TABLE 11.11: Numerical values corresponding to Fig. 11.8.

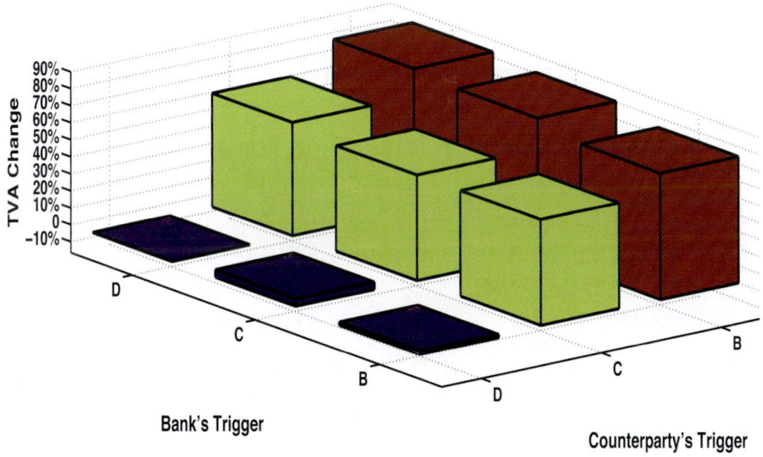

FIGURE 11.9: Mitigation in the TVA of a CDS (in %), $\alpha = 0$, linear collateral rate ρ^i_l.

(B,B)	(B,C)	(C,B)	(C,C)	(B,D)	(D,B)	(C,D)	(D,C)
74.35%	62.23%	80.32%	62.16%	-1.93%	83.25%	4.42%	66.64%

TABLE 11.12: Numerical values corresponding to Fig. 11.9.

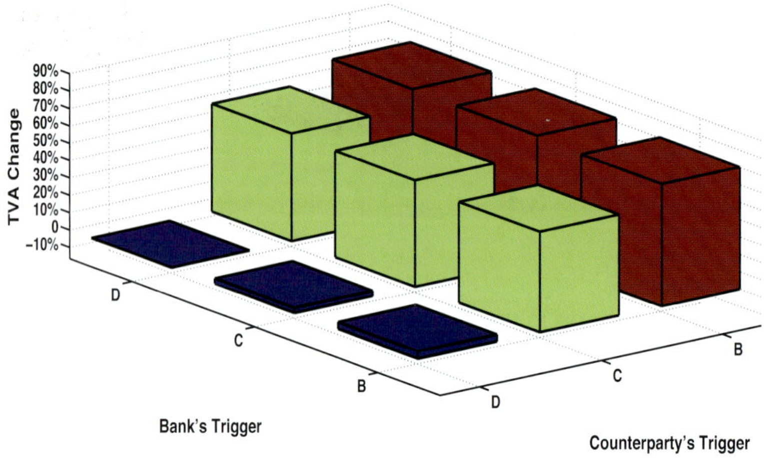

FIGURE 11.10: Mitigation in the TVA of a CDS (in %), $\alpha = 1$, linear collateral rate ρ_l^i.

(B,B)	(B,C)	(C,B)	(C,C)	(B,D)	(D,B)	(C,D)	(D,C)
70.31%	58.06%	72%	61.27%	4.03%	72.37%	3%	62.08%

TABLE 11.13: Numerical values corresponding to Fig. 11.10.

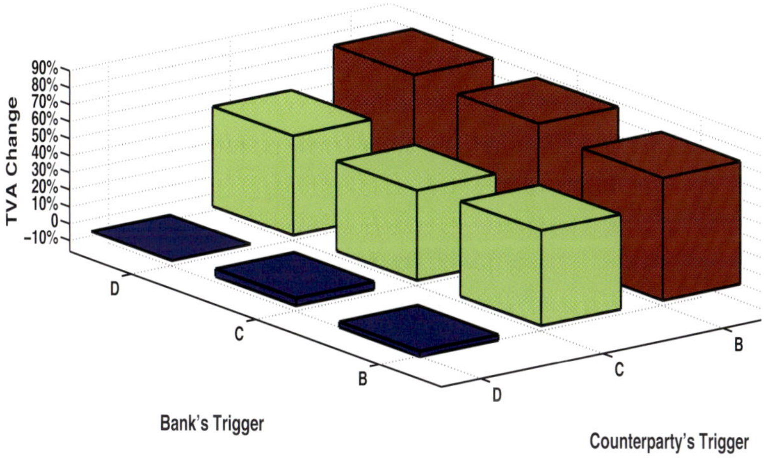

FIGURE 11.11: Mitigation in the TVA of a CDS (in %), $\alpha = 0$, exponential collateral rate ρ_e^i.

(B,B)	(B,C)	(C,B)	(C,C)	(B,D)	(D,B)	(C,D)	(D,C)
72.09%	55.92%	77.42%	52.98%	-3.25%	79.58%	4.53%	58.53%

TABLE 11.14: Numerical values corresponding to Fig. 11.11.

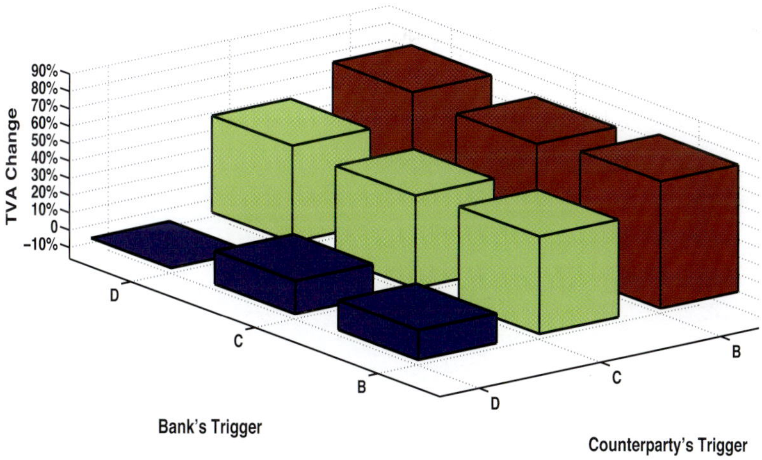

FIGURE 11.12: Mitigation in the TVA of a CDS (in %), $\alpha = 1$, exponential collateral rate ρ_e^i.

(B,B)	(B,C)	(C,B)	(C,C)	(B,D)	(D,B)	(C,D)	(D,C)
73.18%	56.25%	68.31%	53.29%	17.13%	71.55%	19.11%	55.68%

TABLE 11.15: Numerical values corresponding to Fig. 11.12.

Chapter 12

A Unified Perspective

In Part III, a reduced-form counterparty risk modeling approach was presented, under a rather standard immersion hypothesis between the reference filtration and the filtration progressively enlarged by the default times of the two parties. We will refer to this approach as the "basic" (reduced-form counterparty risk modeling) approach.

The present chapter shows how a suitable extension of this basic approach can be devised so that it can be applied in the dynamic copula models of Part IV. Namely, we shall see how the full model filtration $\mathbb{G}$ in these models can be decomposed as $\mathbb{G} = \mathbb{F} \vee \mathbb{K}$, where $\mathbb{F}$ is a reference filtration and $\mathbb{K}$ is a filtration carrying information about the default times of the counterparties, allowing one to develop in these models a reduced-form approach as in Part III. However, this (typically) will be for a default filtration $\mathbb{K}$ larger than the filtration $\mathbb{H}$, generated by τ, used in Part III, and for a suitably chosen new measure $\widetilde{\mathbb{Q}}$, so that $(\mathbb{F}, \widetilde{\mathbb{Q}})$-martingales stopped at time τ are $(\mathbb{G}, \mathbb{Q})$-martingales. Here we only present the basic tenets of the extension and its use. For the complete development we refer to Crépey and Song (2014).

Note that in the present state of the theory this extension only applies to models, such as those of Part IV, that admit only two possible states for each obligor: default and non-default. In more general credit migration setups such as the one of Chapter 11, no analogous extension seems to be possible.

12.1 Introduction

As seen in Part IV, in order to deal with counterparty risk embedded in credit derivatives, one needs a credit portfolio model with the following features. First, as the CVA is an option on the future values of the clean price of the underlyings, it should be a dynamic Markov model, in which future clean prices of credit derivatives can be assessed consistently and computed numerically. Second, the model should calibrate to relevant data sets: CDS data if the targeted application consists of counterparty risk computations on CDS contracts and, additionally, to tranche data for CVA computation on CDO tranches. The model should therefore be a bottom-up model of portfolio credit risk such that individual names are represented in the model (see Bielecki, Crépey, and Jeanblanc (2010)). But, for tractability of the calibration, the model should also enjoy a copula-like separation property between the individual and the dependence model parameters, as well as efficient pricing schemes for vanillas (CDS contracts and/or CDO tranches). One possibility is to use "informationally dynamic copula models", resulting from the introduction of a suitable filtration on top of a static copula model for the default times of the two parties and the reference entities. The simplest example is the dynamic Gaussian copula (DGC) model of Chapter 7, which can be sufficient to deal with CVA on CDS contracts. If CDO tranches are also present in the reference portfolio, then a Gaussian copula dependence structure is not rich enough. Instead, one can use the dynamic Marshall-Olkin (DMO) common-shock model of Chapters 8 through 10.

However, the dynamic copula methodology, as presented in Part IV, will not work in the case of bilateral counterparty risk combined with the related funding issue. Indeed, with nonlinear funding costs, the TVA equation is nonlinear (see Chapters 4 through 6). In the case of credit derivative underlyings, the problem is also very high-dimensional. From the practical point of view, for nonlinear and very high-dimensional problems, any numerical scheme based, even to some extent, on dynamic programming, such as "purely backward" deterministic PDE schemes, but also "hybrid forward/backward" simulation/regression BSDE schemes, are ruled out by the curse of dimensionality (see e.g. Crépey (2013)). In this case the only feasible TVA schemes are "purely forward" simulation schemes, such as the expansions of Fujii and Takahashi (2012) in vanilla cases with explicit clean pricing formulas, or the CVA branching particles of Henry-Labordère (2012) in more exotic cases. But purely forward schemes cannot deal with the implicit terminal condition $\xi_\tau(P_{\tau-} - \Theta_{\tau-})$ of the full TVA equation (4.38) (recalled as equation (12.1) below). To tackle this issue, this chapter shows how a suitable reduced-form approach developed in Crépey and Song (2014) can be applied, beyond the basic immersion setup of Part III, in the context of the dynamic copula models of Part IV.

Outline Sect. 12.2 briefly reviews the extended reduced-form counterparty risk modeling approach of Crépey and Song (2014). Sects. 12.3 and 12.4 show how this approach can be applied in the DGC and in the DMO model, respectively.

12.2 Marked Default Time Reduced-Form Modeling

We consider our usual contract with all the related setup and notation of Chapter 4, including its clean price process P (using a risk-neutral discount factor β at an OIS-riskless rate r_t), exposure at default $\xi_\tau(\pi)$ and funding coefficient g_t, all defined with respect to the pricing stochastic basis $(\Omega, \mathcal{G}_{\bar{T}}, \mathbb{G}, \mathbb{Q})$. For notational simplicity, as in Chapter 6, we only consider the most common situation where the hedge is securely funded, entirely composed of positions either swapped or traded via repo markets. One thus has the case where $\xi_t(\pi, \varsigma) = \xi_t(\pi)$ and $g_t(\pi, \varsigma) = g_t(\pi)$ do not depend on a hedge argument ς (the real number π represents the wealth of the bank's hedging, collateralization and funding portfolio). Extension to hedge-dependent data ξ and g could be considered as in Chapter 5, with the same complication as there that in the Markov case, there is then no longer "intrinsic" Markov predefault TVA PIDE (like (12.7) or (5.42) in Chapter 5), and the predefault TVA PIDE depends on one's hedging criterion (see the remark 5.3.8).

Given the data $\xi_t(\pi)$ and $g_t(\pi)$, it is shown in Chapter 4 how the TVA process Θ can consistently be modeled on $[0,\bar{\tau}]$ as a solution, provided it exists, to the "full BSDE" (4.38) over the random time interval $[0,\bar{\tau}]$, i.e. (in integral form)

$$\beta_t\Theta_t = \mathbb{E}_t\Big[\beta_{\bar{\tau}}\mathbb{1}_{\tau<T}\xi_\tau(P_{\tau-}-\Theta_{\tau-}) + \int_t^{\bar{\tau}}\beta_s g_s(P_s-\Theta_s)ds\Big],\ t\in[0,\bar{\tau}]. \tag{12.1}$$

The object of this chapter is to show how a suitable extension of the basic reduced-form approach of Part III can be applied so to solve equation (12.1) in the context of the dynamic copula models of Chapters 7 through 11.

12.2.1 Pre-default Setup

We assume that $\tau = \min_{e\in E}\tau_e$, where, for each "mark" e in a finite set E, τ_e is a stopping time with intensity λ^e_t and $\mathbb{Q}(\tau_e\neq\tau_{e'}) = 1$ (for $e'\neq e$). We also assume that for any mark e, there exists a family of left-continuous versions of semimartingales parameterized by the real number π, say $\widetilde{\xi}^e_t(\pi)$, such that

$$\xi_\tau(\pi) = \widetilde{\xi}^e_\tau(\pi) \tag{12.2}$$

on the event $\{\tau=\tau_e\}$. We write $\lambda_t\cdot\chi_t = \sum_E\lambda^e_t\chi^e_t$, for any parameterized process χ^e_t.

Lemma 12.2.1 *For every $(\mathbb{G},\mathbb{Q})$- real valued semimartingale Y, the process defined below for $t\in[0,\bar{\tau}]$ is a (local) $(\mathbb{G},\mathbb{Q})$-martingale:*

$$d\mu^Y_t = \xi_t(Y_{t-})dJ_t + \lambda_t\cdot\widetilde{\xi}_t(Y_t)dt. \tag{12.3}$$

***Proof*.** Denoting by J^e the survival process of τ_e, we have $dJ=\sum_{e\in E}dJ^e$ so that, using also (12.2),

$$d\mu^Y_t = \xi_t(Y_{t-})\sum_{e\in E}dJ^e_t + \lambda_t\cdot\widetilde{\xi}_t(Y_t)dt = \sum_{e\in E}\widetilde{\xi}_t(Y_{t-})\left(dJ^e_t+\lambda^e_t dt\right),$$

where each summand is a (local) martingale as the stochastic integral of the left-continuous version of a semimartingale (hence a predictable and locally bounded integrand) against a (local) martingale. □

In particular, $\widetilde{\gamma}_t := \sum_{e\in E}\lambda^e_t = \lambda_t\cdot\mathbb{1}_E$ is the default intensity of τ. For allreal numbers $t\in[0,\bar{\tau}]$ and π, we write

$$\widehat{f}_t(\pi) = g_t(\pi) + \lambda_t\cdot\widetilde{\xi}_t(\pi) - \widetilde{r}_t(P_t-\pi), \tag{12.4}$$

where $\widetilde{r}_t = r_t+\widetilde{\gamma}_t$.

The extended reduced-form methodology presented in this chapter hinges on the following assumption.

Condition (A). There exist:

(A.1) a filtration $\mathbb{F}$ over $[0, T]$ such that $\mathcal{F}_t \subseteq \mathcal{G}_t$ on $[0, \bar{\tau}]$ and $\mathbb{F}$-semimartingales restricted to $[0, \bar{\tau}]$ are $\mathbb{G}$-semimartingales;

(A.2) a probability measure $\widetilde{\mathbb{Q}}$ equivalent to $\mathbb{Q}$ on $\mathcal{F}_{\bar{\tau}}$ such that any $(\mathbb{F}, \widetilde{\mathbb{Q}})$-martingale stopped at $\tau-$ is a $(\mathbb{G}, \mathbb{Q})$-martingale[1];

(A.3) A family of $\mathbb{F}$-progressive processes $f_t(\vartheta)$ (parameterized by the real number ϑ) such that $\int_0^{\cdot} \widehat{f}_t(P_t - \vartheta)dt = \int_0^{\cdot} f_t(\vartheta)dt$ on $[0, \bar{\tau}]$.

The condition (A.3) is a mild technical assumption which is, in particular, satisfied if there exists a pre-default process for $\widehat{f}_t(\pi)$ (cf. Lemma 13.7.2 in the classical immersion setup). The condition (A.1) relates to the $(\mathcal{H}')$-hypothesis between $\mathbb{F}$ and $\mathbb{G}$, which would mean that $\mathbb{F}$-semimartingales are $\mathbb{G}$-semimartingales (see Bielecki, Jeanblanc, and Rutkowski (2009)). The (main) condition (A.2) obviously holds, in particular, if any $(\mathbb{F}, \widetilde{\mathbb{Q}})$-martingale does not jump at τ (then "stopped at $\tau-$" reduces to "stopped at τ" in (A.2)) and any $(\mathbb{F}, \widetilde{\mathbb{Q}})$-martingale stopped at τ is a $(\mathbb{G}, \mathbb{Q})$-martingale. In the case with $\widetilde{\mathbb{Q}} = \mathbb{Q}$, these properties are related to the notion of immersion of $\mathbb{F}$ into $\mathbb{G}$ and to the notion of an $\mathbb{F}$-pseudo stopping time τ, which are recalled in the remark 13.7.4 (see also Sect. 5.2.1). However, even in this "immersion" case where $\widetilde{\mathbb{Q}} = \mathbb{Q}$, in view of the presence of the τ_e, the present setup is richer than the standard reduced-form intensity model of credit risk of Sect. 13.7 (see also Sect. 5.2.1), where the full model filtration $\mathbb{G}$ is given as the reference filtration $\mathbb{F}$ progressively enlarged by τ (or τ_b and τ_c in Part III). Indeed, here the full filtration $\mathbb{G}$ is a priori richer than the reference filtration progressively enlarged by τ or even by τ_b and τ_c (since, in particular, it makes all the τ_e stopping times).

In particular, in the basic setup of Part III, the condition (A) holds with $\mathbb{G} = \mathbb{F} \vee \mathbb{H}$ in (A.1), $\widetilde{\mathbb{Q}} = \mathbb{Q}$ in (A.2) and f defined by (5.7) (compare with (12.4)), assumed to be $\mathbb{F}$-adapted there, in (A.3). But the condition (A) does not exclude a jump of an $\mathbb{F}$-adapted càdlàg process at time τ (see Crépey and Song (2014) for more detailed illustration in the common-shock model of Sect. 12.4), whereas, as seen in Lemma 5.2.1(ii) (cf. also Lemma 13.7.3(ii)), such a jump can't happen in the basic setup of Part III. We will show in Sects. 12.3 and 12.4 that the condition (A) holds in the DGC and DMO models of Part IV, using an equivalent measure $\widetilde{\mathbb{Q}} \neq \mathbb{Q}$ in the first one, and $\widetilde{\mathbb{Q}} = \mathbb{Q}$ (but a well chosen subfiltration $\mathbb{F}$) in the second one.

Let $\widetilde{\mathbb{E}}_t$ stand for the conditional expectation given $\mathcal{F}_t$ under $\widetilde{\mathbb{Q}}$. The following result is shown in Crépey and Song (2014).

Proposition 12.2.2 *Under the condition (A), if an* $(\mathbb{F}, \widetilde{\mathbb{Q}})$*-semimartingale* $\widetilde{\Theta}$ *satisfies the following equation over* $[0, T]$:

$$\widetilde{\Theta}_t = \widetilde{\mathbb{E}}_t \int_t^T f_s(\widetilde{\Theta}_s)ds, \;\; t \in [0, T], \tag{12.5}$$

then the $(\mathbb{G}, \mathbb{Q})$*-semimartingale* Θ *defined as* $\Theta = \widetilde{\Theta}$ *on* $[0, \bar{\tau})$ *and* $\Theta_{\bar{\tau}} = \mathbb{1}_{\tau<T}\xi_\tau(P_{\tau-} - \widetilde{\Theta}_{\tau-})$ *satisfies the full TVA equation* (12.1) *on* $[0, \bar{\tau}]$.

[1]By a process stopped at $\tau-$ we mean a process whose values at and after time τ are equal to the left-limit of the process at time τ.

Note that the equation (12.5) is a BSDE for the process $\widetilde{\Theta}_t$ (in integral form), referred to henceforth as the pre-default TVA BSDE. Proposition 12.2.2 extends Corollary 5.2.11 to the more general setup of the present chapter. Moreover, as in Sect. 5.3, by classical results (see e.g. Crépey (2013)), in the pre-default Markov case where there exists an $\mathbb{R}^q$-valued $(\mathbb{F}, \widetilde{\mathbb{Q}})$-Markov factor process X such that, over $[0, T]$,

$$f_t(\vartheta) = f(t, X_t, \vartheta) \tag{12.6}$$

for some function $f(t, x, \vartheta)$, then the solution $\widetilde{\Theta}_t$ to (12.5) can be represented as $\widetilde{\Theta}(t, X_t)$, for some function $\widetilde{\Theta}(t, x)$ satisfying the following semilinear PIDE:

$$\begin{cases} \widetilde{\Theta}(T, x) = 0, \ x \in \mathbb{R}^q, \\ (\partial_t + \mathcal{X})\,\widetilde{\Theta}(t, x) + f(t, x, \widetilde{\Theta}(t, x)) = 0 \text{ on } [0, T) \times \mathbb{R}^q, \end{cases} \tag{12.7}$$

where $\mathcal{X}$ is the $(\mathbb{F}, \widetilde{\mathbb{Q}})$-generator of the Markov process X_t. Any BSDE/PIDE numerical scheme for (12.7) can then be used for computing the TVA $\widetilde{\Theta}$.

The practical conclusion of the above is that, in cases where a Markov specification of the condition (A) holds, one can model a TVA process Θ as sketched above in terms of a solution $\widetilde{\Theta}_t = \widetilde{\Theta}(t, X_t)$, whenever it exists (for which we refer the reader to Crépey (2012b), Crépey (2013) and the references there), to the corresponding pre-default TVA BSDE/PIDE (12.5)/(12.7). In the next two sections, we show that Markov specifications of the condition (A) indeed hold in the credit portfolio TVA models of Part III (dynamic Gaussian copula DGC model of Chapter 7 and dynamic Marshall-Olkin DMO common-shock model of Chapters 8 through 10).

Let $N = \{-1, 0, 1, \ldots, n\}$, $N^\star = \{1, \ldots, n\}$, and recall that the methodology of Part III hinged on "informationally dynamizing" a copula model for the $(\tau_i)_{i \in N}$ by introduction of a suitable filtration, where $\tau_{-1} = \tau_b$ and $\tau_0 = \tau_c$ were used to model the default times of the bank and its counterparty, whereas the $(\tau_i)_{i \in N^\star}$ represented the default times of a pool of reference entities underlying a credit portfolio between the two parties.

12.3 Dynamic Gaussian Copula TVA Model

12.3.1 Model of Default Times

First, let us briefly revisit the DGC model of Chapter 7. We consider a multivariate Brownian motion $\mathbf{B} = (B^i)_{i \in N}$ with pairwise correlation ϱ, in its own filtration $\mathbb{F}^{\mathbf{B}}$, under a pricing probability measure $\mathbb{Q}$. For any $i \in N$, let h_i be a differentiable increasing function from $\mathbb{R}_+$ to $\mathbb{R}$ with $\lim_0 h_i(s) = -\infty$ and $\lim_{+\infty} h_i(s) = +\infty$. For every $i \in N$, we define a random time

$$\tau_i = h_i^{-1}\Big(\int_0^{+\infty} \varsigma(u) dB_u^i \Big), \tag{12.8}$$

where $\varsigma(\cdot)$ is a square integrable function with unit L^2-norm. So the τ_i jointly follow the standard (static) Gaussian copula model of Li (2000), with correlation parameter ϱ and with marginal distribution function $\Phi \circ h_i$ of τ_i, where Φ is the the standard normal distribution function. In order to make the model dynamic as required by counterparty risk applications, we introduce a model filtration $\mathbb{G}$ given as $\mathbb{F}^{\mathbf{B}}$ progressively enlarged by the τ_i so, for any t,

$$\mathcal{G}_t = \mathcal{F}_t^{\mathbf{B}} \vee \bigvee_{i \in N} \mathcal{H}_t^i.$$

Let

$$\mathbf{m}_t = (m_t^i)_{i \in N}, \text{ with } m_t^i = \int_0^t \varsigma(u) dB_u^i; \ \boldsymbol{\theta}_t = (\theta_t^i)_{i \in N}, \text{ with } \theta_t^i = \tau_i \mathbb{1}_{\{\tau_i \le t\}}.$$

We saw in Chapter 7 that, for processes γ_t^i and λ_t^i of the form

$$\gamma_t^i, \lambda_t^i = \gamma_i, \lambda_i\,(t, \mathbf{m}_t, \boldsymbol{\theta}_t), \tag{12.9}$$

the $\mathbb{G}$-Brownian motions $dW_t^i = dB_t^i - \gamma_t^i dt$ and the $\mathbb{G}$-compensated default indicator processes $dM_t^i = d\mathbb{1}_{\tau_i \le t} - \lambda_t^i dt$ have the $\mathbb{G}$-martingale representation property. Moreover, the process $(\mathbf{m}_t, \boldsymbol{\theta}_t)$ is $\mathbb{G}$-Markov, with a generator $\mathcal{A}$ acting on every function $u = u(t, \mathbf{m}, \boldsymbol{\theta})$ as

$$\mathcal{A}u = \partial_t u + \mathcal{A}_{\mathbf{m}} u + \sum_{i \in N} \lambda_i \delta_i u, \tag{12.10}$$

where

$$\mathcal{A}_{\mathbf{m}} u = \varsigma \sum_{i \in N} \gamma_i \partial_{m_i} u + \frac{\varsigma^2}{2} \left(\sum_{i \in N} \partial^2_{m_i^2} u + \varrho \sum_{i \neq j \in N} \partial^2_{m_i, m_j} u \right) \tag{12.11}$$

and

$$\delta_i u(t, \mathbf{m}, \boldsymbol{\theta}) = u(t, \mathbf{m}, \boldsymbol{\theta}^{i,t}) - u(t, \mathbf{m}, \boldsymbol{\theta}),$$

in which $\boldsymbol{\theta}^{i,t}$ stands for $\boldsymbol{\theta}$ with component i replaced by t.

12.3.2 Pre-default TVA Model

A reduced DGC setup will now be used as a pre-default TVA model applied to credit derivatives, for a mark space of τ given as

$$E = \{-1, 0\},$$

under the related assumption (12.2) that for $i = -1, 0$, there exists a continuous function $\widetilde{\xi}_i$ satisfying

$$\xi_\tau(\pi) = \widetilde{\xi}_i(\tau, \mathbf{m}_\tau, \boldsymbol{\theta}_{\tau-}; \pi) \tag{12.12}$$

on the event $\{\tau = \tau_i\}$. This assumption is discussed in Crépey and Song (2014), where it is shown to hold for any portfolio of CDS contracts and CDO tranches (depending also on the nature of the CSA value and collateral processes Q and Γ that sit in $\xi_t(\pi)$).

By reduced DGC model we mean $X_t = (\mathbf{m}_t, \widetilde{\boldsymbol{\theta}}_t)$, where $\widetilde{\boldsymbol{\theta}}_t = (\theta_t^i)_{i\in N^\star}$, considered relative to the reference filtration $\mathbb{F} = (\mathcal{F}_t)$ generated by the Brownian motion $\mathbf{B}$ and the default times of references names $i \in N^\star$, i.e.

$$\mathcal{F}_t = \mathcal{F}_t^{\mathbf{B}} \vee \bigvee_{i\in N^\star} \mathcal{H}_t^i.$$

The reduced model has Markov and martingale properties analogous to the ones seen above in the full DGC model, subject to obvious modifications. In particular, the relevant fundamental $(\mathbb{F}, \mathbb{Q})$-martingales are now given as

$$d\overline{W}_t^i = dB_t^i - \bar{\gamma}_t^i dt,\ i \in N;\ d\overline{M}_t^i = d\mathbb{1}_{\tau_i \le t} - \bar{\lambda}_t^i dt,\ i \in N^\star, \tag{12.13}$$

where

$$\bar{\gamma}_t^i = \bar{\gamma}_i(t, \mathbf{m}_t, \widetilde{\boldsymbol{\theta}}_t),\ \ \bar{\lambda}_t^i = \bar{\lambda}_i(t, \mathbf{m}_t, \widetilde{\boldsymbol{\theta}}_t). \tag{12.14}$$

Let us change the measure to $\widetilde{\mathbb{Q}}$ such that the "$(\mathbb{F}, \widetilde{\mathbb{Q}})$-intensities" of the $B_t^i, i \in N$ and of the $\mathbb{1}_{\tau_i \le t}, i \in N^\star$ are respectively equal, letting $\boldsymbol{\theta} = (0, 0, \widetilde{\boldsymbol{\theta}})$ for any $\widetilde{\boldsymbol{\theta}} \in \mathbb{R}_+^n$, to

$$\begin{aligned} \widetilde{\gamma}_t^i &= \widetilde{\gamma}_i(t, \mathbf{m}_t, \widetilde{\boldsymbol{\theta}}_t) := \gamma_i(t, \mathbf{m}_t, \boldsymbol{\theta}_t), \\ \widetilde{\lambda}_t^i &= \widetilde{\lambda}_i(t, \mathbf{m}_t, \widetilde{\boldsymbol{\theta}}_t) := \lambda_i(t, \mathbf{m}_t, \boldsymbol{\theta}_t). \end{aligned} \tag{12.15}$$

Thus, the fundamental $(\mathbb{F}, \widetilde{\mathbb{Q}})$-martingales are written as

$$d\widetilde{W}_t^i = dB_t^i - \widetilde{\gamma}_t^i dt,\ i \in N;\ d\widetilde{M}_t^i = d\mathbb{1}_{\tau_i \le t} - \widetilde{\lambda}_t^i dt,\ i \in N^\star. \tag{12.16}$$

Note that, before τ, we have:

$$\gamma_t^i = \widetilde{\gamma}_t^i,\ \ \lambda_t^i = \widetilde{\lambda}_t^i$$

and therefore, until τ, we have:

$$W_t^i = \widetilde{W}_t^i,\ \ M_t^i = \widetilde{M}_t^i.$$

The $(\mathbb{F}, \widetilde{\mathbb{Q}})$-generator of X_t acts on every function $u = u(t, \mathbf{m}, \widetilde{\boldsymbol{\theta}})$ as

$$\mathcal{X}u = \partial_t u + \mathcal{X}_{\mathbf{m}} u + \sum_{i\in N^\star} \widetilde{\lambda}_i \delta_i u, \tag{12.17}$$

where (compare (12.11)):

$$\mathcal{X}_{\mathbf{m}} u = \varsigma \sum_{i\in N} \widetilde{\gamma}_i \partial_{m_i} u + \frac{\varsigma^2}{2}\left(\sum_{i\in N} \partial^2_{m_i^2} u + \varrho \sum_{i\ne j\in N} \partial^2_{m_i, m_j} u\right). \tag{12.18}$$

Before τ, every dependence in $\boldsymbol{\theta}_t$ reduces to a dependence in $\widetilde{\boldsymbol{\theta}}_t$. Under the additional assumption that the processes $r_t, g_t(\pi)$ (for any real number π) and P_t are given before τ as continuous functions of (t, X_t) (functions denoted below by the same letters as the related processes), then in view of (12.4) we have:

$$\widehat{f}_t(P_t - \vartheta)dt = f(t, X_t, \vartheta)dt, \ t \in [0, \bar{\tau}],$$

for the function f such that, with $\boldsymbol{\theta} = (0, 0, \widetilde{\boldsymbol{\theta}})$ for any $\widetilde{\boldsymbol{\theta}} \in \mathbb{R}^n_+$:

$$\begin{aligned} f(t, \mathbf{m}, \widetilde{\boldsymbol{\theta}}, \vartheta) + r(t, \mathbf{m}, \boldsymbol{\theta})\vartheta &= g(t, \mathbf{m}, \boldsymbol{\theta}, P(t, \mathbf{m}, \boldsymbol{\theta}) - \vartheta) \\ &+ \lambda_{-1}(t, \mathbf{m}, \boldsymbol{\theta})\widetilde{\xi}_{-1}(t, \mathbf{m}, \boldsymbol{\theta}; P(t, \mathbf{m}, \boldsymbol{\theta}) - \vartheta) \\ &+ \lambda_0(t, \mathbf{m}, \boldsymbol{\theta})\widetilde{\xi}_0(t, \mathbf{m}, \boldsymbol{\theta}; P(t, \mathbf{m}, \boldsymbol{\theta}) - \vartheta) \\ &- \gamma(t, \mathbf{m}, \boldsymbol{\theta})\vartheta. \end{aligned} \tag{12.19}$$

To summarize, the condition (A) holds for the measure $\widetilde{\mathbb{Q}}$ defined above (12.15) in (A.2) and for the Markov specification of the coefficient $f_t(\vartheta) := f(t, X_t, \vartheta)$ in (A.3), where the function $f = f(t, \mathbf{m}, \widetilde{\boldsymbol{\theta}}, \vartheta)$ is given by (12.19).

12.4 Dynamic Marshall-Olkin Copula TVA Model

The above dynamic Gaussian copula (DGC) model can be sufficient to deal with CVA on portfolios of CDS contracts. If CDO tranches are also present in the reference portfolio, a Gaussian copula dependence structure is not rich enough. Instead, one can use the dynamic Marshall-Olkin (DMO) common-shock model of Chapters 8 through 10.

12.4.1 Model of Default Times

Let's briefly revisit the model. We define a certain number m (typically small: a few units) of groups $I_j \subseteq N$ of obligors who are likely to default simultaneously, for $l = 1, \ldots, m$. The idea is that at every time t, there is a positive probability that the survivors of the group of obligors I_j (obligors of group I_j still alive at time t) default simultaneously. Let $\mathcal{I} = \{I_1, \ldots, I_m\}$, $\mathcal{Y} = \{\{-1\}, \{0\}, \{1\}, \ldots, \{n\}, I_1, \ldots, I_m\}$. Let shock intensity processes X^Y be given in the form of extended CIR processes as, for $Y \in \mathcal{Y}$,

$$dX_t^Y = a(b_Y(t) - X_t^Y)dt + c\sqrt{X_t^Y}dW_t^Y, \tag{12.20}$$

for non-negative constants a, c, non-negative functions $b_Y(t)$ and independent Brownian motions W^Y (in their own filtration $\mathcal{F}^{\mathbf{W}}$), under a pricing measure $\mathbb{Q}$.

Remark 12.4.1 The case of deterministic intensities $X_t^Y = b_Y(t)$ can be embedded in this framework as the limiting case of an "infinite mean-reversion speed" a (see Bielecki, Cousin, Crépey, and Herbertsson (2013b)).

For every $Y \in \mathcal{Y}$, we define

$$\eta_Y = \inf\{t > 0; \int_0^t X_s^Y ds > \mathcal{E}_Y\},$$

where the $\mathcal{E}_Y$ are i.i.d. standard exponential random variables. Then, for any obligor $i \in N$, we set

$$\tau_i = \min_{\{Y \in \mathcal{Y}; i \in Y\}} \eta_Y, \quad H_t^i = \mathbb{1}_{\tau_i \le t}. \tag{12.21}$$

We consider the dynamic model $(\mathbf{X}_t, \mathbf{H}_t) = ((X_t^Y)_{Y \in \mathcal{Y}}, (H_t^i)_{i \in N})$ relative to the filtration $\mathbb{G}$ such that, for $t \ge 0$,

$$\mathcal{G}_t = \mathcal{F}_t^{\mathbf{W}} \vee \bigvee_{i \in N} \mathcal{H}_t^i.$$

As developed in Bielecki, Cousin, Crépey, and Herbertsson (2013a) Part I, $(\mathbf{X}, \mathbf{H})$ is a Markov process with infinitesimal generator $\mathcal{A}$ acting as follows on every function $u = u(t, \mathbf{x}, \mathbf{k})$ with $t \in \mathbb{R}_+, \mathbf{x} = (x_Y)_{Y \in \mathcal{Y}}$ and $\mathbf{k} = (k_i)_{i \in N}$:

$$\mathcal{A}u(t, \mathbf{x}, \mathbf{k}) = \mathcal{A}_x u(t, \mathbf{x}, \mathbf{k}) + \sum_{Y \in \mathcal{Y}} x_Y \delta_Y u(t, \mathbf{x}, \mathbf{k}) \tag{12.22}$$

with

$$\mathcal{A}_x u = \sum_{Y \in \mathcal{Y}} \Big(a(b_Y(t) - x_Y)\partial_{x_Y} u + \frac{1}{2} c^2 \, x_Y \partial^2_{x_Y^2} \Big) u$$

and where we denote, for $Y \in \mathcal{Y}$,

$$\delta_Y u(t, \mathbf{x}, \mathbf{k}) = u(t, \mathbf{x}, \mathbf{k}^Y) - u(t, \mathbf{x}, \mathbf{k}),$$

in which $\mathbf{k}^Y$ denotes the vector obtained from $\mathbf{k} = (k_i)_{i \in N}$ by replacing the components $k_i, i \in Y$ by ones. For every $Z \subseteq N$, we have the following expression for the predictable intensity λ_t^Z of the indicator process H^Z of the event of a joint default of names in set Z and only in Z:

$$\lambda_t^Z = \lambda_Z(t, \mathbf{X}_t, \mathbf{H}_{t-}) = \sum_{Y \in \mathcal{Y};\, Y_t = Z} X_t^Y, \tag{12.23}$$

where Y_t stands for the set of survivors of set Y "right before time t", for any $Y \in \mathcal{Y}$. Thus $Y_t = Y \cap \operatorname{supp}^c(\mathbf{H}_{t-})$. One denotes by M^Z the corresponding compensated set-event martingale so that for $t \in [0, T]$,

$$dM_t^Z = dH_t^Z - \lambda_t^Z dt. \tag{12.24}$$

The W^Y and the M^Z have the $(\mathbb{G}, \mathbb{Q})$-martingale representation property.

12.4.2 TVA Model

A DMO setup can also be used as a credit portfolio TVA model, for a mark space of τ given as

$$E = \{Z \subseteq N;\ -1 \text{ or } 0 \in Z\},$$

under the assumption, enforcing (12.2) in the present setup, that there exists a function $\widetilde{\xi}$ satisfying

$$\xi_\tau(\pi) = \widetilde{\xi}(\tau, \mathbf{X}_\tau, \mathbf{H}_{\tau-}, \mathbf{H}^Z_{\tau-}; \pi) =: \widetilde{\xi}^Z_\tau(\pi) \tag{12.25}$$

on every event $\{\tau = \tau_Z\}$. Note that, in order to meet (12.27) below, we really need

$$\text{“}\widetilde{\xi}(\tau, \mathbf{X}_\tau, \mathbf{H}_{\tau-}, \mathbf{H}^Z_{\tau-}; \pi)\text{”}$$

in (12.25), as opposed to what would be a less specific

$$\text{“}\widetilde{\xi}_Z(\tau, \mathbf{X}_\tau, \mathbf{H}_{\tau-}; \pi)\text{”}$$

(compare (12.12)). But this is fine since, as discussed in Crépey and Song (2014),

$$\text{“}\widetilde{\xi}(\tau, \mathbf{X}_\tau, \mathbf{H}_{\tau-}, \mathbf{H}^Z_{\tau-}; \pi)\text{”}$$

holds in applications.

Also let

$$\mathcal{Z}_t = \{Z \subseteq N; Z = Y_t \text{ for at least one } Y \in \mathcal{Y}\} \setminus \emptyset$$

denote the set of all nonempty sets of survivors of groups Y in $\mathcal{Y}$ right before time t and let

$$\mathcal{Z}^\bullet_t = \{Z \in \mathcal{Z}_t;\ -1 \text{ or } 0 \in Z\} = \mathcal{Z}_t \cap E.$$

Observe that λ^Z_t in (12.23) vanishes for $Z \notin \mathcal{Z}_t$. Therefore the coefficient $\widehat{f}_t(P_t - \vartheta)$ of (12.4) is given, for $t \in [0, \bar{\tau}]$, by:

$$\widehat{f}_t(P_t - \vartheta) + r_t\vartheta = g_t(P_t - \vartheta) + \sum_{Z \in \mathcal{Z}^\bullet_t} \lambda^Z_t \widetilde{\xi}^Z_t(P_t - \vartheta) - \gamma_t\vartheta. \tag{12.26}$$

Moreover, letting

$$\mathcal{Y}^\bullet_t = \{Y \in \mathcal{Y};\ -1 \text{ or } 0 \in Y_t\}$$

for any process of the form

$$V(\tau, \mathbf{X}_\tau, \mathbf{H}_{\tau-}, \mathbf{H}^Z_{\tau-}), \tag{12.27}$$

we have by (12.23):

$$\sum_{Z \in \mathcal{Z}^\bullet_t} \lambda^Z_t V(t, \mathbf{X}_t, \mathbf{H}_{t-}, \mathbf{H}^Z_{t-}) = \sum_{Z \in \mathcal{Z}^\bullet_t} \left(\sum_{Y \in \mathcal{Y}^\bullet_t; Y_t = Z} X^Y_t \right) V(t, \mathbf{X}_t, \mathbf{H}_{t-}, \mathbf{H}^Z_{t-})$$
$$= \sum_{Z \in \mathcal{Z}^\bullet_t} \left(\sum_{Y \in \mathcal{Y}^\bullet_t; Y_t = Z} X^Y_t V(t, \mathbf{X}_t, \mathbf{H}_{t-}, \mathbf{H}^Y_{t-}) \right) = \sum_{Y \in \mathcal{Y}^\bullet_t} X^Y_t V(t, \mathbf{X}_t, \mathbf{H}_{t-}, \mathbf{H}^Y_{t-}),$$

where the middle equality follows from the fact that $\mathbf{H}^Z_{t-} = \mathbf{H}^Y_{t-}$ for any $Y \in \mathcal{Y}$ such that $Y_t = Z$. We have, in particular (in view of (12.25), which fits (12.27), regarding the second equality),

$$\gamma_t = \sum_{Z \in \mathcal{Z}_t^\bullet} \lambda_t^Z = \sum_{Y \in \mathcal{Y}_t^\bullet} X_t^Y, \ \sum_{\mathcal{Z}_t^\bullet} \lambda_t^Z \widetilde{\xi}_t^Z(\pi) = \sum_{Y \in \mathcal{Y}_t^\bullet} X_t^Y \widetilde{\xi}_t^Y(\pi), \tag{12.28}$$

so that, by (12.26):

$$\widehat{f}_t(P_t - \vartheta) + r_t\vartheta = g_t(P_t - \vartheta) + \sum_{Y \in \mathcal{Y}_t^\bullet} X_t^Y \left(\widetilde{\xi}_t^Y (P_t - \vartheta) - \vartheta \right). \tag{12.29}$$

12.4.3 Reduced-Form TVA Approach

Next, we want to use a reduced DMO model in a pre-default Markov TVA approach. Let $\widetilde{\mathcal{Y}} = \{\{1\}, \ldots, \{n\}\} \cup \widetilde{\mathcal{I}}$, where $\widetilde{\mathcal{I}}$ consists of those I_j in $\mathcal{I}$ which do not contain -1 or 0. Defining, for any obligor $i \in N^\star$,

$$\widetilde{\tau}_i = \min_{\{Y \in \widetilde{\mathcal{Y}};\, i \in Y\}} \eta_Y, \ \widetilde{H}_t^i = \mathbb{1}_{\tau_i \le t},$$

we will use the reduced model $X_t = (\mathbf{X}_t, \widetilde{\mathbf{H}}_t)$, where $\widetilde{\mathbf{H}} = (\widetilde{H}^i)_{i \in N^\star}$ with filtration $(\widetilde{\mathcal{H}}_t^i)$ of $\widetilde{H}^i$, relative to the filtration $\mathbb{F}$ such that, for any t,

$$\mathcal{F}_t = \mathcal{F}_t^{\mathbf{W}} \vee \bigvee_{i \in N^\star} \widetilde{\mathcal{H}}_t^i$$

and to the unchanged probability measure $\widetilde{\mathbb{Q}} = \mathbb{Q}$. By virtue of the Markov copula DMO features (see Bielecki, Cousin, Crépey, and Herbertsson (2013a), Part I), the reduced model is $(\mathbb{F}, \mathbb{Q})$- (in fact, even $(\mathbb{G}, \mathbb{Q})$-) Markov, with the following generator in a notation similar to (12.22):

$$\mathcal{X}u(t, \mathbf{x}, \widetilde{\mathbf{k}}) = \mathcal{A}_x u(t, \mathbf{x}, \widetilde{\mathbf{k}}) + \sum_{Y \in \widetilde{\mathcal{Y}}} x_Y \delta_Y u(t, \mathbf{x}, \widetilde{\mathbf{k}}). \tag{12.30}$$

Moreover let, for $Z \subseteq N^\star$,

$$d\widetilde{M}_t^Z = dH_t^Z - \widetilde{\lambda}_t^Z dt \text{ with } \widetilde{\lambda}_t^Z = \sum_{Y \in \widetilde{\mathcal{Y}};\, Y_t = Z} X_t^Y. \tag{12.31}$$

The $W^Y, Y \in \mathcal{Y}$ and the $\widetilde{M}^Z, Z \subseteq N^\star$, have the $(\mathbb{F}, \mathbb{Q})$-martingale representation property. Also note that, for $Z \subseteq N^\star$, we have $\widetilde{M}_t^Z = M_t^Z$ on $\{t \le \bar{\tau}\}$ since, by (12.23) and (12.31):

$$\mathbb{1}_{\{t \le \bar{\tau}\}} \lambda_t^Z dt = \mathbb{1}_{\{t \le \bar{\tau}\}} \widetilde{\lambda}_t^Z dt.$$

Of course, before τ every dependence in $\mathbf{H}_t$ reduces to a dependence in $\widetilde{\mathbf{H}}_t$. Let

$$\mathcal{Y}^\bullet = \{Y \in \mathcal{Y};\ -1 \text{ or } 0 \in Y\}.$$

Under the additional assumption that the processes $r_t, g_t(\pi)$ (for any real number π) and P_t are given before τ as continuous functions of (t, X_t) (functions denoted below by the same letters as the related processes) then, in view of (12.29), we have:

$$\widehat{f}_t(P_t - \vartheta)dt = f(t, X_t, \vartheta)dt,\ \ t \in [0, \bar{\tau}],$$

for the function f given, letting $\mathbf{k} = (0, 0, \widetilde{\mathbf{k}})$ for any $\widetilde{\mathbf{k}} \in \{0,1\}^n$, by:

$$\begin{aligned} f(t, \mathbf{x}, \widetilde{\mathbf{k}}, \vartheta) + r(t, \mathbf{x}, \mathbf{k})\vartheta &= g(t, \mathbf{x}, \mathbf{k}, P(s, \mathbf{m}, \mathbf{k}) - \vartheta) \\ &+ \sum_{Y \in \mathcal{Y}^\bullet} x_Y \left(\widetilde{\xi}(t, \mathbf{x}, \mathbf{k}, \mathbf{k}^Y; P(t, \mathbf{x}, \mathbf{k}) - \vartheta) - \vartheta\right). \end{aligned} \tag{12.32}$$

Note that an $\mathbb{F}$-martingale does not jump at τ and an $\mathbb{F}$-martingale stopped at τ is a $\mathbb{G}$-martingale, by the martingale representation properties that are valid in the full and in the reduced DMO models (all under $\mathbb{Q}$). So the condition (A) holds with $\widetilde{\mathbb{Q}} = \mathbb{Q}$ in (A.2) and for the Markov specification of the coefficient $f_t(\vartheta) := f(t, X_t, \vartheta)$ in (A.3), where the function $f = f(t, \mathbf{x}, \widetilde{\mathbf{k}}, \vartheta)$ is given by (12.32).

Conclusion

In the dynamic Gaussian copula (DGC) model, as in the dynamic Marshall-Olkin (DMO) common-shock model, one ends up with a pre-default TVA equation over $[0, T]$ amenable to purely forward (nonlinear as necessary) simulation schemes. Using our usual specifications for the exposure at default $\xi_t(\pi)$ (see (4.36)) and for the funding coefficient g_t, one can then develop (12.4) and the ensuing expressions (12.19) (in the DGC model) or (12.32) (in the DMO model) in the form of CVA, DVA, LVA and RC decompositions in the manner of (6.15). See Crépey and Song (2014) for further developments in this direction, using default times with marks as a unifying modeling tool allowing one not only to deal with counterparty risk and funding on credit derivatives as shown above, but more broadly to account for various possible wrong-way and gap risk scenarios and features.

Part VI

Mathematical Appendix

Chapter 13

Stochastic Analysis Prerequisites

This is a survey about material of stochastic analysis useful for this book. The main references are Cont and Tankov (2004), Ikeda and Watanabe (1989), Protter (2004), Kunita (2010) and Crépey (2013).

Setup

Let there be given a finite time horizon $T > 0$, a probability space $(\Omega, \mathcal{G}, \mathbb{Q})$ and a filtration $\mathbb{G} = (\mathcal{G}_t)_{t\in[0,T]}$ satisfying the usual conditions of completeness and right-continuity. Let there also be given a measured mark space[1] $(E, \mathcal{B}_E, c)$, where c is a nonnegative σ-finite measure c on $(E, \mathcal{B}_E)$.

By default, any random variable has to be $\mathcal{G}$-measurable, any process is defined on the time interval $[0, T]$, and $\mathbb{G}$-adapted and any (nonnegative) random measure λ is defined on $[0, T] \times E$, such that process $\lambda_t(O) = \lambda([0, t] \times O)$ is $\mathbb{G}$-adapted, for any $O \in \mathcal{B}_E$. All inequalities between random quantities are to be understood $d\mathbb{Q}$-almost surely, $dt \otimes d\mathbb{Q}$-almost everywhere or $dt \otimes d\mathbb{Q} \otimes c(de)$-almost everywhere, as suitable. For simplicity we omit any dependence on ω in the notation.

13.1 Stochastic Integration

In this section we give a very brief account of the semimartingale and random measure integration theories.

13.1.1 Semimartingales

As emphasized in Protter (2004), semimartingales constitute a class of integrator-processes giving rise to the most flexible theory of stochastic integration. In mathematical finance, another motivation for modeling prices of traded assets as semimartingales is

[1]Typically an Euclidean set endowed with its Borel σ-field or a finite mark space endowed with its powerset as σ-field.

that price processes outside this class give rise to arbitrages, unless rather strong conditions are imposed on the trading strategies. In a filtration satisfying the usual conditions, every semimartingale can be considered as càdlàg, a French acronym for "left limited and right continuous". All semimartingales considered in this book are thus understood to be càdlàg. In one of several equivalent characterizations, a semimartingale X can be given as the sum of a local martingale M and of a finite variation process D, where:

- a local martingale M admits an increasing and diverging sequence of stopping times T_n such that every stopped process $M_{\cdot \wedge T_n}$ is a (true) martingale, and
- a finite variation process is a difference between two nondecreasing processes starting from 0.

Recall that a martingale is a local martingale and that a bounded local martingale is a martingale. However, in general a local martingale is not a martingale: e.g. a driftless diffusion process is a local martingale, but not necessarily a martingale.

Any representation $X = D + M$ as above is called a Doob-Meyer decomposition of X. A Doob-Meyer decomposition is not unique in general. However, if there exists a Doob-Meyer decomposition of X with D predictable, then such a decomposition is unique and it is called the canonical Doob-Meyer decomposition of the so called special semimartingale X. Regarding predictability, it is enough in this book to know that any (adapted) left-continuous process, or any deterministic process (Borel function of time), is predictable. Uniqueness for the canonical Doob-Meyer decomposition of a special semimartingale can equivalently be expressed as follows.

Proposition 13.1.1 *The only predictable local martingale of finite variation (in particular, the only time-differentiable local martingale) is the null process.*

Moreover, the local martingale component of a special semimartingale can in turn be uniquely decomposed into the sum of a continuous local martingale and a purely discontinuous local martingale (or compensated sum of jumps, starting from 0).

The stochastic integral of a predictable and locally bounded[2] process Z with respect to a semimartingale X is then defined as

$$Y_t = \int_0^t Z_s dX_s = \int_0^t Z_s dD_s + \int_0^t Z_s dM_s, \qquad (13.1)$$

where $X = D + M$ is a Doob-Meyer decomposition of X and $\int_0^t Z_s dM_s$ is defined by localization of M. A remarkable fact is that the corresponding notion of stochastic integral is independent of the Doob-Meyer decomposition of X which is used in (13.1). Predictable and locally bounded processes notably include all left-limiting processes Z of the form $Z = \widetilde{Z}_-$, where $\widetilde{Z}$ is a càdlàg process (e.g. a semimartingale).

[2]With almost all paths bounded, in particular, and a bit more, since it should be noted that locally bounded actually means "locally uniformly bounded".

Proposition 13.1.2 *In case X is a local martingale, the integral process Y is again a local martingale.*

In the case of a continuous integrator X, it is possible to define the stochastic integral Y for a class of integrands larger than the predictable and locally bounded processes, namely for progressively measurable integrands Z subject to suitable integrability conditions. If the left-limit process Z_- of Z exists, then

$$\int_0^t Z_s dX_s = \int_0^t Z_{s-} dX_s. \tag{13.2}$$

13.1.2 Random Measures Integration Theory

Random measures are useful for the description of infinite activity jump processes. The corresponding integration theory allows one to deal with stochastic differential equations involving such processes. We refer the reader to Kunita (2010), Ikeda and Watanabe (1989), Cont and Tankov (2004) for more developments in this regard.

Definition 13.1.3 Given a nonnegative σ finite (deterministic) measure ρ on $\big([0,T] \times E, \mathcal{B}([0,T]) \otimes \mathcal{B}_E\big)$, a Poisson random measure $n = (n(dt, de))_{t \in [0,T], e \in E}$ with intensity measure ρ on $\big([0,T] \times E, \mathcal{B}([0,T]) \otimes \mathcal{B}_E\big)$ compensating n is such that, denoting $\mathcal{B}^\star = \{B \in \mathcal{B}([0,T]) \otimes \mathcal{B}_E; 0 < \rho(B) < \infty\}$,

0. for every ω, $n(\cdot, \omega)$ is a counting measure on $[0,T] \times E$, i.e. an $\mathbb{N} \cup \{\infty\}$-valued measure on $[0,T] \times E$ such that $n(\{t\} \times E, \omega) \leq 1$, for $t \in [0,T]$,

1. for every disjoint $B_1, \dots, B_n$ in $\mathcal{B}^\star$, the random variables $n(B_1), \dots, n(B_n)$ are independent,

2. for every $B \in \mathcal{B}^\star$, $n(B)$ is Poisson distributed with mean $\rho(B)$.

The standing example of a Poisson random measure n is the measure defined, for any $s < t$ and $O \in \mathcal{B}_E$, by:

$$n((s,t] \times O) = \sharp\{r \in (s,t]; \Delta X_r \in O\} \tag{13.3}$$

for some additive process X, where:

Definition 13.1.4 A $\mathbb{G}$-adapted, real-valued stochastic process $(X_t)_{t \in [0,T]}$, starting from zero, is called an additive (or time-inhomogeneous Lévy) process if it satisfies the following conditions:

0. X is stochastically continuous, i.e. for any $0 \leq t \leq T$ and $\epsilon > 0$,

$$\lim_{s \to t} \mathbb{P}(|X_t - X_s| > \epsilon) = 0,$$

1. X has independent increments with respect to $\mathbb{F}$, i.e. for any $0 \leq s \leq t \leq T$ the random variable $(X_t - X_s)$ is independent from $\mathcal{F}_s$ (i.e. from any $\mathcal{F}_s$-measurable random variable).

One can show that an additive process X is a semimartingale, the càdlàg version of which is considered. Note that the first condition in the above definition does not imply pathwise continuity of X. It is, for instance, satisfied by the Poisson process (which, together with the Brownian motion, is the standing example of a Lévy process). This condition is only intended to avoid unwanted "calendar effects", such as X jumping at fixed times. Actually one can show that an additive process X is quasi-left continuous, which for a càdlàg process is equivalent to continuity at predictable stopping times. This implies the existence of a sequence of totally inaccessible stopping times T_n (denumerable but possibly not orderable) exhausting the jumps of X, so that (13.3) can equivalently be stated in differential form as

$$n(dt, de) = \sum_n \boldsymbol{\delta}_{T_n, \Delta X_{T_n}}(dt, de),$$

where ΔX_{T_n} is the jump of X at T_n and $\boldsymbol{\delta}_{T_n, \Delta X_{T_n}}$ is a Dirac measure at the (random) point $(T_n, \Delta X_{T_n})$.

Remark 13.1.5 Since an additive process has independent increments with respect to the filtration $\mathbb{G}$, it is a $\mathbb{G}$-Markov process.

We henceforth assume that the compensator ρ of n can be disintegrated in the form $\rho(dt, de) = c(de)dt$, for a σ-finite (deterministic) jump measure c on $(E, \mathcal{B}_E)$. This is, for instance, the case for the counting measure n of a Lévy process X, where:

Definition 13.1.6 A Lévy process X is an additive process that is time-homogeneous, meaning that it satisfies the following condition in addition to those of the definition 13.1.4:

2. X has stationary increments, i.e. for any $0 \leq s \leq t \leq T$ the distribution of the random variable $X_{t+s} - X_t$ does not depend on t.

The measure c thus associated with a Lévy process is called a Lévy measure; it satisfies the following integrability condition:

$$\int_E \frac{|e|^2}{1 + |e|^2} c(de) < \infty,$$

often written as

$$\int_E (1 \wedge |e|^2) c(de) < \infty.$$

By a predictable integrand $V = V_t(e)$, we mean a real valued random function $V_t(\omega, e)$, measurable with respect to the product of the predictable σ-algebra on $[0, T] \times \Omega$ by $\mathcal{B}_E$.

Since $n(dt, de)$ and $\rho(dt, de) = c(de)dt$ are nonnegative σ-finite measures, we can define the integrals $V_t \cdot dn_t$ and $V_t \cdot c\,dt$ of V with respect to n and ρ in the Lebesgue-Stieltjes pathwise sense as, respectively,

$$\begin{aligned} \int_0^{\cdot} V_t \cdot dn_t &= \int_0^{\cdot}\int_E V_t^+(e)n(dt,de) - \int_0^{\cdot}\int_E V_t^-(e)n(dt,de), \\ \int_0^{\cdot} V_t \cdot c\,dt &= \int_0^{\cdot}\int_E V_t^+(e)c(de)dt - \int_0^{\cdot}\int_E V_t^-(e)c(de)dt, \end{aligned} \tag{13.4}$$

where the terms in the right-hand sides are well defined whenever

$$\int_0^{\cdot}\int_E |V_t(e)|n(dt,de) < \infty, \text{ respectively } \int_0^{\cdot}\int_E |V_t(e)|c(de)dt < \infty.$$

Moreover, one can show that if the second condition holds almost surely ("$V \in \mathcal{P}^1_{loc}$" in the notation to be introduced below), so does the first one.

In case $V_t \cdot c\,dt$ and $V_t \cdot dn_t$ fail to be well defined separately, it may still be possible to define a stochastic integral $V_t \cdot dm_t$ of V with respect to the compensated Poisson measure $m(dt, de) = n(dt, de) - \rho(dt, de)$. We refer the reader to Kunita (2010) (see also Ikeda and Watanabe (1989)) for the construction of this stochastic integral, using the properties of a Poisson random measure in the definition 13.1.3, such that

$$V_t \cdot dm_t = V_t \cdot dn_t - V_t \cdot c\,dt \tag{13.5}$$

holds for any $V \in \mathcal{P}^1_{loc}$.

We introduce the following spaces of predictable integrands:

- $\mathcal{P}$, respectively $\mathcal{P}_{loc}$, the space of predictable integrands V such that

$$\mathbb{E}\int_0^T \frac{|V_t|^2}{1+|V_t|} \cdot c\,dt < \infty, \text{ respectively } \int_0^T \frac{|V_t|^2}{1+|V_t|} \cdot c\,dt < \infty \text{ a.s.,} \tag{13.6}$$

- $\mathcal{P}^1$, respectively $\mathcal{P}^1_{loc}$, the space of predictable integrands V such that

$$\mathbb{E}\int_0^T |V_t| \cdot c\,dt < \infty, \text{ respectively } \int_0^T |V_t| \cdot c\,dt < \infty \text{ a.s.,} \tag{13.7}$$

- $\mathcal{P}^2$, respectively $\mathcal{P}^2_{loc}$, the space of predictable integrands V such that $|V|^2 \in \mathcal{P}^1$, respectively $\mathcal{P}^1_{loc}$.

For any of these spaces, denoted by $\mathcal{Q}$, we have that $V \in \mathcal{Q}$ if $|V| \leq |U|$ for some $U \in \mathcal{Q}$. In particular, in the case of a Lévy measure $c(de)$ integrating $1 \wedge |e|^2$ (so that $1 \wedge |e| \in \mathcal{P}_{loc}$), we have $V \in \mathcal{P}_{loc}$ as soon as $|V_t(e)| \leq C(1 \wedge |e|)$. Also note that $\mathcal{P}^i \subseteq \mathcal{P}^i_{loc} \subseteq \mathcal{P}_{loc}$ for $i = 1, 2$. In the case of a finite jump measure c with $c(E) < \infty$, we have that

$$\mathcal{P}^2 \subseteq \mathcal{P}^1 = \mathcal{P},\ \mathcal{P}^2_{loc} \subseteq \mathcal{P}^1_{loc} = \mathcal{P}_{loc}.$$

All the above (concepts, spaces and notation) is extended componentwise to the case of vector-valued integrands.

The stochastic integral $V_t \cdot dm_t$ yields a well-defined local martingale, respectively martingale, respectively locally square integrable local martingale, respectively square integrable martingale, provided V lies in $\mathcal{P}_{loc}$, respectively $\mathcal{P}$, respectively $\mathcal{P}^2_{loc}$, respectively $\mathcal{P}^2$. For a real valued $V \in \mathcal{P}^2$, we have the following "Poisson measure isometry property" , which plays a central role in the construction of the stochastic integral $\int_0^{\cdot} V_t \cdot dm_t$:

$$\mathbb{V}\mathrm{ar}(\int_0^T V_t \cdot dm_t) = \mathbb{E}(\int_0^T V_t^2 \cdot c\, dt)$$

(consistent with the well known identities $\mathbb{V}\mathrm{ar} N_T = \lambda T = \mathbb{E} N_T$, in the case of a Poisson process N_t with parameter λ). See Kunita (2010) or Ikeda and Watanabe (1989) for more details and properties.

13.2 Itô Processes

In addition to the above Poisson random measure n with disintegrated intensity measure $c\, dt$ and compensated measure $m(dt, de) = n(dt, de) - c(de)dt$, let $W = (W_t)_{t \in [0,T]}$ be a standard q-dimensional Brownian motion, assumed to exist on the same stochastic basis. We introduce the following spaces of progressively measurable processes:

- $\mathcal{M}^1$, respectively $\mathcal{M}^1_{loc}$, the space of progressively measurable real integrands Z such that

$$\mathbb{E}\int_0^T |Z_t| dt < \infty, \text{ respectively } \int_0^T |Z_t| dt < \infty \text{ a.s.}, \tag{13.8}$$

- $\mathcal{M}^2$, respectively $\mathcal{M}^2_{loc}$, the space of progressively measurable real integrands Z such that $|Z|^2 \in \mathcal{M}^1$, respectively $\mathcal{M}^1_{loc}$,

- $\mathcal{S}^p$, for any real number $p \geq 2$, the space of progressively measurable real processes Y such that

$$\|Y\|_{\mathcal{S}^p} = \left(\mathbb{E}\left\{\sup_{t \in [0,T]} |Y_t|^p\right\}\right)^{\frac{1}{p}} < \infty.$$

We also denote by the same letters the componentwise extensions of these spaces to vector or matrix cases. Given a q-dimensional random vector b_t, matrix σ_t and function $j_t(e)$, we consider a q-variate Itô process X such that

$$dX_t = b_t\, dt + \sigma_t\, dW_t + j_t \cdot dm_t, \tag{13.9}$$

where it is assumed for well-definiteness of the successive terms on the right-hand side that

$$b \in \mathcal{M}^1_{loc}, \ \sigma \in \mathcal{M}^2_{loc}, \ j \in \mathcal{P}_{loc}. \tag{13.10}$$

In particular, an Itô process is a special semimartingale, with canonical Doob-Meyer decomposition given by (13.9). The second and third terms in this decomposition are the continuous local martingale and the purely discontinuous local martingale components of X. If $\sigma \in \mathcal{M}^2$ and $j \in \mathcal{P}$, the corresponding local martingale terms in (13.9) are true martingales (and a square integrable martingale, in $\mathcal{S}^2$, in the case of the second term). Alternatively, strengthening the condition $j \in \mathcal{P}_{loc}$ to $j \in \mathcal{P}^2_{loc}$, (respectively $\sigma \in \mathcal{M}^2_{loc}$ into $\sigma \in \mathcal{M}^2$ and $j \in \mathcal{P}_{loc}$ into $j \in \mathcal{P}^2$) in (13.10), we will speak of an Itô process of locally integrable quadratic variation (respectively of integrable quadratic variation). In this case the local martingale terms in (13.9) are both local square integrable martingales (respectively square integrable martingales). The "quadratic variation" terminology refers to the notion of quadratic variation (or square bracket) of a process X that will be introduced in Sect. 13.2.2.

13.2.1 Finite Variation Jumps

We first assume that $j \in \mathcal{P}^1_{loc}$. This situation can be shown to correspond to the "finite variation case", where the purely discontinuous martingale component $j_t \cdot dm_t$ of X has finite variation (see e.g. Cont and Tankov (2004)). In other words this corresponds to the case where all the three stochastic integrals (in differential notation) are well defined and have finite variation in

$$j_t \cdot dm_t = j_t \cdot dn_t - j_t \cdot c\, dt$$

In this case, the integrals on the right-hand side are well defined in the Lebesgue-Stieltjes pathwise sense recalled in the first line of (13.4), so that the stochastic integral $j_t \cdot dm_t$ can also be understood pathwise.

The dynamics (13.9) of X can then be rewritten in the following equivalent form:

$$dX_t = b^\circ_t\, dt + \sigma_t\, dW_t + j_t \cdot dn_t, \tag{13.11}$$

with

$$b^\circ_t = b_t - j_t \cdot c. \tag{13.12}$$

Let there be given a real valued function $u = u(t, x)$ of class $\mathcal{C}^{1,2}$ on $[0, T] \times \mathbb{R}^d$ forming an Itô pair with X, in the sense that

$$\delta u_t \in \mathcal{P}^1_{loc}, \tag{13.13}$$

where

$$\delta u_t(e) = u(t, X_{t-} + j_t(e)) - u(t, X_{t-}). \tag{13.14}$$

Note that, given $j \in \mathcal{P}^1_{loc}$, we have:

Remark 13.2.1 a $\mathcal{C}^{1,2}$-function u forms an Itô pair with X whenever (there are no jumps in X or)

(i) $|j_t(e)| \le \widehat{j}_t$ for a nonnegative predictable and locally bounded process $\widehat{j}_t$ (observe that this condition is only about X, not u) , or

(ii) u has linear growth, i.e. $|u(t,x)| \le C(1+|x|)$, or

(iii) u has polynomial growth, i.e. $|u(t,x)| \le C(1+|x|^n)$ for some nonnegative integer n, and c integrates $|j_t(e)|^n \mathbb{1}_{|j_t(e)|\ge 1}$ (against $dt \otimes d\mathbb{Q} \otimes c(de)$).

Knowing the Itô formula for a continuous Itô process, the following result is then apparent from (13.11).

Proposition 13.2.2 (Itô formula for an Itô process with finite variation jumps) If $j \in \mathcal{P}^1_{loc}$, for any function $u = u(t,x)$ forming an Itô pair with X, we have:

$$\begin{aligned} du(t,X_t) &= \partial_t u(t,X_t)dt + \partial u(t,X_t)b^\circ_t dt + \partial u(t,X_t)\sigma_t\, dW_t \\ &\quad + \frac{1}{2}\partial^2 u(t,X_t) : a_t dt + \delta u_t \cdot dn_t, \end{aligned} \tag{13.15}$$

in which:

- $\partial u(t,x)$ and $\partial^2 u(t,x)$ denote the row-gradient and the Hessian of u with respect to x,
- $\partial^2 u(t,x) : a_t$ stands for the trace (sum of the diagonal elements) of the product of the Hessian matrix $\partial^2 u(t,x)$ with the covariance matrix $a_t = \sigma_t \sigma_t^{\mathsf{T}}$ of X, and
- $\delta u_t(e)$ was defined in (13.14).

Or, equivalent to (13.15):

$$\begin{aligned} du(t,X_t) &= \left(\partial_t u(t,X_t) + \partial u(t,X_t)b^\circ_t + \frac{1}{2}\partial^2 u(t,X_t) : a_t + \delta u_t \cdot c\right) dt \\ &\quad + \partial u(t,X_t)\sigma_t\, dW_t + \delta u_t \cdot dm_t. \end{aligned} \tag{13.16}$$

Given (13.13), the process $\int_0^{\cdot} \delta u_t \cdot dm_t$ is a local martingale, the purely discontinuous local martingale component of the special semimartingale $u(t,X_t)$ (the continuous local martingale component being given as $\partial u(t,X_t)\sigma_t\, dW_t$).

13.2.2 General Case

Now, an important fact is that in the general case $j \in \mathcal{P}_{loc}$ of a possibly infinite variation compensated sum of jumps (purely discontinuous martingale component) of X, the Itô formula in the special semimartingale form (13.16) we just obtained is still valid, rewritten in the form (13.18) below, under the Itô pair integrability condition modified into

$$\delta u_t - \partial u(t,X_{t-})j_t \in \mathcal{P}^1_{loc}, \quad \delta u_t \in \mathcal{P}_{loc}. \tag{13.17}$$

Remark 13.2.3 Moreover, the remark 13.2.1 holds verbatim in this setting. Note that X_{t-} is locally bounded, hence bounded on its trajectories on $[0, T]$, as is therefore also $\partial u(t, X_{t-})$ (since ∂u is a continuous function). Consequently, in the special case $j \in \mathcal{P}^1_{loc}$ (case of finite variation jumps), we also have that $\partial u(t, X_{t-}) j_t \in \mathcal{P}^1_{loc}$, so that (13.17) reduces to (13.13).

Therefore:

Proposition 13.2.4 (Itô formula in special semimartingale form) Given an Itô pair (X, u) composed of an Itô process X and a $\mathcal{C}^{1,2}$-function u satisfying (13.17) we have, for $t \in [0, T]$,

$$du(t, X_t) = (\partial_t u(t, X_t) + \mathcal{A}u_t)\,dt + \partial u(t, X_t)\sigma_t\,dW_t + \delta u_t \cdot dm_t, \quad (13.18)$$

for the (random) generator

$$\mathcal{A}u_t = \frac{1}{2}\partial^2 u(t, X_t) : a_t + \partial u(t, X_t) b_t + \big(\delta u_t - \partial u(t, X_t) j_t\big) \cdot c. \quad (13.19)$$

Therefore:

$$\begin{aligned}(dt)^{-1}\mathbb{E}\left(du(t, X_t)\middle|\mathcal{G}_t\right) &:= \lim_{h \to 0} h^{-1}\mathbb{E}\left(u(t+h, X_{t+h}) - u(t, X_t)\middle|\mathcal{G}_t\right)\\ &= \partial_t u(t, X_t) + \mathcal{A}u_t,\end{aligned}$$

where, by the left-hand side on the first line and in analogous expressions in the sequel, we mean the corresponding limit when h goes to 0 in the right hand side.

Remark 13.2.5 (Truncation functions) An Itô process X satysfying (13.9)-(13.10) can equivalently be represented in the following form:

$$dX_t = b^\bullet_t\,dt + \sigma_t\,dW_t + j^\circ_t \cdot dn_t + j^\bullet_t \cdot dm_t, \quad (13.20)$$

where j° and $j^\bullet$ respectively correspond to the "large" and the "small" jumps of X as ruled by the (arbitrary) cutoff level one, i.e.

$$j^\circ_t(e) = j_t(e)\mathbb{1}_{|j_t(e)| \geq 1},\ j^\bullet_t(e) = j_t(e)\mathbb{1}_{|j_t(e)| < 1} = j_t(e) - j^\circ_t(e) \quad (13.21)$$

and where, by identification with (13.9):

$$b_t = b^\bullet_t + j^\circ_t \cdot c. \quad (13.22)$$

Observe that, since $j \in \mathcal{P}_{loc}$, it follows that $c(de)dt$ integrates $j^\circ_t(e)$, so that the jump term $j^\circ_t \cdot dn_t$ in (13.20) is well defined and of finite variation.

Remark 13.2.6 Since we consider a semimartingale X in a càdlàg version, the jump times of X of size ≥ 1 almost surely form a finite sequence over $[0, T]$. Therefore $j^\circ_t \cdot dn_t$ actually reduces to a (randomly) finite sum over the jump times of X of size ≥ 1.

Now, every Lévy (not necessarily Itô in our sense) process X admits a representation of the form (13.20) with $b^\bullet$ and σ constant, $j_t(e) = e$ and $c(de)$ integrating $1 \wedge |e|^2$. In the context of Lévy processes, the semimartingale decomposition (13.20) is more general than the "fully compensated', special semimartingale, decomposition (13.9). Namely, every Lévy process X admits a semimartingale decomposition of the form (13.20). But it is only for $c(de)$ integrating $e\mathbb{1}_{|e|\geq 1}$ that a special semimartingale decomposition (13.9) of X follows. In other words, not all Lévy processes are special semimartingales (or, again, Itô processes in our sense).

Remark 13.2.7 In line with the existence of the general semimartingale Itô formula which will be alluded to in the remark 13.2.10, it is not difficult to build upon the Itô formula in special semimartingale form (13.18), for an Itô pair (X, u), in order to derive a general Itô formula for Itô processes, valid for any function $u = u(t, x)$ of class $C^{1,2}$ and any Itô process X (not necessarily satisfying the Itô pair integrability conditions (13.17)). The notation is as in (13.21)-(13.22) where, for $j \in \mathcal{P}_{loc}$, $c(de)dt$ almost surely integrates $j_t^\circ(e)$. One may thus rewrite (13.9) in the form (13.20), or

$$dX_t = dX_t^\bullet + j_t^\circ \cdot dn_t, \tag{13.23}$$

with

$$dX_t^\bullet = b_t^\bullet\, dt + \sigma_t\, dW_t + j_t^\bullet \cdot dm_t,$$

Since $|j^\bullet| \leq 1$, we have an Itô pair $(X^\bullet, u)$ for any $C^{1,2}$ function u (see the remarks 13.2.3 and 13.2.1(i)). One can thus apply the Itô formula in special semimartingale form (13.18) to the pair $(X^\bullet, u)$, which yields in the end, returning to the process X, the following "general Itô process Itô formula":

$$\begin{aligned} du(t, X_t) &= (\partial_t u(t, X_t) + \mathcal{A}^\bullet u_t)\, dt + \delta_t^\circ u(X_{t-}) \cdot dn_t \\ &\quad + \partial u(t, X_t)\sigma_t\, dW_t + \delta^\bullet u_t \cdot dm_t, \end{aligned} \tag{13.24}$$

in which

$$\mathcal{A}^\bullet u_t = \frac{1}{2}\partial^2 u(t, X_t) : a_t + \partial u(t, X_t) b_t^\bullet + \left(\delta^\bullet u_t - \partial u(t, X_t) j_t^\bullet\right) \cdot c \tag{13.25}$$

and where

$$\delta^\circ u_t \cdot dn_t$$

reduces to an almost surely finite sum over the jump times of X of size ≥ 1. However, (13.25) does not provide a special semimartingale decomposition of $u(t, X_t)$ and the process $u(t, X_t)$ may fail to be an Itô process in our sense. In particular, (13.24) cannot be used to conclude that $\partial_t u(t, X_t) + \mathcal{A}^\bullet u_t$ should be zero, should $u(t, X_t)$ be a local martingale (cf. Sect. 13.4).

13.2.2.1 Brackets

We now consider the case of an Itô process X of locally integrable quadratic variation. In addition, we strengthen to $\delta u_t \in \mathcal{P}^2_{loc}$ the second integrability condition in the Itô pair requirements (13.17). We then speak of an "Itô pair of locally integrable quadratic variation". Note that, in the general setup of this subsection, if X is of locally integrable quadratic variation, then $\delta u_t \in \mathcal{P}^2_{loc}$ holds under any of the additional conditions stated at the remark 13.2.1 (or of course if there are no jumps in X).

The local martingale terms are then local square integrable martingales in the Itô formula (13.18). Given another Itô pair of locally integrable quadratic variation (X, v) for the same process X and some other real valued function v, this implies that

$$\begin{aligned}&(dt)^{-1}\mathbb{C}\mathrm{ov}\left(du(t,X_t), dv(t,X_t)\middle|\mathcal{G}_t\right)\\&\quad= \partial u(t,X_t)a_t(\partial v(t,X_t))^\mathsf{T} + \gamma(u,v)_t \cdot c = \mathcal{C}(u,v)_t,\end{aligned} \tag{13.26}$$

where $(u,v) \mapsto \gamma(u,v)_t$ and $(u,v) \mapsto \mathcal{C}(u,v)_t$ are the bilinear *carré du champ* (random) operators associated with the linear (random) operators $u \mapsto \delta u_t$ and $u \mapsto \mathcal{A}u_t$, so

$$\gamma(u,v)_t - \delta(uv)_t \quad u(t,X_{t-})\delta v_t - v(t,X_{t-})\delta u_t = \delta u_t \delta v_t$$

and

$$\mathcal{C}(u,v)_t = \mathcal{A}(uv)_t - u\mathcal{A}v_t - v\mathcal{A}u_t.$$

Letting $Y_t = u(t,X_t)$ and $Z_t = v(t,X_t)$, the process $\mathbb{C}\mathrm{ov}\left(dY_t, dZ_t\middle|\mathcal{G}_t\right)$ corresponds to the so called sharp bracket process $d\langle Y, Z\rangle_t$ through

$$\mathcal{C}(u,v)_t = \frac{d\langle Y, Z\rangle_t}{dt}. \tag{13.27}$$

To summarize:

Proposition 13.2.8 *The (random) generator $u \mapsto \mathcal{A}u_t$ of X and its carré du champ $(u,v) \mapsto \mathcal{C}(u,v)_t$ are such that, letting $Y_t = u(t,X_t)$ and $Z_t = v(t,X_t)$ for any Itô pairs of locally integrable quadratic variation (X,u) and (X,v):*

$$\begin{aligned}(dt)^{-1}\mathbb{E}\left(du(t,X_t)\middle|\mathcal{G}_t\right) &= \partial_t u(t,X_t) + \mathcal{A}u_t\\(dt)^{-1}\mathbb{C}\mathrm{ov}\left(du(t,X_t), dv(t,X_t)\middle|\mathcal{G}_t\right) &= \mathcal{C}(u,v)_t,\\= \frac{d\langle Y,Z\rangle_t}{dt} &= \partial u(t,X_t)a_t(\partial v(t,X_t))^\mathsf{T} + (\delta u_t \delta v_t)\cdot c.\end{aligned} \tag{13.28}$$

In particular,

$$\begin{aligned}(dt)^{-1}\mathbb{V}\mathrm{ar}\left(du(t,X_t)\middle|\mathcal{G}_t\right) &= \mathcal{C}(u,u)_t\\= \frac{d\langle Y\rangle_t}{dt} &= \partial u(t,X_t)a_t(\partial u(t,X_t))^\mathsf{T} + (\delta u_t)^2\cdot c.\end{aligned} \tag{13.29}$$

By letting u and v range over the various coordinate mappings of X, we obtain:

Corollary 13.2.9 *Denoting in matrix form* $< X >= (< X^i, X^j >)_i^j$, *we have*

$$(dt)^{-1}\mathbb{C}\text{ov}\left(dX_t \middle| \mathcal{G}_t\right) = \frac{d\langle X\rangle_t}{dt} = a_t + (j_t j_t^{\mathsf{T}}) \cdot c \tag{13.30}$$

Returning to $Y_t = u(t, X_t)$ and $Z_t = v(t, X_t)$, observe that the sharp brackets compensate the corresponding square brackets (quadratic covariation and variation) defined by $[Y, Z]_0 = 0$ and

$$d[Y, Z]_t = \partial u(t, X_t) a_t (\partial v(t, X_t))^{\mathsf{T}} dt + \left(\delta u_t \delta v_t\right) \cdot dn_t,$$

and thus $[Y, Y]_0 = 0$ and

$$d[Y, Y]_t = \partial u(t, X_t) a_t (\partial u(t, X_t))^{\mathsf{T}} dt + \left(\delta u_t\right)^2 \cdot dn_t.$$

Notably, if X is a continuous Itô process, the above sharp and square brackets (exist and) coincide.

Remark 13.2.10 The square brackets can equivalently be defined as limits in probability[3] of the realized covariance and variance processes over vanishing mesh sizes. They can be so defined and are finite variation processes for any semimartingales Y, Z. They are key in the following ***semimartingale integration by parts formulas*** (in differential form):

$$d(Y_t Z_t) = Y_{t-} dZ_t + Z_{t-} dY_t + d[Y, Z]_t. \tag{13.31}$$

They can also be used for stating a general ***semimartingale Itô formula***, valid for any function $u = u(t, x)$ of class $C^{1,2}$ and any semimartingale X_t.

13.3 Jump-Diffusions

By a jump-diffusion we mean hereafter an Itô process X in the sense of Sect. 13.2, but for a Markov SDE (13.9), meaning that the random coefficients b_t, σ_t and $j_t(e)$ of (13.9) are now given deterministically in terms of X_{t-}[4] as

$$b_t = b(t, X_t),\ \sigma_t = \sigma(t, X_t),\ j_t(e) = j(t, X_{t-}, e). \tag{13.32}$$

Endowed with an initial condition $X_0 = x$ (a constant, say, for simplicity), equation (13.9) is thus now an SDE in X. We assume the following linear growth and Lipschitz conditions on the coefficients:

$$\begin{aligned} |b(t,x)|,\ |\sigma(t,x)| &\le C(1 + |x|), \\ |b(t,x) - b(t,y)|,\ |\sigma(t,x) - \sigma(t,y)| &\le C|x - y| \end{aligned} \tag{13.33}$$

[3]Or almost sure limits in the case of nested meshes.

[4]Or X_t in the case of b, σ, which in view of (13.2) makes no difference in (13.9), by continuity of t and W_t.

and

$$\begin{aligned} |j(t,x,e)| &\le \Gamma(e)(1+|x|), \\ |j(t,x,e)-j(t,y,e)| &\le \Gamma(e)|x-y|, \end{aligned} \tag{13.34}$$

with $\Gamma^2 \cdot c < \infty$ almost surely. This entails the following a priori estimate on any solution[5] X to (13.32), for $p \ge 2$:

$$\|X\|^p_{\mathcal{S}^p} \le C_p(1+|x|^p), \tag{13.35}$$

as well as error estimates implying uniqueness of a solution (if any). Note that, as a consequence of (13.35), any solution X to (13.9), (13.32) satisfies

$$b(t,X_t) \in \mathcal{M}^2, \;\; \sigma(t,X_t) \in \mathcal{M}^2, \;\; j(t,X_{t-},e) \in \mathcal{P}^2 \tag{13.36}$$

and is therefore a square integrable Itô process with square integrable martingale terms. Given this observation, existence (and uniqueness) of a solution X can be established by Picard iteration. A notable feature of X is the so-called ***Markov property***, meaning that

$$\mathbb{E}(\phi(X_s, s\in[t,T]) \,|\, \mathcal{G}_t) = \mathbb{E}(\phi(X_s, s\in[t,T]) \,|\, X_t),$$

for any (possibly path dependent) functional ϕ of X that makes sense on both sides of the equality. So "the past of X does not influence its future", the present of X gathering all the relevant information.

Given a real valued function $u = u(t,x)$ resulting in an Itô pair (X,u), we have by (13.18) the following ***Itô formula for a jump-diffusion***:

$$du(t,X_t) = (\partial_t + \mathcal{A})\, u(t,X_t)dt + \partial u(t,X_t)\sigma_t\, dW_t + \delta u(t,X_{t-}) \cdot dm_t, \tag{13.37}$$

where, for $(t,x,y) \in [0,T] \times \mathbb{R}^d \times E$,

$$\delta u(t,x,e) = u(t,x+j(t,x,e)) - u(t,x)$$

and where the infinitesimal generator $\mathcal{A}$ of X acts on u at time t as

$$\begin{aligned} (\mathcal{A}u)(t,x) = \frac{1}{2}\partial^2 u(t,x) : a(t,x) + \partial u(t,x) b(t,x) + \\ \big(\delta u(t,x) - \partial u(t,x) j(t,x)\big) \cdot c. \end{aligned} \tag{13.38}$$

Remark 13.3.1 The martingale part in (13.37) can also be abbreviated $\mathcal{B}u(t,X_{t-}) \cdot dM_t$, where $\mathcal{B}u(t,x) = (\partial u(t,x) b(t,x), \delta u(t,x))$ and where M_t refers to the set of fundamental martingales, W and m, that drive X.

[5] In the strong sense to which we will limit ourselves in this book.

In the case of Itô pairs of locally integrable quadratic variation (X, u) and (X, v), we have (cf. (13.28)):

$$\begin{aligned}&\mathbb{E}\left(du(t, X_t)\middle|\mathcal{G}_t\right) = (\partial_t + \mathcal{A})u(t, X_t)dt,\\&\mathbb{C}\mathrm{ov}\left(du(t, X_t), dv(t, X_t)\middle|\mathcal{G}_t\right) = \mathcal{C}(u, v)(t, X_t)dt,\end{aligned} \tag{13.39}$$

where

$$\mathcal{C}(u, v)(t, x) = \big(\mathcal{A}(uv) - u\mathcal{A}v - v\mathcal{A}u\big)(t, x). \tag{13.40}$$

Thus, writing $Y_t = u(t, X_t)$ and $Z_t = v(t, X_t)$:

$$\begin{aligned}\mathcal{C}(u, v)(t, X_t) &= \frac{d\langle Y, Z\rangle_t}{dt}\\&= \partial u(t, X_t)a_t(\partial v(t, X_t))^{\mathsf{T}} + \big(\delta u \delta v\big)(t, X_t) \cdot c(t, X_t).\end{aligned} \tag{13.41}$$

Also, the conditionings with respect to $\mathcal{G}_t$ in (13.39) can be replaced by conditionings with respect to X_t, by the Markov property of X.

13.4 Feynman-Kac Formula

Let X be given as a jump-diffusion and let $u = u(t, x)$ be a function of class $\mathcal{C}^{1,2}$ forming an Itô pair with X. If $u(t, X_t)$ is a local martingale then, in view of the Itô formula (13.37), we conclude from Proposition 13.1.1 that the time-differentiable local martingale

$$(\partial_t + \mathcal{A})\, u(t, X_t)dt = du(t, X_t) - \partial u(t, X_t)\sigma_t\, dW_t - \delta u(t, X_{t-}) \cdot dm_t$$

is constant. Using for instance BSDE techniques to be mentioned in Sect. 13.5 (see also Crépey (2013)), this in turn translates into the following partial integro-differential equation (deterministic PIDE to be satisfied by the function u:

$$(\partial_t + \mathcal{A})\, u(t, x) = 0,\ x \in \mathbb{R}^d. \tag{13.42}$$

The fundamental situation of this kind corresponds to a Doob-martingale

$$u(t, X_t) = \mathbb{E}\left(\phi(X_T)\middle|X_t\right) = \mathbb{E}\left(\phi(X_T)\middle|\mathcal{G}_t\right)$$

for an integrable terminal condition $\phi(X_T)$. The second equality, which grounds the martingale property of $u(t, X_t)$, holds in virtue of the Markov property of X. In this case the function u can typically be characterized as the unique solution to the PIDE (13.42), along with a terminal condition $u = \phi$ at time T.

More generally, given suitable running and terminal cost functions f and ϕ and a discount rate function r, let

$$u(t, X_t) = \mathbb{E}\left(\int_t^T e^{-\int_t^s r(\zeta, X_\zeta)d\zeta} f(s, X_s)ds + e^{-\int_t^T r(s,X_s)ds}\phi(X_T)\Big|X_t\right)$$

$$= \mathbb{E}\left(\int_t^T e^{-\int_t^s r(\zeta, X_\zeta)d\zeta} f(s, X_s)ds + e^{-\int_t^T r(s,X_s)ds}\phi(X_T)\Big|\mathcal{G}_t\right), \tag{13.43}$$

the second equality resulting from the Markov property of X. Immediate extension of the previous computations yield:

- on the one hand, the following local martingale that arises from the Itô formula applied to $u(t, X_t)$ (assuming an Itô pair (X, u)):

$$du(t, X_t) - (\partial_t u + \mathcal{A}u)(t, X_t)dt = \partial u(t, X_t)\sigma_t\, dW_t + \delta u(t, X_{t-}) \cdot dm_t; \tag{13.44}$$

- on the other hand, the following Doob-martingale (conditional expectation of an integrable terminal condition) that arises from (13.43):

$$du(t, X_t) + (f(t, X_t) - ru(t, X_t))\, dt. \tag{13.45}$$

Substracting (13.44) from (13.45) yields the local martingale

$$(\partial_t u + \mathcal{A}u + f - ru)(t, X_t)dt,$$

which is therefore constant as a time-differentiable local martingale (Proposition 13.1.1). Also, accounting for the terminal condition $u = \phi$ at time T, this translates into the following PIDE for the function u:

$$\begin{cases} u(T, x) = \phi(x), x \in \mathbb{R}^d, \\ (\partial_t u + \mathcal{A}u + f - ru)(t, x) = 0, \; t < T, x \in \mathbb{R}^d. \end{cases} \tag{13.46}$$

The function u can then be characterized and computed (including numerically if needed/possible) as the unique solution, in some sense, to (13.46).

13.4.1 An Affine Formula

Let X be an extended CIR process with dynamics

$$dX_t = a(b(t) - X_t)dt + c\sqrt{X_t}dW_t, \tag{13.47}$$

where a and c are positive constants and $b(\cdot)$ is a nonnegative deterministic function. Let Φ_y and Ψ_y satisfy the following Riccati system of ODEs, parameterized by a real number y :

$$\begin{cases} \dot{\Phi}_y(t) = -a\Phi_y(t) - \frac{c^2}{2}\Phi_y^2(t) + 1, \; \Phi_y(0) = y, \\ \dot{\Psi}_y(t) = a\Phi_y(t), \; \Psi_y(0) = 0. \end{cases} \tag{13.48}$$

By a classical result, these ODEs are solved explicitly as

$$\begin{aligned}\Phi_y(t &= \frac{1 + C_y e^{-D_y t}}{A + B_y e^{-D_y t}},\\ \Psi_y(t) &= \frac{a}{A}\Big\{\frac{B_y - AC_y}{D_y B_y}\ln\frac{A + B_y e^{-D_y t}}{A + B_y} + t\Big\},\end{aligned} \tag{13.49}$$

where A, B_y, C_y and D_y are given by

$$A = \frac{1}{2}\big(a + \sqrt{a^2 + 2c^2}\big), \quad B_y = (1 - Ay)\frac{a + c^2 y - \sqrt{a^2 + 2c^2}}{2ay + c^2 y - 2},$$

$$C_y = (A + B_y)y - 1, \quad D_y = \frac{-B_y(2A - a) + C_y(c^2 + aA)}{AC_y - B_y}.$$

Proposition 13.4.1 *For any $s \geq t$ and $y \geq 0$, we have:*

$$\mathbb{E}\Big(e^{-\int_t^s X_u du - yX_s} | X_t = x\Big) = e^{-I_{s,y}(t,x)}, \tag{13.50}$$

where

$$I_{s,y}(t,x) = x\Phi_y(s - t) + a\int_t^s \Phi_y(s - u)b(u)du. \tag{13.51}$$

Also, we have:

$$\mathbb{E}\Big(X_s e^{-\int_t^s X_u du} | X_t = x\Big) = \partial_s I_{s,0}(t,x)e^{-I_{s,0}(t,x)}, \tag{13.52}$$

where the function $\dot{\Phi}$ implicit in $\partial_s I$ in (13.52) *can be computed explicitly via the first line in (13.49).*

If $b(\cdot)$ is piecewise-constant, such that $b(t) = b_k$ on every interval $[T_{k-1}, T_k)$ of a time-grid (T_k), and if $i \leq j$, such that $t \in [T_{i-1}, T_i)$ and $s \in [T_{j-1}, T_j)$, the second term in (13.51) *reduces to*

$$\begin{aligned} a\int_t^s \Phi_y(s - u)b(u)du &= \left(\Psi_y(s - t) - \Psi_y(s - T_i)\right) b_i\\ &+ \sum_{k=i+1}^{j-1}\left(\Psi_y(s - T_{k-1}) - \Psi_y(s - T_k)\right) b_k + \Psi_y(s - T_{j-1})b_j\end{aligned} \tag{13.53}$$

if $i < j$; otherwise $a\int_t^s \Phi_y(s - u)b(u)du = \Psi_y(s - t)b_i$.

Proof. Formula (13.52) follows, by differentiating in y and evaluating at $y = 0$, from (13.50), itself a classical result in the theory of (time-inhomogenous) affine processes. This formula (13.50) can also be verified by checking that $v(t,x) := e^{-I_{s,y}(t,x)}$ satisfies the following PDE, which characterizes the left-hand side in (13.50), viewed as a function of t, x, for any fixed s, y:

$$\partial_t v(t,x) + \mathcal{A}v(t,x) - xv(t,x) = 0, \;\; v(s,x) = e^{-xy},$$

where

$$\mathcal{A}v(t,x) = a(b(t) - x)\partial_x v(t,x) + \frac{1}{2}c^2 x \partial^2_{x^2} v(t,x)$$

is the infinitesimal generator of the affine process X in (13.47). Finally, in the case of a piecewise-constant $b(\cdot)$, formula (13.53) immediately follows from (13.51), in view of the second line in (13.49). □

13.5 Backward Stochastic Differential Equations

In view of (13.44) and (13.46), we can interpret the triplet of processes (parameterized by $e \in E$, in the case of V)

$$(Y_t, Z_t, V_t(e)) = (u(t,X_t), \partial u(t,X_t)\sigma(t,X_t), \delta u(t,X_{t-},e)) \tag{13.54}$$

as a solution to the following backward SDE (see e.g. Crépey (2013)):

$$\begin{cases} Y_T = \phi(X_T) \text{ and, for } t < T, \\ -dY_t = \big(f(t,X_t) - r(t,X_t)Y_t\big)dt - Z_t dW_t - V_t \cdot dm_t. \end{cases} \tag{13.55}$$

In the case of a diffusion X (no jumps, i.e. $c = m = n = 0$), there is no component V involved in the solution (or, equivalently, $V = 0$).

The Feynman-Kac formula (13.43) for the solution u to (13.46), written in the equivalent form (as obvious from (13.55))

$$Y_t = \mathbb{E}\left(\int_t^T \big(f(s,X_s) - r(s,X_s)Y_s\big)ds + \phi(X_T)\Big|\mathcal{G}_t\right), \tag{13.56}$$

can thus also be regarded as the Feynman-Kac representation of the solution (Y, Z, V) to the BSDE (13.55). In fact, the simplest way to rigorously solve (13.46) for a function u satisfying (13.43) is to go the other way around, i.e. first solving (13.55) for a triplet of processes $(Y_t, Z_t, V_t(e))$ and then redoing the above computations in the reverse order to establish that the function u subsequently defined via $u(t, X_t) = Y_t$ solves (13.46) and satisfies (13.43).

Note that the "intrinsic" (non discounted) form (13.56) of the Feynman-Kac representation (13.43) is implicit, meaning that the right-hand side in (13.56) depends on Y. In this case this is not a real issue, however, as revealed by the equivalent explicit discounted representation (13.43). Now, the power of BSDEs lies precisely in the fact that this theory allows one to solve more general problems than the linear equations (13.46), (13.55), namely nonlinear problems in which the BSDE coefficient, $g(t, X_t, Y_t) = f(t, X_t) - rY_t$ in the case of (13.46), (13.55), depends nonlinearly on Y and possibly also on Z and V. Let us thus consider the following BSDE, to be solved for a triplet of processes $(Y_t, Z_t, V_t(e))$ in $\mathcal{S}^2 \times \mathcal{M}^2 \times \mathcal{P}^2$:

$$\begin{cases} Y_T = \phi(X_T) \text{ and, for } t < T, \\ -dY_t = g(t, X_t, Y_t, Z_t, \widehat{V}_t)dt - Z_t dW_t - V_t \cdot dm_t, \end{cases} \tag{13.57}$$

where

$$\widehat{V}_t = \big(V_t \eta(t, X_t)\big) \cdot c(t, X_t) = \int_E V_t(e)\eta(t, X_t, e)c(t, X_t, de),$$

for a (possibly vector-valued) integration kernel $\eta \in \mathcal{P}^2$ (so that $V\eta \in \mathcal{P}^1$). Now let a function $u = u(t, x)$ solve the following semilinear PIDE:

$$\begin{cases} u(T, x) = \phi(x),\ x \in \mathbb{R}^d, \\ \partial_t u(t, x) + \mathcal{A}u(t, x) + g\left(t, x, u(t, x), \partial u(t, x)\sigma(t, x), \widehat{\delta} u(t, x)\right) = 0,\ t < T, x \in \mathbb{R}^d, \end{cases} \tag{13.58}$$

where

$$\widehat{\delta} u(t, x) = \big(\delta u(t, x)\eta(t, x)\big) \cdot c(t, x) = \int_E \delta u(t, x, e)\eta(t, x, e)c(t, x, de).$$

Straightforward extensions of the computations that led from (13.46) to (13.55) show that the triplet $(Y, Z, V(e))$, given in terms of u by the formula (13.54), solves the nonlinear BSDE (13.55). For this reason the formula (13.54) is sometimes called a nonlinear Feynman-Kac formula.

13.6 Measure Changes and Random Intensity of Jumps

By measure changes based on Girsanov transforms, we can extend the previous developments to models with a random jump intensity measure of the form $c_t(\omega, de)dt = \kappa_t(\omega, e)c(de)dt$ (or a more specific Markov form $c(t, X_t, de)dt = \kappa(t, X_t, e)c(de)dt$) for a suitable nonnegative bounded kernel κ : formally replace $c(de)$ by $c_t(de)$, $c(t, X_t, de)$ or $c(t, x, de)$ as suitable, in all the equations above. This allows one to design models with a stochastic jump intensity measure $c_t(de)$ ($c(t, X_t, de)$ in the Markov case), relevant in many applications.

More precisely, let a change of measure function ψ be defined as the exponential of a function of class $\mathcal{C}^{1,2}$ with compact support on $[0, T] \times \mathbb{R}^d$ (with ψ, $1/\psi$ and $\partial\psi$ bounded in particular). We write $\psi(t, X_{t-}) = \psi_t$. Let there be given an Itô pair of integrable quadratic variation, i.e. a pair (X, ψ) for X of the form, (13.9) with

$$\sigma_t \in \mathcal{M}^2,\ j_t \in \mathcal{P}^2,\ \delta\psi_t \in \mathcal{P}^2$$

(and therefore also, since ψ is positively bounded, $\delta\psi_t/\psi_t \in \mathcal{P}^2$). Let a $\mathbb{Q}$-local martingale Ψ be defined as the following stochastic exponential process: $\Psi_0 = 1$ and, for $t \in [0,T]$,

$$d\Psi_t/\Psi_{t-} = (\delta\psi_t/\psi_t) \cdot dm_t. \tag{13.59}$$

Standard a priori SDE estimates yield that Ψ is a square integrable martingale. In particular, Ψ is a positive $\mathbb{Q}$-martingale with $\Psi_0 = 1$, so that it can be used to define an equivalent probability measure $\mathbb{Q}^\psi$ on $(\Omega, \mathcal{G})$ via the Radon-Nikodym density

$$\frac{d\mathbb{Q}^\psi}{d\mathbb{Q}} = \Psi_T.$$

Since $\delta\psi_t/\psi_t \in \mathcal{P}^2$, c integrates $j_t\delta\psi_t/\psi_t$, which allows one to rewrite the equation (13.9) for X as

$$dX_t = b_t^\star dt + \sigma_t dW_t + (j_t \cdot (dn_t - c\,dt) - (j_t\delta\psi_t/\psi_t \cdot c\,dt), \tag{13.60}$$

where we let

$$b_t^\star = b_t + (j_t\delta\psi_t)/\psi_t \cdot c\,dt.$$

Moreover, standard computations show that

$$\eta_t \cdot (dn_t - c\,dt) - (\eta_t\delta\psi_t/\psi_t) \cdot c\,dt \tag{13.61}$$

is a $\mathbb{Q}^\psi$-local martingale, for any η in $\mathcal{P}^2_{loc}$. Let $c_t(de) = \kappa_t(e)c(de)$, with

$$\kappa_t(e) = 1 + \delta\psi_t/\psi_t = \psi(t, X_{t-} + j_t(e))/\psi(t, X_{t-}).$$

Defining pathwise by (13.61) the $\mathbb{Q}^\psi$-stochastic integral of η in $\mathcal{P}^2_{loc}$ against the random measure $(dn_t - c_t dt)$, one then extends in the standard way (see e.g. Kunita (2010)) this notion of a stochastic integral such that $\eta_t \cdot (dn_t - c_t dt)$ yields, for any $\eta \in \mathcal{P}_{loc}$, a well-defined $\mathbb{Q}^\psi$-local martingale. Note that the definitions of $\mathcal{P}^2_{loc}$ and $\mathcal{P}_{loc}$ are independent of a reference probability measure, such as $\mathbb{Q}^\psi$, equivalent to $\mathbb{Q}$. Rewriting (13.60) in terms of this $\mathbb{Q}^\psi$ stochastic integral against $(dn_t - c_t dt)$, one obtains in the end the following $\mathbb{Q}^\psi$-special semimartingale decomposition of X:

$$dX_t = b_t^\star dt + \sigma_t dW_t + j_t \cdot (dn_t - c_t dt),$$

with

$$b^\star \in \mathcal{M}^1_{loc},\ \sigma \in \mathcal{M}^2_{loc},\ j \in \mathcal{P}^2_{loc}.$$

We have thus constructed a model X with a random disintegrated jump intensity $c_t(de)dt$ relative to the probability measure $\mathbb{Q}^\psi$ (where, in this perspective, the original measure $\mathbb{Q}$ was purely instrumental).

13.7 Reduction of Filtration and Hazard Intensity Pre-Default Credit Risk Modeling

This section gives mathematical tools underlying the so-called reduced-form intensity approach in credit risk modeling, based on reduction of filtration (see Crépey (2013)). Given a $[0, T] \cup \{+\infty\}$-valued stopping time τ without atom on $[0, T]$, let $J_t = \mathbb{1}_{\{\tau > t\}}$ denote the related survival indicator process and let $\bar{\tau} = \tau \wedge T$. We assume further that $\mathbb{G} = \mathbb{G}^\tau := \mathbb{F} \vee \mathbb{H}$, where the filtration $\mathbb{H}$ is generated by the process J and where $\mathbb{F}$ is some reference filtration. The Azéma supermartingale associated with τ is the process A defined, for $t \in [0, T]$, by:

$$A_t = \mathbb{Q}(\tau > t \,|\, \mathcal{F}_t) = \mathbb{E}(J_t \,|\, \mathcal{F}_t). \tag{13.62}$$

Assuming a positive $A_t =: e^{-\Gamma_t}$, where Γ is called the hazard process, we have the following "key lemma" of single-name credit risk (see e.g. page 143 of Bielecki and Rutkowski (2001)).

Lemma 13.7.1 *If ξ is an (integrable) random variable, then*

$$J_t \mathbb{E}[\xi | \mathcal{G}_t] = J_t \frac{\mathbb{E}(\xi J_t \,|\, \mathcal{F}_t)}{\mathbb{Q}(\tau > t \,|\, \mathcal{F}_t)} = J_t e^{\Gamma_t} \mathbb{E}(\xi J_t \,|\, \mathcal{F}_t). \tag{13.63}$$

For ξ of the form $J_s \chi$, for some $\mathcal{F}_s$-measurable χ with $s \geq t$, we have:

$$\mathbb{E}[J_s \chi | \mathcal{G}_t] = J_t e^{\Gamma_t} \mathbb{E}(\chi J_s \,|\, \mathcal{F}_t) = J_t \mathbb{E}(\chi e^{-(\Gamma_s - \Gamma_t)} \,|\, \mathcal{F}_t). \tag{13.64}$$

Proof. The left-hand side in (13.63) (where the right-hand side is only notational) results from the fact that, on $\{\tau > t\}$, the σ-field $\mathcal{G}_t$ is generated by $\mathcal{F}_t$ and the random variable $\{\tau > t\}$. In (13.64), the left-hand side follows by an application of (13.63) to $\xi = J_s \chi$; the right-hand side then results from the tower law by taking an inner conditional expectation with respect to $\mathcal{F}_s$. □

In particular, (13.64) with $\chi = e^{\Gamma_s}$ proves that the process $\mathfrak{E}_t = J_t e^{\Gamma_t}$ is a $\mathbb{G}$-martingale, since for $s \geq t$:

$$\mathbb{E}[J_s e^{\Gamma_s} | \mathcal{G}_t] = J_t \mathbb{E}(e^{\Gamma_s} e^{-(\Gamma_s - \Gamma_t)} | \mathcal{F}_t] = J_t e^{\Gamma_t}.$$

Lemma 13.7.2 *For any $\mathbb{G}$-adapted, respectively $\mathbb{G}$-predictable, process Y, there exists a unique $\mathbb{F}$-adapted, respectively $\mathbb{F}$-predictable, process $\widetilde{Y}$, called the pre-default value process of Y, such that $JY = J\widetilde{Y}$, respectively $J_- Y = J_- \widetilde{Y}$.*

Proof. In view of (13.63), we can take, in the adapted case, $\widetilde{Y}_t = e^{\Gamma_t} \mathbb{E}(Y_t J_t \,|\, \mathcal{F}_t)$. For the predictable case see §75, page 186 in Dellacherie, Maisonneuve, and Meyer (1992) and Proposition 9.12 in (Nikeghbali 2006). □

Additionally, assuming a continuous and nonincreasing process A_t and letting $\mathfrak{M}_t = -(J_t + \Gamma_{t \wedge \tau})$, we have that $d\mathfrak{E}_t = -\mathfrak{E}_{t-} d\mathfrak{M}_t$ and therefore $d\mathfrak{M}_t = -e^{-\Gamma_t} d\mathfrak{E}_t$, so that $\mathfrak{M}_t$ also is a $\mathbb{G}$-martingale. Moreover:

Lemma 13.7.3 **(i)** *An $\mathbb{F}$-martingale stopped at τ is a $\mathbb{G}$-martingale.*
(ii) *An $\mathbb{F}$-adapted càdlàg process cannot jump at τ. We thus have that $\Delta X_\tau = 0$, almost surely, for any $\mathbb{F}$-adapted càdlàg process X.*

Proof. **(i)** Since τ has a positive, continuous and nonincreasing Azéma supermartingale, it is known from Elliot, Jeanblanc, and Yor (2000) that an $\mathbb{F}$-martingale stopped at τ is a $\mathbb{G}$-martingale.
(ii) As A is continuous, τ avoids $\mathbb{F}$-stopping times, i.e. $\mathbb{Q}(\tau = \sigma) = 0$ for any $\mathbb{F}$-stopping time σ (see for instance Coculescu and Nikeghbali (2012)). Now, by Theorem 4.1, page 120 in He, Wang, and Yan (1992), there exists a sequence of $\mathbb{F}$-stopping times exhausting the jump times of an $\mathbb{F}$-adapted càdlàg process. □

Remark 13.7.4 Our assumptions on A exclude that τ could be an $\mathbb{F}$-stopping time (otherwise one would have $A = J$, which jumps at τ). However, they imply that τ is an $\mathbb{F}$-pseudo stopping time, meaning that an $\mathbb{F}$-local martingale stopped at τ is a $\mathbb{G}$-local martingale. This is a slight relaxation of the immersion or $(\mathcal{H})$-hypothesis which would mean that an $\mathbb{F}$-local martingale is a $\mathbb{G}$-local martingale (see the remark 5.2.2). Note that the "key lemma" 13.7.1 is true regardless whether immersion holds, so from this perspective immersion or not is irrelevant.

Letting $\beta_t = e^{-\int_0^t r_s ds}$ denote the discount factor at some $\mathbb{F}$-progressively measurable risk-free rate r_t, we model the cumulative discounted future cash flows of a defaultable claim in the form of the following $\mathcal{G}_{\bar{\tau}}$-measurable random variable, assumed to be well-defined for any $t \in [0, \bar{\tau}]$:

$$\beta_t \pi^t = \int_t^{\bar{\tau}} \beta_s f_s ds + \beta_{\bar{\tau}} \Big(\mathbb{1}_{\{t<\tau<T\}} R_\tau + \mathbb{1}_{\tau>T} \xi \Big), \tag{13.65}$$

for some $\mathbb{F}$-progressively measurable dividend rate process f, some $\mathbb{F}$-predictable recovery process R and some $\mathcal{F}_T$-measurable payment at maturity (random variable) ξ. Note that the assumption that the data r_t, f, R and ξ are in $\mathbb{F}$ is not restrictive in view of Lemma 13.7.2.

Now, assuming A_t time-differentiable, we define the hazard intensity γ and the credit-risk-adjusted-discount-factor α by:

$$\gamma_t = -\frac{d \ln A_t}{dt} = \frac{d\Gamma_t}{dt}, \ \alpha_t = \beta_t \exp(-\int_0^t \gamma_s ds) = \exp(-\int_0^t (r_s + \gamma_s) ds).$$

The next result shows that the computation of conditional expectations of cash flows π^t with respect to $\mathcal{G}_t$ can be reduced to the computation of conditional expectations of "$\mathbb{F}$-equivalent" cash flows $\widetilde{\pi}^t$ with respect to $\mathcal{F}_t$.

Lemma 13.7.5 *We have*

$$\mathbb{E}\big(\pi^t \,\big|\, \mathcal{G}_t\big) = J_t \, \mathbb{E}\big(\widetilde{\pi}^t \,\big|\, \mathcal{F}_t\big),$$

where $\widetilde{\pi}^t$ is given, with $g = f + \gamma R$, by:

$$\alpha_t \widetilde{\pi}^t = \int_t^T \alpha_s g_s ds + \alpha_T \xi. \tag{13.66}$$

Proof. Since $\mathfrak{M}_t = -(J_t + \Gamma_{t \wedge \tau})$ is a $\mathbb{G}$-martingale,

$$\mathbb{E}[\mathbb{1}_{\{t<\tau<T\}} \beta_\tau R_\tau | \mathcal{G}_t] = -\mathbb{E}[\int_t^T \beta_s R_s dJ_s | \mathcal{G}_t] = \int_t^T \mathbb{E}[J_s \gamma_s \beta_s R_s | \mathcal{G}_t] ds.$$

The proof is concluded by repeated applications of (13.64). □

In order to compute the left-hand side in (13.66), we can apply a pre-default BSDEs/PDEs modeling approach to the $\mathbb{E}(\widetilde{\pi}^t \mid \mathcal{F}_t)$-term in the right-hand side, following the lines of Sects. 13.4-13.5 relative to an $\mathbb{F}$-jump-diffusion (pre-default factor process) X. In this approach, the valuation of defaultable claims is handled in essentially the same way as default-free claims, provided the default-free discount factor process β is replaced by a credit risk adjusted discount factor α and a fictitious dividend continuously paid at rate γ is introduced to account for recovery on the claim upon default (note that a "default-free" discount factor β can itself be interpreted in terms of a default risk with "intensity" r_t). This approach can also be applied to the hedging issue by decomposing the $\mathbb{G}$-martingale of the hedging error of a specific hedging scheme as the sum of an $\mathbb{F}$-martingale stopped at τ (hence a $\mathbb{G}$-martingale) and a jump-to-default $\mathbb{G}$-compensated martingale (see Chapter 5 and Crépey (2013)).

13.7.1 Portfolio Credit Risk

We label by $i \in N = \{1, \ldots, n\}$ the n names of a credit pool. We denote by τ_i the default time of name i. Given a reference filtration $\mathbb{F} = (\mathcal{F}_t)_{t \in [0,T]}$ satisfying the usual conditions, the full model filtration $\mathbb{G} = (\mathcal{G}_t)_{t \in [0,T]}$ is defined as the progressive enlargement of $\mathbb{F}$ by the τ_i, i.e.

$$\mathcal{G}_t = \mathcal{F}_t \vee \bigvee_{i \in N} \mathcal{H}_t^i,$$

where $(\mathcal{H}_t^i)$ is the natural filtration of the indicator process H^i of τ_i, so that $\mathcal{H}_t^i = (\tau_i \wedge t) \vee (\tau_i > t)$. The filtration $\mathbb{G}$ can be shown to be right-continuous by a combination of the arguments of Amendinger (1999) and of the appendix[6] of Bélanger, Shreve, and Wong (2001).

Moreover, for $I \subseteq N$, we define the filtration $\mathbb{G}^I = (\mathcal{G}_t^I, t \geq 0)$ as the initial enlargement of $\mathbb{F}$ by the τ_i for $i \in I$, i.e.

$$\mathcal{G}_t^I = \mathcal{F}_t \vee \bigvee_{i \in I} \tau_i.$$

[6] Available online, not present in the Mathematical Finance published version of the paper.

By a straightforward multidefault extension of the analysis in Amendinger (1999), this filtration can be shown to be right-continuous. The result stated in the following lemma can be seen as a multidimensional counterpart of the key lemma 13.7.1 of single-name credit risk. Let

$$\mathfrak{J}_t = \{i \in N \,|\, \tau_i \le t\}, \text{ respectively } J_t = N \setminus \mathfrak{J}_t,$$

denote the random set of the indices of the obligor in default, respectively alive, at time t, i.e. for any $I \subseteq N$:

$$\{\mathfrak{J}_t = I\} = \{\tau_i \le t,\ i \in I;\ \tau_j > t,\ j \in J\}. \tag{13.67}$$

We also write, for any random function f_t of I:

$$\mathbb{E}(f_t(\mathfrak{J}_t) \,|\, \mathcal{G}_t^{\mathfrak{J}_t}) = \sum_{I \subseteq N} \mathbb{1}_{\{\mathfrak{J}_t = I\}} \, \mathbb{E}(f_t(I) \,|\, \mathcal{G}_t^I). \tag{13.68}$$

Lemma 13.7.6 *For every integrable random variable X,*

$$\mathbb{E}[X \,|\, \mathcal{G}_t] - \frac{\mathbb{E}(X \mathbb{1}_{\{\tau_j > t,\ j \in J_t\}} \,|\, \mathcal{G}_t^{\mathfrak{J}_t})}{\mathbb{P}(\tau_j > t, j \in J_t \,|\, \mathcal{G}_t^{\mathfrak{J}_t})}. \tag{13.69}$$

Proof. On the set $\{\mathfrak{J}_t = I\}$, any $\mathcal{G}_t$-measurable random variable is equal to a $\mathcal{G}_t^I$-measurable random variable X_t^I, so that

$$\mathbb{E}[\mathbb{1}_{\{\mathfrak{J}_t = I\}} X \,|\, \mathcal{G}_t] = \mathbb{1}_{\{\mathfrak{J}_t = I\}} \mathbb{E}[X \,|\, \mathcal{G}_t] = \mathbb{1}_{\{\mathfrak{J}_t = I\}} X_t^I. \tag{13.70}$$

Taking conditional expectations given $\mathcal{G}_t^I$ on both sides and multiplying by $\{\mathfrak{J}_t = I\}$ yields that

$$\mathbb{1}_{\{\mathfrak{J}_t = I\}} \mathbb{E}[\mathbb{1}_{\{\mathfrak{J}_t = I\}} X \,|\, \mathcal{G}_t^I] \;=\; \mathbb{1}_{\{\mathfrak{J}_t = I\}} X_t^I \mathbb{E}[\mathbb{1}_{\{\mathfrak{J}_t = I\}} \,|\, \mathcal{G}_t^I]. \tag{13.71}$$

In view of (13.67), where $\{\tau_i \le t,\ i \in I\}$ is $\mathcal{G}_t^I$-measurable, (13.71) reduces to

$$\mathbb{1}_{\{\mathfrak{J}_t = I\}} \mathbb{E}[\mathbb{1}_{\{\tau_j > t,\ j \in J\}} X \,|\, \mathcal{G}_t^I] \;=\; \mathbb{1}_{\{\mathfrak{J}_t = I\}} X_t^I \mathbb{Q}(\tau_j > t,\ j \in J \,|\, \mathcal{G}_t^I).$$

We saw in (7.5) that $\mathbb{Q}(\tau_j > t,\ j \in J \,|\, \mathcal{G}_t^I) > 0$. Thus, we can substitute

$$\mathbb{1}_{\{\mathfrak{J}_t = I\}} \frac{\mathbb{E}[\mathbb{1}_{\{\tau_j > t,\ j \in J\}} X \,|\, \mathcal{G}_t^I]}{\mathbb{Q}(\tau_j > t,\ j \in J \,|\, \mathcal{G}_t^I)}$$

for $\mathbb{1}_{\{\mathfrak{J}_t = I\}} X_t^I$ in (13.70), which yields the formula (13.69) (in the notation of (13.68)).
□

Chapter 14

Markov Consistency and Markov Copulas

14.1 Introduction

In this chapter we provide a brief account of the theory of Markov consistency and the theory of Markov copulas, keeping the technical detail to a minimum. We refer to the existing literature (cf. Bielecki, Jakubowski, and Nieweglowski (2012, 2013) and the references there) for a more comprehensive exposition. The main objective of the theory of Markov copulas can be summarized as follows:

- Consider Y^i, $i = 1, \ldots, n$, a family of univariate Markov processes, defined on an underlying probability space $(\Omega, \mathcal{G}, \mathbb{Q})$, taking values in $\mathbb{R}^1$ (more generally, in some locally compact separable topological space).

- Goal: To construct a copula for Markov processes Y^i, $i = 1, \ldots, n$. Namely, to construct a multivariate Markov process, say $X = (X^i, i = 1, \ldots, n)$, defined on a (possibly different) probability space $(\widetilde{\Omega}, \widetilde{\mathcal{G}}, \widetilde{\mathbb{Q}})$ and taking values in $\mathbb{R}^n$ (more generally – in the appropriate product space) in such a way that each component X^i is Markov with respect to some filtration and with the same law as Y^i, for $i = 1, \ldots, n$.

The starting point for studying Markov copulas is the study of so called Markov consistency property for a multivariate Markov process. Two concepts of Markov consistency property are studied in the literature: the strong Markov consistency and the weak Markov consistency. Here, we will only discuss the former concept (that was historically the first to be introduced), i.e. the concept of a strong Markov copula. Henceforth then, the terms "Markov consistency" and "Markov copula" will refer to "strong Markov consistency" and "strong Markov copula", respectively.

In what follows we introduce the class of consistent Markov processes. Next, we define and construct Markov copulas using the so called generator approach. For an alternative approach (the so called symbolic approach), see Bielecki, Jakubowski, and Nieweglowski (2012).

14.2 Consistent Markov Processes

Let $E = \times_{i=1}^{n} E_i$, where E_i are locally compact separable topological spaces. For any index set $I \subset \{1, \ldots, n\}$, we denote by J its complementary set and we write $E_I = \times_{i \in I} E_i$. For $x \in E$ and a process X with values in E, we use the notation $x_I = (x_i,\, i \in I)$ and $X^I = (X^i,\, i \in I)$, respectively. Let $X = (X^1, \ldots, X^n)$ be an $(\Omega, \mathbb{G}, \mathbb{Q})$-Markov process taking values in E. In general, the components of a vector $\mathbb{G}$-Markov process are not $\mathbb{G}$-Markov themselves. In fact, requiring that a $\mathbb{G}$-Markov process X has $\mathbb{G}$-Markov component X^I is a stringent requirement. However, if the components of a multivariate Markov process X are themselves Markov, then we can apply the rich analytical apparatus of Markov processes to the analysis of both X and its components. This observation motivates the following definition.

Definition 14.2.1 We say that a Markov process X has the Markov consistency property for X^I (or briefly consistency property if X^I is predetermined) if

$$\mathbb{E}(\phi(X^I_{t+s}) \mid \mathcal{G}_t) = \mathbb{E}(\phi(X^I_{t+s}) \mid X^I_t), \tag{14.1}$$

for all appropriately measurable and bounded functions ϕ.

If, in addition, the law of X^I agrees with the law of a given Markov process Y taking values in E_I and defined on some probability space $(\widetilde{\Omega}, \widetilde{\mathcal{G}}, \widetilde{\mathbb{Q}})$, i.e. for any positive integer d, any $t_1, t_2, \ldots, t_d \geq 0$ and any measurable subsets of E_I, $A_1, A_2, \ldots, A_d$,

$$\mathbb{Q}(X^I_{t_l} \in A_l,\, l = 1, 2, \ldots, d) = \widetilde{\mathbb{Q}}(Y_{t_l} \in A_l,\, l = 1, 2, \ldots, d), \tag{14.2}$$

then we say that X has the Markov consistency property for (X^I, Y).

Remark 14.2.2 Let $\mathbb{X}^I$ be the natural filtration of the process X^I and let $\bar{\mathbb{G}}$ be any filtration satisfying $\mathbb{X}^I \subseteq \bar{\mathbb{G}} \subseteq \mathbb{G}$. It is an immediate consequence of (14.1) and the chain rule for conditional expectation that X^I remains a Markov process with respect to $\mathbb{G}$. In other words, Markov consistency also implies the Markov property for the component in its own filtration and in any intermediate filtration between $\mathbb{X}^I$ and $\bar{\mathbb{G}}$.

Let $\mathcal{A}$ denote the infinitesimal generator of X. As seen in Chapter 13, $\mathcal{A}f(x)$ determines the expected infinitesimal evolution of the process $f(X_t)$, given the initial state $X_t = x$. Intuitively, for X^I to have the Markov property in the filtration $\mathbb{G}$, its infinitesimal probabilistic behavior should not depend on the state of the components X^J. In terms of the infinitesimal generator, this means that, for a function f which is a constant function of x_J, $\mathcal{A}f(x)$ should only depend on the variables x_I. This observation allows providing sufficient conditions for the Markov consistency property in terms of the generator $\mathcal{A}$. We refer to Vidozzi (2009) for details.

14.3 Markov Copulas

The question of Markov consistency asks whether a vector-valued $\mathbb{G}$-Markov process is such that its given group of components of X is also Markov. Now we consider the problem from the opposite perspective. For simplicity we only look at univariate components. Let $Y = \{Y^i, i = 1, \ldots, n\}$, where Y^i is E_i-valued, $i = 1, \ldots, n$, be a given collection of Markov processes.

Definition 14.3.1 Any process $X = (X^1, X^2, \ldots, X^n)$ with values in $E = E_1 \times E_2 \times \cdots \times E_n$ that is Markov with respect to its natural filtration, say $\mathbb{G}$, and has the Markov consistency property for $(X^i, Y^i),\ i = 1, \ldots, n$, is called a Markov copula for Y.

The above definition can be rephrased in terms of corresponding infinitesimal generators. Roughly speaking, if we denote by $\{\mathcal{A}^i, i = 1, \ldots, n\}$ the collection of infinitesimal generators for the processes in the family $\{Y^i, i = 1, \ldots, n\}$, then we say that the generator $\mathcal{A}$ of X is a Markov copula for $\{\mathcal{A}^i, i = 1, \ldots, n\}$ iff X is a Markov copula for $\{Y^i, i = 1, \ldots, n\}$.

Clearly one can always take independent processes $X^1, X^2, \ldots, X^n$ such that their laws coincide with the laws of $Y^i, i = 1, \ldots, n$; such a collection constitutes what we call an independence Markov copula. In order to find other Markov copulas, one needs to look for appropriate solutions to an operator equation. Sometimes, a solution to such an equation can be guessed, constructed in fact. This way of proceeding falls into the technique that we call the operator approach to Markov copulas.

Remark 14.3.2 In Bielecki, Jakubowski, and Nieweglowski (2012) symbolic Markov copulas are discussed. The question of computing of such copulas corresponds to the question of computing of solutions to certain functional equations.

Note that the discussion presented here regards time-homogenous Markov processes. However, the time variable can always be added as one of the states of the process.

14.4 Examples

In this section we provide some examples of Markov copulas.

14.4.1 Diffusions

We consider a collection of n operators,

$$\mathcal{A}_i f(x_i) \quad = \quad b_i(x_i)\partial_{x_i} f(x_i) + \frac{1}{2}\sigma_i(x_i)^2 \partial^2_{x_i^2} f(x_i), \qquad (14.3)$$

where the coefficients $b_i(x_i)$ and $\sigma_i(x_i)$ are given functions. In what follows we use the shorthand notation $\bar{\mathcal{A}}_i$ for $I^1 \otimes \cdots \otimes \mathcal{A}_i \otimes \cdots \otimes I^n$, where I^m is the identity operator.

Proposition 14.4.1 *Let $\mathcal{A}_i$ be as in (14.3) and define a linear operator $\mathcal{A}$ by*

$$\mathcal{A}\phi(x) := \sum_{i=1}^{n} \bar{\mathcal{A}}_i\phi(x) + \frac{1}{2}\sum_{i,j=1,i\neq j}^{n} a_{i,j}(x_i,x_j)\partial^2_{x_i^2}\phi(x), \tag{14.4}$$

where $a_{i,j}(x_i,x_j)$ are such that $a_{i,i}(x_i,x_i) = \sigma_i^2(x_i)$ and the (diffusion) matrix $\Sigma(x) = [a_{i,j}(x_i,x_j)]$ is symmetric nonnegative definite and admits a square root $[\sigma_{i,j}] = \Sigma^{\frac{1}{2}}$. Then the operator $\mathcal{A}$ is a Markov copula for $\{\mathcal{A}_i, i = 1,\ldots,n\}$.

Remark 14.4.2 Note that dependence between the components X^i is entirely characterized by the functions $a_{i,j}(\cdot,\cdot)$, $i \neq j$. Therefore, any diffusion copula can be associated with a particular choice of the functions $a_{i,j}(\cdot,\cdot)$.

14.4.2 Jump-Diffusions

In this example we assume that $E_i \subset \mathbb{R}$ are compact sets for all i. For $i = 1,\ldots,n$, we consider a family of operators given by

$$\mathcal{A}_i\phi(x_i) = \lambda_i(x_i)\int_{E_i}(\phi(z_i) - \phi(x_i))\, w_i(x_i,dz_i), \tag{14.5}$$

where $\lambda_i(x_i)$ are continuous functions, $w_i(x_i,dz_i)$ are probability measures on E and the mappings $x_i \to w_i(x_i,B)$ are continuous for all i and $B \subset E_i$. We write

$$\mathcal{S} = 2^{\{1,\ldots,d\}}\setminus\emptyset,\ \ \mathcal{S}_n = \{S\in\mathcal{S}; |S|\geq n\}.$$

Proposition 14.4.3 *Let $\mathcal{A}_i$ be as in (14.5) and define an operator $\mathcal{A}$ as*

$$\begin{aligned}\mathcal{A}\phi(x) &= \sum_{i=1}^{n}\bar{\mathcal{A}}_i\phi(x)\\ &+ \sum_{S\in\mathcal{S}_1}\lambda_S(x)\int_E(\phi(z)-\phi(x))\,w_S(x,dz),\end{aligned} \tag{14.6}$$

where:

(i) *$w_S(x,dz)$ is a probability measure on E defined for $S\in\mathcal{S}$ as*

$$w_S(x,dz) = \otimes_{i\in S}w_i(x_i,dz_i)\otimes_{j\notin S}\boldsymbol{\delta}_{x_j}(dz_j),$$

(ii) *for any $S\in\mathcal{S}_2$, the functions λ^S are nonnegative, continuous and*

$$\sum_{S\in\mathcal{S}_2:i\in S}\lambda_S(x)\leq\lambda_i(x_i)\ \textit{for}\ x\in E\ \textit{and}\ i=1,\ldots,n, \tag{14.7}$$

(iii) *$\lambda_{\{i\}}(x) = \lambda_i(x_i) - \sum_{S\in\mathcal{S}_2:i\in S}\lambda_S(x)$ for all i (so that, in particular, $\sum_{S\in\mathcal{S}_1:i\in S}\lambda_S(x) = \lambda_i(x_i)$).*

Then the operator $\mathcal{A}$ is a Markov copula for $\{\mathcal{A}^i, i=1,\ldots,n\}$.

14.4.2.1 Finite Markov Chains

As a special case of Markov copulas pertaining to Markov jump processes, we present here some core results regarding Markov copulas with regard to finite Markov chains. Again, to simplify the notation, we only consider time-homogeneous Markov chains.

Thus we consider here a collection of finite sets E_i, $i = 1, \ldots, n$, representing state spaces for the coordinate processes. For $i = 1, \ldots, n$, we consider a family of operators on $\mathbb{R}^{E_i}$ (space of real valued functions on E_i) given by

$$\mathcal{A}_i\phi(x_i) = \sum_{y_i \in E_i, y_i \neq x_i} \lambda_i(x_i, y_i)\left(\phi(y_i) - \phi(x_i)\right), \tag{14.8}$$

where $\lambda_i(x_i, y_i) \geq 0$. This can be written as (cf. (14.5))

$$\mathcal{A}_i\phi(x_i) = \lambda_i(x_i)\int_{E_i} \left(\phi(z_i) - \phi(x_i)\right) w_i(x_i, dz_i), \tag{14.9}$$

where $\lambda_i(x_i) = \sum_{y_i \in E_i, y_i \neq x_i} \lambda_i(x_i, y_i)$ and $w_i(x_i, dz_i) = \sum_{y_i \in E_i, y_i \neq x_i} \frac{\lambda_i(x_i, y_i)}{\lambda_i(x_i)} \boldsymbol{\delta}_{y_i}(dz_i)$.

As it is customary in the theory of Markov chains, we identify operators $\mathcal{A}_i$ with matrices

$$A^i = [\lambda_i(x_i, y_i)]_{x_i, y_i \in E_i},$$

where

$$\lambda_i(x_i, x_i) = -\lambda_i(x_i).$$

Now let us consider a system of n algebraic equations in unknowns λ^x_y, where $x = (x^1, x^2, \ldots, x^n)$, $y = (y^1, y^2, \ldots, y^n) \in E = E_1 \times \cdots \times E_n$ and $x \neq y$:

$$\lambda_i(x_i, y_i) = \sum_{y_j \in E_j, j \neq i} \lambda^{y_1, \ldots, y_i, \ldots, y_n}_{x_1, \ldots, x_i, \ldots, x_n}, \quad \forall x_j \in E_j,\ j \neq i, \quad \forall x_i,\ y_i \in E_i,\ y_i \neq x_i,$$
$$i = 1, 2, \ldots, d. \tag{14.10}$$

Note that any solution λ^y_x to (14.10) satisfies the following property (known as condition (M), cf. Bielecki, Jakubowski, Vidozzi, and Vidozzi (2008) and Bielecki, Jakubowski, and Nieweglowski (2012)):

$$\sum_{y_j \in E_j, j \neq i} \lambda^{y_1, \ldots, y_i, \ldots, y_n}_{x'_1, \ldots, x'_{i-1}, x_i, x'_{i+1}, \ldots, x'_n} = \sum_{y_j \in E_j, j \neq i} \lambda^{y_1, \ldots, y_i, \ldots, y_n}_{x_1, \ldots, x_i, \ldots, x_n},$$
$$\forall x'_j, x_j \in E_j,\ j \neq i, \quad \forall x_i,\ y_i \in E_i,\ x_i \neq y_i,$$
$$i = 1, 2, \ldots, n. \tag{14.11}$$

Thus (see Bielecki, Jakubowski, Vidozzi, and Vidozzi (2008) and Bielecki, Jakubowski, and Nieweglowski (2012)), if λ^y_x is a solution to (14.10) that produces (after appropriately defining the diagonal elements) a valid generator matrix $A = [\lambda^y_x]_{x, y \in E}$, A is a Markov copula for $\{A^i, i = 1, \ldots, n\}$.

Remark 14.4.4 It is known (cf. e.g. Bielecki, Jakubowski, Vidozzi, and Vidozzi (2008)) that the above system admits at least one solution λ_y^x such that, defining

$$\lambda_x^x = -\sum_{y\in E, y\neq x} \lambda_x^y, \tag{14.12}$$

then the matrix

$$A = [\lambda_x^y]_{x,y\in E}$$

is a generator of a Markov chain, say $X = (X^1, \ldots, X^n)$, with state space E.

In fact, the elements of the matrix A defined as (I below denotes n-dimensional identity matrix)

$$\begin{aligned} A &= A^1 \otimes I \otimes I \cdots \otimes I + I \otimes A^2 \otimes I \cdots \otimes I \\ &\quad + I \otimes I \cdots \otimes I \otimes A^n \end{aligned} \tag{14.13}$$

satisfy conditions (14.10) and (14.12), and this matrix A is a generator matrix. This, indeed, is the independence Markov copula for finite Markov chains.

14.4.3 Diffusion Modulated Markov Jump Processes.

Let Y be a diffusion process in $\mathbb{R}^n$, with infinitesimal generator $\mathcal{A}_y$ given by (cf. Proposition 13.2.2 for the meaning of the trace operator " : ")

$$\mathcal{A}_y\phi(y) = \partial\phi(y)b(y) + \frac{1}{2}\partial^2\phi(y) : a(y),$$

where $b(\cdot)$ and $a(\cdot)$ are sufficiently regular. Using $\mathcal{A}_y$, we define a collection of n operators

$$\mathcal{A}_i\phi(y, x_i) = (\mathcal{A}_y \otimes I^i)\phi(y, x_i) + \widetilde{\mathcal{A}}_i\phi(y, x_i), \tag{14.14}$$

where I^i is the identity operator on E_i (a compact subset of $\mathbb{R}$) and

$$\widetilde{\mathcal{A}}_i\phi(y, x_i) = \lambda_i(y, x_i)\int_{E_i} (\phi(y, z_i) - \phi(y, x_i))\, w_i(y, x_i, dz_i), \tag{14.15}$$

where $\lambda_i(\cdot,\cdot)$ is a continuous and bounded function of both arguments and $w_i(y, x_i, dz_i)$ is a probability measure for any y, x_i, such that, for any measurable set B in E_i the map $y, x_i \to w_i(y, x_i, B)$ is continuous and bounded for $i = 1, 2, \ldots, n$. Also let $\widehat{\mathcal{A}}_i = I^1 \otimes \cdots \otimes \widetilde{\mathcal{A}}_i \otimes \cdots \otimes I^n$.

Proposition 14.4.5 *Define an operator $\mathcal{A}$ by*

$$\begin{aligned} \mathcal{A}f(x, y) &= (\mathcal{A}_y \otimes I)f(y, x) + \sum_{i=1}^n \widehat{\mathcal{A}}_i f(y, x) \\ &\quad + \sum_{S\in\mathcal{S}_1} \lambda_S(y, x)\int_E (f(y, z) - f(y, x))\, w_S(y, x, dz), \end{aligned} \tag{14.16}$$

where:

(i) $w_S(y, x, dz)$ *is a probability measure on* E *defined as*

$$w_S(y, x, dz) = \otimes_{j \in S} w_i(y, x_i, dz_i) \otimes_{j \notin S} \boldsymbol{\delta}_{x_j}(dz_j),$$

(ii) *the nonnegative continuous functions* $\lambda^S(y, x)$ *are such that*

$$\sum_{S \in \mathcal{S}_2 : i \in S} \lambda_S(y, x) \leq \lambda_i(y, x_i), \quad \forall (x, y) \in E \times \mathbb{R}^n,\, \forall i \in \{1, \ldots, n\},$$

(iii) $\lambda_{\{i\}}(y, x) = \lambda_i(x_i) - \sum_{S \in \mathcal{S}_2 : i \in S} \lambda_S(y, x)$ *for all* i *(so that, in particular,* $\sum_{S \in \mathcal{S}_1 : i \in S} \lambda_S(y, x) = \lambda_i(y, x_i)$*).*

Then $\mathcal{A}$ *is a Markov copula for* $\mathcal{A}_i, i = 1, \ldots, n$.

This proposition can then be specialized to diffusion modulated Markov chains following the approach of Sect. 14.4.2.1.

Bibliography

Albanese, C., T. Bellaj, G. Gimonet, and G. Pietronero (2011). Coherent global market simulations and securitization measures for counterparty credit risk. *Quantitative Finance 11*(1), 1–20.

Albanese, C., D. Brigo, and F. Oertel (2013). Restructuring counterparty credit risk. *International Journal of Theoretical and Applied Finance 16*(2), 1350010 (29 pages).

Albanese, C. and S. Iabichino (2013). The FVA-DVA puzzle: completing markets with collateral trading strategies.

Amendinger, J. (1999). *Initial enlargement of filtrations and additional information in financial markets*. Ph. D. thesis, Technische Universität Berlin.

Andersen, L. and J. Sidenius (2004). Extensions to the Gaussian copula: random recovery and random factor loadings. *Journal of Credit Risk 1*(1), 29–70.

Assefa, S., T. R. Bielecki, S. Crépey, and M. Jeanblanc (2011). CVA computation for counterparty risk assessment in credit portfolios. In T. Bielecki, D. Brigo, and F. Patras (Eds.), *Credit Risk Frontiers*, Chapter 12, pp. 397–436. Wiley.

Basel Committee on Banking Supervision (2004). International convergence of capital measurement and capital standards. http://www.oenb.at/en/img/bcbs107a_tcm16-13376.pdf.

Basel Committee on Banking Supervision (2006). International convergence of capital measurement and capital standards – A revised framework comprehensive version. http://www.bis.org/publ/bcbs128.pdf.

Basel Committee on Banking Supervision (2011a). Basel III: A global regulatory framework for more resilient banks and banking systems. http://www.bis.org/publ/bcbs189.pdf.

Basel Committee on Banking Supervision (2011b). Revisions to the Basel II market risk framework. http://www.bis.org/publ/bcbs193.pdf.

Basel Committee on Banking Supervision (2012). Application of own credit risk adjustments to derivatives. http://www.bis.org/publ/bcbs214.pdf.

Basel Committee on Banking Supervision and Board of the International Organization of Securities Commissions (2012, July). Margin requirements for non-centrally-cleared derivatives. Consultative Document, http://www.bis.org/publ/bcbs226.pdf.

Basel Committee on Banking Supervision and Board of the International Organization of Securities Commissions (2013, February). Margin requirements for non-centrally-cleared derivatives. Second Consultative Document, http://www.bis.org/publ/bcbs242.pdf.

Bean, C. (2007). An indicative decomposition of Libor spreads. *Quaterly Bulletin of the The Bank of England 47*(4), 498–99.

Bélanger, A., S. Shreve, and D. Wong (2001). A unified model for credit derivatives. Long preprint version of the *Mathematical Finance* 2004 paper "A general framework for pricing credit risk".

Beumee, J., D. Brigo, D. Schiemert, and G. Stoyle (2010). Charting a course through the CDS big bang. In: Wigan, D. (ed.), Credit Derivatives: The March to Maturity, Thomson Reuters.

Beylkin, G. and L. Monzon (2005). On approximation of functions by exponential sums. *Journal of Applied and Computational Harmonic Analysis 19*(1), 17–48.

Bianchetti, M. (2010). Two curves, one price. *Risk Magazine* (August), 74–80.

Bielecki, T., I. Cialenco, and I. Iyigunler (2013). Collateralized CVA valuation with rating triggers and credit migrations. *International Journal of Theoretical and Applied Finance 16*, 1350009 (32 pages).

Bielecki, T., S. Crépey, M. Jeanblanc, and B. Zargari (2012). Valuation and hedging of CDS counterparty exposure in a Markov copula model. *International Journal of Theoretical and Applied Finance 15*(1), 1250004 (39 pages).

Bielecki, T., A. Vidozzi, and L. Vidozzi (2008a). A Markov copulae approach to pricing and hedging of credit index derivatives and ratings triggered step-up bonds. *Journal of Credit Risk 4*(1), 47–76.

Bielecki, T. R., A. Cousin, S. Crépey, and A. Herbertsson (2013a). A bottom-up dynamic model of portfolio credit risk – Part I: Markov copula perspective & Part II: Common-shock interpretation, calibration and hedging issues. In A. Takahashi, Y. Muromachi, and T. Shibata (Eds.), *Recent Advances in Financial Engineering 2012*. pp. 25–74, World Scientific.

Bielecki, T. R., A. Cousin, S. Crépey, and A. Herbertsson (2013b). A bottom-up dynamic model of portfolio credit risk with stochastic intensities and random recoveries. *Communications in Statistics–Theory and Methods. 43*(7), 1362–1389.

Bielecki, T. R., A. Cousin, S. Crépey, and A. Herbertsson (2013c). Dynamic hedging of portfolio credit risk in a Markov copula model. *Journal of Optimization Theory and Applications*. Forthcoming, DOI 10.1007/s10957-013-0318-4.

Bielecki, T. R. and S. Crépey (2013). Dynamic hedging of counterparty exposure. In Y. Kabanov, M. Rutkowski, and T. Zariphopoulou (Eds.), *Inspired by Finance*. Springer Berlin.

Bielecki, T. R., S. Crépey, and M. Jeanblanc (2010). Up and down credit risk. *Quantitative Finance 10*(10), 1469–7696.

Bielecki, T. R., S. Crépey, M. Jeanblanc, and M. Rutkowski (2011). Convertible bonds in a defaultable diffusion model. In A. Kohatsu-Higa, N. Privault, and S. Sheu (Eds.), *Stochastic Analysis with Financial Applications*, pp. 255–298. Birkhäuser.

Bielecki, T. R., J. Jakubowski, and M. Nieweglowski (2012). Study of dependence for some stochastic processes: symbolic Markov copulae. *Stochastic Processes and their Applications 122 (3)*, 930–951.

Bielecki, T. R., J. Jakubowski, and M. Nieweglowski (2013). Intricacies of dependence between components of multivariate Markov chains: weak Markov consistency and weak Markov copulae. *Electronic Journal of Probability 18*(45), 1–21.

Bielecki, T. R., J. Jakubowski, A. Vidozzi, and L. Vidozzi (2008). Study of dependence for some stochastic processes. *Stochastic Analysis and Applications 26*, 903–924.

Bielecki, T. R., M. Jeanblanc, and M. Rutkowski (2008). Pricing and trading credit default swaps. *Annals of Applied Probability 18*, 2495–2529.

Bielecki, T. R., M. Jeanblanc, and M. Rutkowski (2009). *Credit Risk Modeling*. Osaka University CSFI Lecture Notes Series 2. Osaka University Press.

Bielecki, T. R. and M. Rutkowski (2001). *Credit Risk: Modeling, Valuation and Hedging*. Springer.

Bielecki, T. R. and M. Rutkowski (2013). Valuation and hedging of OTC contracts with funding costs, collateralization and counterparty credit risk: Part 1. *Working paper*. http://math.iit.edu/ bielecki/publication/BR_1_July_5.pdf.

Bielecki, T. R., A. Vidozzi, and L. Vidozzi (2008b). A Markov copulae approach to pricing and hedging of credit index derivatives and ratings triggered step–up bonds. *Journal of Credit Risk 4*(1), 47–76.

Blanchet-Scalliet, C. and F. Patras (2008). Counterparty risk valuation for CDS. http://arxiv.org/abs/0807.0309.

Blanchet-Scalliet, C. and F. Patras (2011). Structural counterparty risk valuation for CDS. In T. Bielecki, D. Brigo, and F. Patras (Eds.), *Credit Risk Frontiers: Subprime crisis, Pricing and Hedging, CVA, MBS, Ratings and Liquidity*. Wiley.

Blundell-Wignall, A. and P. Atkinson (2010). Thinking beyond Basel III: Necessary solutions for capital and liquidity. *OECD Journal: Financial Market Trends 2*(1), 9–33. http://www.oecd.org/dataoecd/42/58/45314422.pdf.

Brigo, D. (2005). Market models for CDS options and callable floaters. *Risk Magazine* (January), 89–94.

Brigo, D., C. Buescu, and M. Morini (2012). Counterparty risk pricing: impact of closeout and first-to-default times. *International Journal of Theoretical and Applied Finance 15*(6), 1250039.

Brigo, D. and A. Capponi (2008a). Bilateral counterparty risk valuation with stochastic dynamical models and application to credit default swaps. Working paper available at http://arxiv.org/abs/0812.3705 . Short updated version in *Risk*, March 2010 issue.

Brigo, D. and A. Capponi (2008b). Bilateral counterparty risk valuation with stochastic dynamical models and application to credit default swaps. Working paper.

Brigo, D., A. Capponi, and A. Pallavicini (2014). Arbitrage-free bilateral counterparty risk valuation under collateralization and application to credit default swaps. *Mathematical Finance 24*(1), 125–146.

Brigo, D., A. Capponi, A. Pallavicini, and V. Papatheodorou (2011). Collateral margining in arbitrage-free counterparty valuation adjustment including re-hypotecation and netting. Working paper, available at http://arxiv.org/abs/1101.3926.

Brigo, D. and K. Chourdakis (2008). Counterparty risk for credit default swaps: impact of spread volatility and default correlation. *International Journal of Theoretical and Applied Finance 12*(7), 1007–1026.

Brigo, D., K. Chourdakis, and I. Bakkar (2008). Counterparty risk valuation for energy-commodities swaps. *Energy Risk* (March).

Brigo, D. and M. Masetti (2005). Risk neutral pricing of counterparty risk. In M. Pykhtin (Ed.), *Counterparty Credit Risk Modelling: Risk Management, Pricing and Regulation*. Risk Books.

Brigo, D. and F. Mercurio (2006). *Interest Rate Models – Theory and Practice* (2nd ed.). Springer.

Brigo, D. and M. Morini (2010a). Dangers of bilateral counterparty risk: the fundamental impact of closeout conventions. Preprint available at ssrn.com or at arxiv.org.

Brigo, D. and M. Morini (2010b). Rethinking counterparty default. *Creditflux Newsletter Analysis 114*, 18–19.

Brigo, D. and M. Morini (2011). Close-out convention tensions. *Risk Magazine* (December), 86–90.

Brigo, D., M. Morini, and A. Pallavicini (2013). *Counterparty Credit Risk, Collateral and Funding with pricing cases for all asset classes*. Wiley.

Brigo, D., M. Morini, and M. Tarenghi (2011). Equity return swap valuation under counterparty risk. In T. Bielecki, D. Brigo, and F. Patras (Eds.), *Credit Risk Frontiers*, pp. 457–484. Wiley.

Brigo, D. and A. Pallavicini (2007). Counterparty risk under correlation between default and interest rates. In J. Miller, D. Edelman, and J. Appleby (Eds.), *Numerical Methods for Finance*. Chapman Hall.

Brigo, D. and A. Pallavicini (2008). Counterparty Risk and contingent CDS under correlation between interest-rates and default. *Risk Magazine* (February), 84–88.

Brigo, D., A. Pallavicini, and V. Papatheodorou (2011). Arbitrage-free valuation of bilateral counterparty risk for interest-rate products: Impact of volatilities and correlations. *International Journal of Theoretical and Applied Finance 14*(6), 773–802.

Brigo, D., A. Pallavicini, and R. Torresetti (2007). Cluster-based extension of the generalized Poisson loss dynamics and consistency with single names. *International Journal of Theoretical and Applied Finance 10*(4), 607–632.

Brigo, D., A. Pallavicini, and R. Torresetti (2010). *Credit Models and the Crisis: a Journey into CDOs, Copulas, Correlations and Dynamic Models*. Wiley.

Brigo, D. and M. Tarenghi (2004). Credit default swap calibration and equity swap valuation under counterparty risk with a tractable structural model. Working Paper, available at www.damianobrigo.it/cdsstructural.pdf. Reduced version in *Proceedings of the FEA Conference at MIT, Cambridge, Massachusetts, November* and in *Proceedings of the Counterparty Credit Risk 2005 C.R.E.D.I.T. conference*, Venice, Vol 1.

Brigo, D. and M. Tarenghi (2005). Credit default swap calibration and counterparty risk valuation with a scenario based first passage model. Working Paper, available at www.damianobrigo.it/cdsscenario1p.pdf Also in: *Proceedings of the Counterparty Credit Risk C.R.E.D.I.T. conference*, Venice, Vol 1.

Brunnermeier, M. and L. Pedersen (2009). Market liquidity and funding liquidity. *Review of Financial Studies 22*(6), 2201–2238.

Burgard, C. and M. Kjaer (2011a). In the balance. *Risk Magazine* (October), 72–75.

Burgard, C. and M. Kjaer (2011b). PDE representations of options with bilateral counterparty risk and funding costs. *Journal of Credit Risk 7*(3), 1–19.

Burgard, C. and M. Kjaer (2012). The FVA debate: in theory and practice. http://ssrn.com/abstract=2157634.

Canabarro, E. and D. Duffie (2004). Measuring and marking counterparty risk. Proceedings of the Counterparty Credit Risk 2005 C.R.E.D.I.T. conference, Venice, Vol 1.

Capponi, A. (2013). Pricing and mitigation of counterparty credit exposures. In J.-P. Fouque and J.Langsam (Eds.), *Handbook of Systemic Risk*, pp. 41–55. Cambridge University Press.

Carver, L. (2013). Introducing the XVA desk – a treasurer's nightmare. *Risk Magazine* (August). http://www.risk.net/risk-magazine/feature/2291080/introducing-the-xva-desk-a-treasurers-nightmare.

Castagna, A. (2011). Funding, liquidity, credit and counterparty risk: Links and implications. http://ssrn.com/abstract=1855028.

Cesari, G., J. Aquilina, N. Charpillon, Z. Filipovic, G. Lee, and I. Manda (2010). *Modelling, Pricing and Hedging Counterparty Credit Exposure: A Technical Guide*. Springer.

Coculescu, D. and A. Nikeghbali (2012). Hazard processes and martingale hazard processes. *Mathematical Finance 22*, 519–537.

Cont, R. and Y. H. Kan (2011). Dynamic hedging of portfolio credit derivatives. *SIAM Journal on Financial Mathematics 2*(1), 112–140.

Cont, R. and T. Kokholm (2012). Central clearing of OTC derivatives: bilateral vs multilateral netting. Working paper.

Cont, R. and A. Minca (2013). Recovering portfolio default intensities implied by CDO quotes. *Mathematical Finance 23*(1), 94–121.

Cont, R., R. Mondescu, and Y. Yu (2011). Central clearing of interest rate swaps: a comparison of offerings. SSRN eLibrary.

Cont, R., E. Santos, and A. Moussa (2013). Network structure and systemic risk in banking systems. In J.-P. Fouque and J. Langsam (Eds.), *Handbook of Systemic Risk*. Cambridge University Press.

Cont, R. and P. Tankov (2004). *Financial Modelling with Jump Processes*. Chapman & Hall.

Cousin, A., S. Crépey, and Y. H. Kan (2012). Delta-hedging correlation risk? *Review of Derivatives Research 15*(1), 25–56.

Cousin, A., M. Jeanblanc, and J.-P. Laurent (2011). Hedging CDO tranches in a Markovian environment. In *Paris-Princeton Lectures in Mathematical Finance 2010*, Lecture Notes in Mathematics, pp. 1–61. Springer.

Crépey, S. (2011). A BSDE approach to counterparty risk under funding constraints. LAP Preprint n° 326, June 2011, available at http://www.maths.univ-evry.fr/prepubli/index.html.

Crépey, S. (2012a). Bilateral Counterparty risk under funding constraints – Part I: Pricing. *Mathematical Finance*. Forthcoming, DOI 10.1111/mafi.12004.

Crépey, S. (2012b). Bilateral Counterparty risk under funding constraints – Part II: CVA. *Mathematical Finance*. Forthcoming, DOI 10.1111/mafi.12005.

Crépey, S. (2013). *Financial Modeling: A Backward Stochastic Differential Equations Perspective*. Springer.

Crépey, S. (2014). Reduced-form modeling of counterparty risk on credit derivatives. In C. Hillairet, M. Jeanblanc, and Y. Jiao (Eds.), *Arbitrage, Credit and Informational Risks*. World Scientific. Proceedings of the Sino-French Research Program in Financial Mathematics Conference, Beijing June 2013, forthcoming.

Crépey, S., T. R. Bielecki, and D. Brigo (2014). *Counterparty Risk and Funding – A Tale of Two Puzzles*. Taylor & Francis.

Crépey, S. and R. Douady (2013a). LOIS: credit and liquidity. *Risk Magazine* (June), 82–86.

Crépey, S. and R. Douady (2013b). The whys of the LOIS: Credit skew and funding spreads volatility. *Bloomberg Brief / Risk* (May), 6–7.

Crépey, S., R. Gerboud, Z. Grbac, and N. Ngor (2013). Counterparty risk and funding: The four wings of the TVA. *International Journal of Theoretical and Applied Finance 16*(2), 1350006 (31 pages).

Crépey, S., Z. Grbac, N. Ngor, and D. Skovmand (2013). A Lévy HJM multiple-curve model with application to CVA computation. Working paper, available at http://ssrn.com/abstract=2334865.

Crépey, S., Z. Grbac, and H. N. Nguyen (2012). A multiple-curve HJM model of interbank risk. *Mathematics and Financial Economics 6 (3)*, 155–190.

Crépey, S., M. Jeanblanc, and D. L. Wu (2013). Informationally dynamized Gaussian copula. *International Journal of Theoretical and Applied Finance 16*(2), 1350008 (29 pages).

Crépey, S., M. Jeanblanc, and B. Zargari (2010). Counterparty risk on a CDS in a Markov chain copula model with joint defaults. In M. Kijima, C. Hara, Y. Muromachi, and K. Tanaka (Eds.), *Recent Advances in Financial Engineering 2009*, pp. 91–126. World Scientific.

Crépey, S. and A. Rahal (2013). Simulation/regression pricing schemes for CVA computations on CDO tranches. *Communications in Statistics – Theory and Methods. 43*(7), 1390–1408.

Crépey, S. and S. Song (2014). Counterparty risk modeling: Beyond immersion. Working paper.

Cvitanic, J. and J. Ma (1996). Hedging Options for a Large Investor and Forward-Backward SDE's. *Annals of Applied Probability 6*(2), 370–398.

De Franco, C., P. Tankov, and X. Warin (2013). Numerical methods for the quadratic hedging problem in Markov models with jumps. http://arxiv.org/abs/1206.5393.

Delbaen, F. and W. Schachermayer (2006). *The Mathematics of Arbitrage*. Springer.

Dellacherie, C., B. Maisonneuve, and P.-A. Meyer (1992). *Probabilités et Potentiel, Chapitres XVII-XXIV.* Hermann.

Dellacherie, C. and P.-A. Meyer (1975). *Probabilité et Potentiel, Vol. I.* Hermann.

Delong, L. and P. Imkeller (2010). Backward stochastic differential equations with time delayed generators – results and counterexamples. *Annals of Applied Probability 20*(4), 1512–1536.

Dong Li, W. (2013). *Density models and applications to counterparty credit risk*. Ph. D. thesis, Université d'Evry Val d'Essonne.

Duffie, D. (2010). *How Big Banks Fail: and What To Do About It*. Princeton University Press.

Duffie, D. and M. Huang (1996). Swap rates and credit quality. *Journal of Finance 51*, 921–950.

Eberlein, E., D. Madan, M. Pistorius, and M. Yor (2013). Bid and ask prices as non-linear continuous time G-expectations based on distortions. Preprint University of Freiburg, available at http://www.stochastik.uni-freiburg.de/ eberlein/papers/NLEMD2.pdf.

Ehlers, P. and P. Schönbucher (2006). The influence of FX risk on credit spreads. Working Paper, ETH Zurich.

Eisenschmidt, J. and J. Tapking (2009). Liquidity risk premia in unsecured interbank money markets. Technical Report 1025, ECB Working Paper Series.

El Karoui, N., M. Jeanblanc, and Y. Jiao (2009). What happens after a default: the conditional density approach. *Stochastic Processes and their Applications 120*, 1011–1032.

El Karoui, N., S. Peng, and M.-C. Quenez (1997). Backward stochastic differential equations in finance. *Mathematical Finance 7*, 1–71.

El Karoui, N. and M.-C. Quenez (1995). Dynamic programming and pricing of contingent claims in an incomplete market. *SIAM Journal of Control and Optimization 33*, 29–66.

Elliot, R., M. Jeanblanc, and M. Yor (2000). On models of default risk. *Mathematical Finance 10*, 179–195.

Elouerkhaoui, Y. (2007). Pricing and hedging in a dynamic credit model. *International Journal of Theoretical and Applied Finance 10*(4), 703–731.

Ethier, H. J. and T. G. Kurtz (1986). *Markov processes: Characterization and convergence*. John Wiley & Sons.

European Central Bank (2009). Credit default swaps and counterparty credit risk. http://www.ecb.europa.eu/pub/pdf/other/ creditdefaultswapsandcounterpartyrisk2009en.pdf.

Fermanian, J.-D. and O. Vigneron (2010). Pricing and hedging basket credit derivatives in the Gaussian copula. *Risk Magazine* (February), 92–96.

Fermanian, J.-D. and O. Vigneron (2013). On break-even correlation: the way to price structured credit derivatives by replication. *Quantitative Finance*. Forthcoming, DOI:10.1080/14697688.2013.812233.

Filipović, D. (2005). Time-inhomogeneous affine processes. *Stochastic Processes and their Applications 115*, 639–659.

Filipović, D. and A. B. Trolle (2013). The term structure of interbank risk. *Journal of Financial Economics 109*(3), 707–733.

Frey, R. and J. Backhaus (2010). Dynamic hedging of synthetic CDO tranches with spread and contagion risk. *Journal of Economic Dynamics and Control 34*(4), 710–724.

Fujii, M., Y. Shimada, and A. Takahashi (2011). A market model of interest rates with dynamic basis spreads in the presence of collateral and multiple currencies. *Wilmott Magazine 54*, 61–73.

Fujii, M. and A. Takahashi (2011a). Choice of collateral currency. *Risk Magazine* (January), 120–125.

Fujii, M. and A. Takahashi (2011b). Derivative pricing under asymmetric and imperfect collateralization and CVA. SSRN eLibrary.

Fujii, M. and A. Takahashi (2012). Perturbative expansion technique for non-linear FBSDEs with interacting particle method. arXiv:1204.2638.

Gregory, J. (2009). *Counterparty Credit Risk and Credit Value Adjustment: The New Challenge for Global Financial Markets*. Wiley.

Gregory, J. (2012). *Counterparty Credit Risk and Credit Value Adjustment: A Continuing Challenge for Global Financial Markets*. Wiley.

Gupton, G. M., C. C. Finger, and M. Bathia (1997). CreditMetrics. Technical document, available at defaultrisk.com.

Guyon, J. and P. Henry-Labordère (2012). *Nonlinear Pricing Methods in Quantitative Finance*. Chapman & Hall/CRC. Financial Mathematics Series.

Hastie, T., R. Tibshirani, and J. Friedman (2009). *The Elements of Statistical Learning: Data Mining, Inference and Prediction*. Springer.

He, S.-W., J.-G. Wang, and J.-A. Yan (1992). *Semimartingale Theory and Stochastic Calculus*. CRC Press.

Henrard, M. (2007, July). The irony in the derivatives discounting. *Wilmott Magazine 30*, 92–98. http://www.wilmott.com/pdfs/130912_henrard.pdf.

Henrard, M. (2010). The irony in the derivatives discounting part II: the crisis. *Wilmott Magazine 2*(6), 301–316.

Henry-Labordère, P. (2012). Cutting CVA's complexity. *Risk Magazine* (July), 67–73.

Herbertsson, A. (2011). Modelling default contagion using multivariate phase-type distributions. *Review of Derivatives Research 14*(1), 1–36.

Horst, U., Y. Hu, P. Imkeller, A. Reveillac, and J. Zhang (2011). Forward-backward systems for expected utility maximization. Working paper.

Huge, B. and D. Lando (1999). Swap pricing with two-sided default risk in a rating-based model. *European Finance Review 3*, 239–268.

Hull, J. and A. White (2001). Valuing credit default swaps II: Modeling default correlation. *The Journal of Derivatives 8*(3), 12–22.

Hull, J. and A. White (2013a). Collateral and credit issues in derivatives pricing. Working paper available at www.defaultrisk.com.

Hull, J. and A. White (2013b). The FVA debate. *Risk Magazine* (August). http://www.risk.net/risk-magazine/analysis/2188684/risk-25-the-fva-debate.

Hull, J. and A. White (2013c). Libor vs OIS: The derivatives discounting dilemma. Working paper available at www.defaultrisk.com.

Hull, J. and A. White (2013d). Valuing derivatives: Funding value adjustments and fair value. Working paper available at www.defaultrisk.com.

Ikeda, N. and S. Watanabe (1989). *Stochastic Differential Equations and Diffusion Processes* (2nd ed.). North-Holland.

International Swaps and Derivatives Association (2002). ISDA 2002 Master Agreement. Available at http://www.isda.org.

International Swaps and Derivatives Association (2009). AIG and credit default swaps. ISDA regulatory discussion papers. http://www.isda.org/c_and_a/pdf.

International Swaps and Derivatives Association (2010). Market Review of OTC Derivative Bilateral Collateralization Practices. http://www.isda.org/c_and_a/pdf/Collateral-Market-Review.pdf.

Iscoe, I., K. Jackson, A. Kreinin, and X. Ma (2010). On exponential approximation to the hockey stick function. Working Paper, Department of Computer Science, University of Toronto.

Iscoe, I., K. Jackson, A. Kreinin, and X. Ma (2013). Pricing correlation dependent derivatives based on exponential approximations to the hockey stick. *Journal of Computational Finance 16*(3), 127–150.

Jamshidian, F. (2002). Valuation of credit default swap and swaptions. *Finance and Stochastics 8*, 343–371.

Jarrow, R. and F. Yu (2001). Counterparty risk and the pricing of defaultable securities. *Journal of Finance 56*, 1765–1799.

Jeanblanc, M. and Y. Le Cam (2007). Reduced form modelling for credit risk. http://www.defaultrisk.com/pp_model135.htm.

Jeanblanc, M. and Y. Le Cam (2009). Progressive enlargement of filtrations with initial times. *Stochastic Processes and their Applications 119*, 2523–2543.

Jeanblanc, M., M. Yor, and M. Chesney (2010). *Mathematical Methods for Financial Markets*. Springer Finance Textbooks. Springer.

Jorion, P. (2006). *Value at Risk*. McGraw Hill. 3rd edition.

Kenyon, C. (2010). Short-rate pricing after the liquidity and credit shocks: including the basis. *Risk Magazine* (November), 83–87.

Kenyon, C. and R. Stamm (2012). *Discounting, Libor, CVA and Funding*. Palgrave Macmillan.

Kijima, M., K. Tanaka, and T. Wong (2009). A multi-quality model of interest rates. *Quantitative Finance 9*(2), 133–145.

Kunita, H. (2010). Itô's stochastic calculus: Its surprising power for applications. *Stochastic Processes and Applications 120*, 622–652.

Laurent, J.-P., A. Cousin, and Fermanian (2011). Hedging default risks of CDOs in Markovian contagion models. *Quantitative Finance 11*(12), 1773–1791.

Leung, S. and Y. Kwok (2005). CDS valuation with counterparty risk. *The Kyoto Economic Review 74*(1), 24–45.

Levels, A. and J. Capel (2012). Is collateral becoming scarce? Evidence for the euro area. *Journal of Financial Market Infrastructures 1*(1), 29–53.

Li, D. (2000). On default correlation: a copula function approach. *Journal of Fixed Income 9*, 43–54.

Lipton, A. and A. Sepp (2009a). Credit value adjustment for credit default swaps via the structural default model. *The Journal of Credit Risk 5*(2), 123–146.

Lipton, A. and A. Sepp (2009b). Credit value adjustment for credit default swaps via the structural default model. *The Journal of Credit Risk 5*, 123–146.

Lipton, S. and D. Shelton (2012). Credit default swaps with an without counterparty and collateral adjustments. *Stochastics: An International Journal of Probability and Stochastic Processes 84*, 603–624.

Ma, J. and J. Yong (2007). *Forward-Backward Stochastic Differential Equations and their Applications* (3rd ed.), Volume 1702 of *Lecture Notes in Mathematics*. Springer.

Marshall, A. W. and I. Olkin (1967). A multivariate exponential distribution. *Journal of the American Statistical Association 2*, 84–98.

McNeil, A., R. Frey, and P. Embrechts (2005). *Quantitative Risk Management: Concepts, Techniques and Tools*. Princeton University Press.

Mercurio, F. (2010a). Interest rates and the credit crunch: new formulas and market models. Technical report, Bloomberg Portfolio Research Paper No. 2010-01-FRONTIERS.

Mercurio, F. (2010b). A Libor market model with stochastic basis. *Risk Magazine* (December), 84–89.

Mercurio, F., R. Caccia, and M. Cutuli (2012). Downgrade termination costs. *Risk Magazine* (March), 66–71.

Moreni, N. and A. Pallavicini (2013a). Parsimonious HJM modelling for multiple yield-curve dynamics. *Quantitative Finance*. Forthcoming, preprint version available at http://dx.doi.org/10.2139/ssrn.1699300.

Moreni, N. and A. Pallavicini (2013b). Parsimonious multi-curve HJM modelling with stochastic volatility. *Interest Rate Modelling After The Financial Crisis (Risk Books)*. Forthcoming.

Morini, M. (2009). Solving the puzzle in the interest rate market. SSRN eLibrary.

Morini, M. and A. Prampolini (2011). Risky funding with counterparty and liquidity charges. *Risk Magazine* (March), 70–75.

Musiela, M. and M. Rutkowski (2005). *Martingale Methods in Financial Modelling* (2nd ed.). Springer.

Nikeghbali, A. (2006). An essay on the general theory of stochastic processes. *Probability Surveys 3*, 345–412.

Nikeghbali, A. and M. Yor (2005). A definition and some characteristic properties of pseudo-stopping times. *Annals of Probability 33*, 1804–1824.

Pallavicini, A. and D. Brigo (2013). CCP clearing pricing of interest-rate derivatives: funding costs and wrong-way risk. arXiv.org.

Pallavicini, A., D. Perini, and D. Brigo (2011). Funding valuation adjustment consistent with CVA and DVA, wrong way risk, collateral, netting and re-hypotecation. SSRN.com and arXiv.org.

Pallavicini, A., D. Perini, and D. Brigo (2012). Funding, collateral and hedging: Uncovering the mechanics and the subtleties of funding valuation adjustments. http://dx.doi.org/10.2139/ssrn.2161528.

Picoult, E. (2005). Calculating and hedging exposure, credit value adjustment and economic capital for counterparty credit risk. In M. Pykhtin (Ed.), *Counterparty Credit Risk Modelling*. Risk Books.

Piterbarg, V. (2010). Funding beyond discounting: collateral agreements and derivatives pricing. *Risk Magazine* (August), 57–63.

Piterbarg, V. (2012). Cooking with collateral. *Risk Magazine* (July), 58–63.

Pollack, L. (2012a). Barclays visits the securitisation bistro. Financial Times Alphaville Blog, Posted by Lisa Pollack on Jan 17, 11:20.

Pollack, L. (2012b). Big banks seek regulatory capital trades. Financial Times Alphaville Blog, Posted by Lisa Pollack on April 29, 19:27.

Pollack, L. (2012c). The latest in regulation-induced innovation. – Part 2. Financial Times Alphaville Blog, Posted by Lisa Pollack on Apr 11 16:50.

Protter, P. (2004). *Stochastic Integration and Differential Equations* (3rd ed.). Springer.

Pykhtin, M. (2005). In M. Pykhtin (Ed.), *Counterparty Credit Risk Modelling: Risk Management, Pricing and Regulation*. Risk Books.

Redon, C. (2006). Wrong way risk modelling. *Risk Magazine*, April 90–95.

Royer, M. (2006). BSDEs with jumps and related non linear expectations. *Stochastics Processes and their Applications 116*, 1357–1376.

Schweizer, M. (2001). A Guided Tour through Quadratic Hedging Approaches. In J. C. E. Jouini and M. Musiela (Eds.), *Option Pricing, Interest Rates and Risk Management*, Volume 12, pp. 538–574. Cambridge University Press.

Singh, M. (2010). Collateral, netting and systemic risk in the OTC derivatives market. Technical report, IMF working paper WP/10/99.

Singh, M. and J. Aitken (2009). Counterparty risk, impact on collateral flows, and role for central counterparties. Technical report, IMF working paper WP/09/173.

Smith, J. (2010). The term structure of money market spreads during the financial crisis. Preprint.

Torresetti, R., D. Brigo, and A. Pallavicini (2009). Risk-neutral versus objective loss distribution and CDO tranche valuation. *Journal of Risk Management in Financial Institutions* 2(2), 175–192.

Vidozzi, A. (2009). *Two Essays In Mathematical Finance*. Ph. D. thesis, Illinois Institute of Technology.

Watt, M. (2011). Corporates fear CVA charge will make hedging too expensive. *Risk Magazine*, October. http://www.risk.net/risk-magazine/feature/2111823/corporates-fear-cva-charge-hedging-expensive.

Wheatley, M. (2012). The Wheatley review of Libor: final report. Technical report.

Wong, G. (2013). A comprehensive FVA modelling approach and collateral trading strategies. Slides of The 9th WBS Fixed Income Conference, Munich, 18/10/2013.

Yi, C. (2011). Dangerous knowledge: Credit value adjustment with credit triggers. *International Journal of Theoretical and Applied Finance 14 (6)*, 839–865.

Zhou, R. (2013). Counterparty risk subject to additional termination event clauses. *Journal of Credit Risk* 9(1), 39–73.

Ziegel, J. (2013). Coherence and elicitability. Working paper available at http://arxiv.org/abs/1303.1690.

Index